Black Holes and Relativistic Stars

BLACK HOLES AND RELATIVISTIC STARS

Edited by Robert M. Wald

The University of Chicago Press
Chicago and London

ROBERT M. WALD is professor in the Department of Physics, the
Enrico Fermi Institute, and the College, at the University of Chicago.
He is the author of *Space, Time, and Gravity: The Theory of the Big
Bang and Black Holes* (1977; 2d ed. 1992); *General Relativity* (1984);
and *Quantum Field Theory in Curved Spacetime and Black Hole
Thermodynamics* (1994)—all published by the University of Chicago
Press.

The University of Chicago Press, Chicago 60637
The University of Chicago Press, Ltd., London
© 1998 by The University of Chicago
Chapter 3 © 1998 by Kip S. Thorne
All rights reserved. Published 1998
Printed in the United States of America
07 06 05 04 03 02 01 00 99 98   1 2 3 4 5

ISBN: 0-226-87034-0 (cloth)

Library of Congress Cataloging-in-Publication Data

Black holes and relativistic stars / edited by Robert M. Wald.
    p. cm
  Includes bibliographical references.
  ISBN 0-226-87034-0 (cloth : alk. paper)
    1. Black holes (Astronomy) 2. Chandrasekhar, S.
(Subrahmanyan), 1910–95. I. Wald, Robert M.
QB843.B55B585 1988
523.8'8756dc21   97-43102
                                                             CIP

♾ The paper used in this publication meets the minimum
requirements of the American National Standard for Information
Sciences Permanence of Paper for Printed Library Materials,
ANSI Z39.48-1992.

Contents

Contributors — vii

Preface — ix

Part I

1. Gravitational Waves, Stars and Black Holes — 3
   Valeria Ferrari

2. Rotating Relativistic Stars — 23
   John L. Friedman

3. Probing Black Holes and Relativistic Stars with Gravitational Waves — 41
   Kip S. Thorne

4. Astrophysical Evidence for Black Holes — 79
   Martin J. Rees

5. The Question of Cosmic Censorship — 103
   Roger Penrose

6. Black Hole Collisions, Toroidal Black Holes, and Numerical Relativity — 123
   Saul A. Teukolsky

7. The Internal Structure of Black Holes — 137
   Werner Israel

Part II

8 Black Holes and Thermodynamics 155
   Robert M. Wald

9 The Statistical Mechanics of Black Hole Thermodynamics 177
   Rafael D. Sorkin

10 Generalized Quantum Theory in Evaporating Black Hole Spacetimes 195
   James B. Hartle

11 Is Information Lost in Black Holes? 221
   Stephen W. Hawking

12 Quantum States of Black Holes 241
   Gary T. Horowitz

Chandra Remembered

Chandra: A Tribute 269
Kameshwar C. Wali

Our Song 273
Lalitha Chandrasekhar

Contributors

Lalitha Chandrasekhar
5825 S. Dorchester Avenue
Chicago, IL 60637, USA

Valeria Ferrari
Dipartimento di Fisica "G. Marconi"
Università di Roma "La Sapienza" and Sezione INFN ROMA1
p.le A. Moro 5
I-00185 Rome, Italy

John L. Friedman
Department of Physics
University of Wisconsin-Milwaukee
P.O. Box 413
Milwaukee, WI 53201, USA

James B. Hartle
Institute of Theoretical Physics and Department of Physics
University of California
Santa Barbara, CA 93106-4030, USA

Stephen W. Hawking
Department of Applied Mathematics and Theoretical Physics
University of Cambridge
Silver Street
Cambridge CB3 9EW, UK

Gary T. Horowitz
Department of Physics
University of California
Santa Barbara, CA 93106, USA

Werner Israel
Department of Physics and Astronomy
University of Victoria
Victoria, BC, V8W 3P6, Canada

Roger Penrose
  Mathematical Institute
  24–29 St Giles'
  Oxford OX1 3LB, UK
  and
  Center for Gravitational Physics and Geometry
  Pennsylvania State University
  University Park, PA 16801, USA

Martin J. Rees
  Institute of Astronomy
  Madingley Road
  Cambridge CB3 0HA, UK

Rafael D. Sorkin
  Department of Physics
  Syracuse University
  Syracuse, NY 13244-1130, USA

Saul A. Teukolsky
  Departments of Physics and Astronomy
  Cornell University
  Ithaca, NY 14853, USA

Kip S. Thorne
  Department of Physics
  California Institute of Technology
  Pasadena, CA 91125, USA

Robert M. Wald
  Enrico Fermi Institute and Department of Physics
  University of Chicago
  5640 S. Ellis Avenue
  Chicago, IL 60637-1433, USA

Kameshwar C. Wali
  Department of Physics
  Syracuse University
  Syracuse, NY 13244-1130, USA

Preface

A Symposium on Black Holes and Relativistic Stars was held on the University of Chicago campus on the weekend of December 14 and 15, 1996. The symposium was dedicated to the memory of Subrahmanyan Chandrasekhar, who died in August 1995. Chandra had devoted most of the last thirty years of his extraordinary scientific career to research on the theory of black holes and relativistic stars. This volume, which is based upon this symposium, also is dedicated to his memory.

In keeping with wishes that Chandra had expressed to me before his death, the symposium was not a memorial function for him, but rather was a working scientific symposium on recent developments in the theory of black holes and relativistic stars. The inevitability of gravitational collapse for sufficiently massive bodies was demonstrated convincingly by Chandra more than sixty-five years ago, but until the discovery of quasars and pulsars in the 1960s, black holes and neutron stars seem generally to have been viewed as exotica which had no place in serious analyses of physical or astrophysical phenomena. This attitude changed dramatically during the decade from the mid-1960s to the mid-1970s—although some vestiges of the former attitude can still be found today!—mainly as a result both of significant new observational evidence for black holes and neutron stars and of major breakthroughs in the development of the theory of black holes. Today, it is generally recognized that black holes and neutron stars are the key components of many astrophysical systems, and that black holes are central to our understanding of fundamental quantum gravitational phenomena. A considerable amount of progress has been made during the past decade on both the theory and observation of black holes and relativistic stars. Most of this progress is summarized in this volume.

The content of this volume corresponds very closely to the program of the symposium.[1] I have divided this volume into two parts. The first part primarily concerns black holes and relativistic stars as astrophysical objects

---
[1] Edward Witten spoke on "Quantum and Stringy Geometry" at the symposium but did not provide a written version of his talk for this volume. For scheduling reasons Gary Horowitz' talk on "Quantum States of Black Holes" was presented at the Texas Symposium on Relativistic Astrophysics, which was held in Chicago during the following week. The written version of Horowitz' talk is included in this volume.

and/or as objects whose structure and properties can be understood within the framework of classical general relativity. The second part concerns black holes as fundamental objects in quantum gravitational physics, with the main focus being on the thermodynamic properties of black holes and the possibility of the occurrence of information loss in the process of black hole formation and evaporation.

Part I begins with two articles, by Valeria Ferrari and John Friedman, which treat aspects of the theory of relativistic stars. Both concern research with which Chandra was intimately involved. Ferrari reviews her work with Chandra on the reformulation of the theory of nonradial oscillations of stars as a problem in the scattering of gravitational waves. Friedman gives a comprehensive review of recent work on the structure and stability of rapidly rotating relativistic stars. The analysis of the gravitational wave induced, nonaxisymmetric instabilities of rotating stars was pioneered by Chandra more than twenty-five years ago.

Kip Thorne's article provides a comprehensive summary of the opportunities for using gravitational radiation to probe the physics of black holes and neutron stars. The construction of highly sensitive laser interferometers in the LIGO and VIRGO projects is presently underway. With these interferometers, it should be possible to directly detect the gravitational radiation from inspiraling neutron star and/or black hole binaries and, possibly, from other strong field phenomena. High sensitivity to very low frequency gravitational radiation will be achieved if the spaceborne interferometers planned for the LISA project are deployed.

Martin Rees' article reviews the current status of observational evidence for black holes. The evidence for stellar mass black holes in binary X-ray systems continues to grow, but the most dramatic new developments of the past few years concern greatly strengthened evidence for massive black holes at the centers of nearby galaxies, including our own.

Without question, one of the key outstanding issues in classical general relativity is whether the "cosmic censor hypothesis" is valid: Does complete gravitational collapse always produce a black hole, or can a "naked singularity" sometimes result? Roger Penrose's article reviews the various formulations of cosmic censorship and the progress—or lack thereof—that has been made toward a proof (or disproof) of their validity.

Saul Teukolsky's article describes some of the ongoing research in numerical relativity, which is aimed primarily at analyzing the collisions of black holes. Some insights into the structure of the event horizon in black hole formation and in the collision of black holes has already been obtained. It is hoped that in the near future, numerical relativity will aid us in making breakthroughs in our understanding of strong field, dynamical phenomena involving black holes.

The interior structure of black holes is the subject of Werner Israel's article. It has long been argued that the "inner horizon" of the Reissner-

Nordström and Kerr black hole solutions must be unstable to small perturbations and, thus, should be singular in a realistic gravitational collapse. However, the precise character of the resulting singularity within these black holes has not been investigated in detail until relatively recently. Israel summarizes the current status of research in this area.

Part II begins with two articles on the thermodynamic properties of black holes. My own contribution reviews the derivation of the laws of black hole thermodynamics—as well as the laws of ordinary thermodynamics—emphasizing the universal character of these laws. Rafael Sorkin's article focuses on the issues of the "microscopic origin" of black hole entropy and on the validity of the generalized second law of thermodynamics, i.e., the nondecrease of the sum of black hole entropy plus the entropy of matter outside of black holes.

The articles by James Hartle and Stephen Hawking both primarily address aspects of the issue of whether information is lost in the process of black hole formation and evaporation, i.e., whether an initially pure quantum state can evolve to a mixed state. Hartle analyzes this issue from the viewpoint of generalized quantum mechanics and argues that no violation of the fundamental principles of quantum mechanics would occur if pure states evolve to mixed states in this process. Hawking reviews the arguments in favor of information loss and describes some of his recent research on information loss effects produced by virtual black hole pairs.

Part II concludes with an article by Gary Horowitz reviewing recent work on black holes in string theory. There currently is a great deal of excitement in this area as a result of the fact that the ordinary entropy of certain string states in the weak coupling limit of the theory has been calculated and found to agree with the Bekenstein-Hawking entropy of certain corresponding black hole solutions occurring in the strong coupling, low energy limit of the theory. This work has opened up many new avenues of exploration on the quantum properties of black holes.

I have included at the end of this volume the written versions of the talks given at the symposium banquet, held on the evening of December 14, 1996. The first, by Kameshwar Wali, provides some personal reminiscences of Chandra. The second, by Mrs. Lalitha Chandrasekhar, contains many insights concerning their life together, Chandra's relationship with Eddington, and the "cycles" of Chandra's research endeavors.

The symposium was made possible by financial support from the University of Chicago through the Hudnall endowment and other university funds, and by a grant from the National Science Foundation. Many people played key roles in making the symposium a success. Judith Spurgin and Houston Patterson of the University of Chicago Office of Special Events worked tirelessly to make all of the necessary arrangements. Many staff members and graduate students in the Department of Physics helped out with various important tasks during the symposium to ensure that everything ran smoothly. The Organizing Committee provided valuable advice on a number of critical issues. I particularly wish to thank Mrs. Lalitha Chandrasekhar both for her strong support of the symposium and for her sound advice on many matters that arose in the course of its organization.

Ted Jacobson and Bernard Schutz served as (originally anonymous) reviewers of this volume for the University of Chicago Press. Their comments and criticisms were extremely valuable and helped improve the volume in many ways. Finally, I am greatly indebted to James Geddes, Daniel Holz, and Ted Quinn for converting the individual contributions to this volume into a common format and for their assistance with many other aspects of producing this volume.

Robert Wald
August 27, 1997

Part I

1

Gravitational Waves, Stars and Black Holes

Valeria Ferrari

Abstract

The theory of nonradial oscillations of stars can be formulated, in close analogy with the theory of perturbations of black holes, as a problem of resonant scattering of gravitational waves by the potential barrier generated by the spacetime curvature, and it is governed by an energy conservation law.

This approach discloses new relativistic effects, such as the existence of slowly damped axial modes in ultra-compact stars and the coupling of the polar and axial perturbations in slowly rotating stars, induced by the Lense-Thirring effect.

I had the privilege of collaborating with Professor Chandrasekhar for twelve years during which we explored the general theory of relativity and developed a new formulation of the theory of stellar perturbations, the startling complexity and richness of which I will try to describe in this article.

In order to understand the basic ideas underlying our approach, we need to frame the problem from a historical perspective and start with a description of some major results of the theory of perturbations of a Schwarzschild black hole, which is beautifully illustrated in Chandra's book *The Mathematical Theory of Black Holes* [1]. In 1957 T. Regge and J. A. Wheeler [2] derived the equations governing the perturbations of a static, spherically symmetric black hole. The separation of variables was accomplished by expanding the perturbed metric tensor in tensorial spherical harmonics, and since these harmonics behave differently under the angular transformation $\theta \to \pi - \theta$, $\varphi \to \pi + \varphi$, the separated equations split in two sets: the *polar* or *even*, belonging to the parity $(-1)^\ell$, and the *axial* or *odd*, belonging to the parity $(-1)^{(\ell+1)}$. Regge and Wheeler reduced the equations describing the *axial* perturbations to a single Schroedinger-like equation

$$\frac{d^2 Z_\ell^-}{dr_*^2} + \left[\omega^2 - V_\ell^-(r)\right] Z_\ell^- = 0, \qquad (1.1)$$

where
$$V_\ell^-(r) = \frac{1}{r^3}\left(1 - \frac{2M}{r}\right)[\ell(\ell+1)r - 6M].$$

Here, $r_* = r + 2M\log(r/2M - 1)$, M is the black hole mass, ω is the frequency, and the perturbed functions have been Fourier-expanded. The theory of perturbations of black holes was born.

Due to the analytical complexity of the polar equations, only much later, in 1970, was F. Zerilli [3] able to derive for the *polar* perturbations a single Schroedinger-like equation, but with a different potential barrier:

$$\frac{d^2 Z_\ell^+}{dr_*^2} + [\omega^2 - V_\ell^+(r)]Z_\ell^+ = 0, \qquad (1.2)$$

where

$$V_\ell^+(r) = \frac{2(r-2M)}{r^4(nr+3M)^2}[n^2(n+1)r^3 + 3Mn^2r^2 + 9M^2nr + 9M^3],$$
$$n = \frac{1}{2}(\ell+1)(\ell-2).$$

Equations (1.1) and (1.2) show that the curvature generated by a pointlike mass appears in the perturbed equations as a potential barrier which extends throughout spacetime. Consequently, the response of a black hole to a generic perturbation can be studied by investigating the manner in which a gravitational wave incident on the black hole is transmitted, absorbed, and reflected by this barrier, a phenomenon with which we are familiar in elementary quantum theory.

1.1 The quasi-normal modes of a black hole

In 1970 Vishveshwara [4] pointed out that the equations governing the perturbations of a Schwarzschild black hole should allow complex frequency solutions behaving at radial infinity as pure outgoing waves. W. H. Press [5] confirmed this idea by numerically integrating the equations and showing that an arbitrary initial perturbation ends in a ringing tail, which indicates that black holes possess some proper modes of vibration.

Since the oscillations must be damped by the emission of gravitational waves, these modes were called *quasi-normal modes*, and they were defined to be solutions of the perturbed equations belonging to complex eigenfrequencies $\omega = \omega_0 + i\omega_i$ and satisfying the boundary conditions of a pure outgoing wave at infinity and of a pure ingoing wave at the horizon. The first condition identifies physically acceptable modes, i.e. those that damp the star (provided $\omega_i > 0$). The latter is the requirement that nothing can escape from the horizon. It should be noted that in scattering theory these boundary conditions associated to a Schroedinger equation with a

Chapter 1. Gravitational Waves, Stars and Black Holes

	$M\omega + iM\omega_i$		$M\omega + iM\omega_i$
$\ell = 2$	0.3737 + i0.0890	$\ell = 3$	0.5994 + i0.0927
	0.3467 + i0.2739		0.5826 + i0.2813
	0.3011 + i0.4783		0.5517 + i0.4791
	0.2515 + i0.7051		0.5120 + i0.6903

Table 1.1: Complex characteristic frequencies of the quasi-normal modes of a Schwarzschild black hole.

one-dimensional potential barrier identify the singularities of the scattering amplitude.

In 1975 S. Chandrasekhar and S. Detweiler [6] computed the complex eigenfrequencies of the quasi-normal modes of a Schwarzschild black hole. The first few values for $\ell = 2$ and $\ell = 3$ are given in table 1.1.

The real part of the frequency is inversely proportional to the mass, while the damping time is proportional to it. If the black hole mass is $M = nM_\odot$, the oscillation frequency and the damping time of the modes can be computed by the following formulae:

$$\nu_0 = \frac{c(M\omega)}{2\pi n M_\odot} = \frac{32.26}{n}(M\omega)\,\text{kHz}, \tag{1.3}$$

$$\tau = \frac{nM_\odot}{(M\omega)c} = \frac{n \cdot 0.4937 \times 10^{-5}}{(M\omega)}\,\text{s}. \tag{1.4}$$

For example, the lowest $\ell = 2$ quasi-normal modes of a black hole of one solar mass and of a supermassive black hole of $10^6 M_\odot$ belong, respectively, to the following frequencies:

$$\begin{aligned} M &= 1M_\odot, & \nu_0 &= 12.06\,\text{kHz}, & \tau &= 5.55 \times 10^{-5}\,\text{s}, \\ M &= 10^6 M_\odot, & \nu_0 &= 1.21 \times 10^{-2}\,\text{Hz}, & \tau &= 55.5\,\text{s}. \end{aligned} \tag{1.5}$$

The frequencies of oscillation of a black hole depend exclusively on the parameters that identify the spacetime geometry: the mass, and the angular momentum or the charge if the black hole is rotating or charged.

In [6] S. Chandrasekhar and S. Detweiler also showed that the transmission and the reflection coefficients associated respectively to the polar and to the axial potential barriers are equal. This equality can be explained in terms of a transformation theory which clarifies the relations that exist between potential barriers admitting the same reflection and absorption coefficients (this theory is extensively illustrated in [1]).[1] However, the physical reason why this happens is still unclear:

---

[1] The equality of the transmission and reflection coefficients can also be justified by the following considerations. The perturbations of a Schwarzschild black hole can be

In spite of $V^{(+)}$ and $V^{(-)}$ appearing so very different, they are *iso-spectral* in the sense that the reflection and absorption coefficient for incident polar and axial gravitational waves are identically the same for all frequencies. In tracing the origin of this identity, one is led to a "transformation theory" whose significance remains elusive. [8]

Numerical integration of the wave equations (1.1) and (1.2) with different sources (see [9] for an extensive bibliography) has shown that the gravitational signal emitted as a consequence of a generic perturbation will, during the last stages, decay as a superposition of the quasi-normal modes. In addition, a newborn black hole generated either by the gravitational collapse of a massive star or by the coalescence of two compact objects will oscillate and emit gravitational waves until its residual mechanical energy is radiated away, and again the dominant contribution is expected to be due to the quasi-normal modes. Since the axial and polar perturbations are isospectral, the gravitational radiation emitted in these processes will carry a definite signature on the nature of the emitting source; in fact, as we shall later discuss, the axial and polar perturbations of a star are *not* isospectral [10].

1.2 A conservation law for the scattering of gravitational waves by a black hole

One of the major problems in general relativity is that an energy conservation law governing the scattering of gravitational waves by black holes does not exist in the framework of the exact nonlinear theory. However, such a law can be derived in perturbation theory for Schwarzschild, Kerr and Reissner-Nordström black holes. We shall now derive the conservation law for a Schwarzschild black hole, by following the procedure adopted in [1].

Due to the short-range character of the potential barriers of eqs. (1.1) and (1.2), the asymptotic behaviour of the solution Z at $r_* = \pm\infty$ is, in general, a superposition of outgoing and ingoing waves:

$$Z_{\text{out}} \sim e^{-i\omega r_*} \quad \text{and} \quad Z_{\text{in}} \sim e^{+i\omega r_*}. \tag{1.6}$$

Consider two solutions of the wave equations, say Z_1 and Z_2, satisfying

---

described in terms of the Bardeen-Press equation [7] written for the Weyl scalars Ψ_0 and Ψ_4, which represent the ingoing and outgoing radiative parts of the gravitational field. The Bardeen-Press equation admits solutions which satisfy the boundary conditions of the quasi-normal modes, and since both Ψ_0 and Ψ_4 can be expressed as combinations of the Regge-Wheeler and of the Zerilli functions and their first derivatives [1], the axial and the polar perturbations must be isospectral.

Chapter 1. Gravitational Waves, Stars and Black Holes

respectively the following boundary conditions:

$$
\begin{aligned}
r_* &\to +\infty, & Z_1 &\to e^{-i\omega r_*}, & &\text{pure outgoing wave,} \\
r_* &\to -\infty, & Z_2 &\to e^{+i\omega r_*}, & &\text{pure ingoing wave.}
\end{aligned}
\quad (1.7)
$$

The pairs (Z_1, Z_1^*) and (Z_2, Z_2^*), where the $*$ indicates complex conjugation and ω is assumed to be real, will be pairs of independent solutions of the wave equation, since their Wronskians are different from zero. In fact, by a direct evaluation, for example, at $\pm\infty$, one finds[2]

$$
\begin{aligned}
r_* &\to +\infty, & [Z_1, Z_1^*]_{r_*} &= -2i\omega, \\
r_* &\to -\infty, & [Z_2, Z_2^*]_{r_*} &= +2i\omega.
\end{aligned}
\quad (1.8)
$$

Therefore, we can write Z_1 as a linear combination of (Z_2, Z_2^*) and vice versa:

$$
\begin{aligned}
Z_1 &= A(\omega) Z_2 + B(\omega) Z_2^*, \\
Z_2 &= C(\omega) Z_1 + D(\omega) Z_1^*.
\end{aligned}
\quad (1.9)
$$

We now divide Z_2 by $D(\omega)$ and define

$$
Z_R = \frac{Z_2}{D(\omega)} = R_1(\omega) Z_1 + Z_1^*, \quad (1.10)
$$

where $R_1(\omega) = C(\omega)/D(\omega)$, and similarly

$$
Z_L = \frac{Z_1}{B(\omega)} = R_2(\omega) Z_2 + Z_2^*, \quad (1.11)
$$

where $R_2(\omega) = A(\omega)/B(\omega)$. Z_R and Z_L have the following asymptotic behaviour:

$$
\begin{aligned}
Z_R &\to \begin{cases} T_1(\omega) e^{+i\omega r_*}, & r_* \to -\infty \\ e^{+i\omega r_*} + R_1(\omega) e^{-i\omega r_*}, & r_* \to +\infty \end{cases} \\
Z_L &\to \begin{cases} e^{-i\omega r_*} + R_2(\omega) e^{+i\omega r_*}, & r_* \to -\infty \\ T_2(\omega) e^{-i\omega r_*}, & r_* \to +\infty \end{cases}
\end{aligned}
\quad (1.12)
$$

where we have set $T_1(\omega) = 1/D(\omega)$ and $T_2(\omega) = 1/B(\omega)$.

Thus, Z_R represents a wave of unitary amplitude incident on the potential barrier from $+\infty$ which gives rise to a reflected wave of amplitude $R_1(\omega)$ and to a transmitted wave of amplitude $T_1(\omega)$. Conversely, Z_L is a unitary wave incident from $-\infty$ which is partially reflected ($R_2(\omega)$) and partially transmitted ($T_2(\omega)$). Furthermore, by computing the Wronskian of the two solutions at $\pm\infty$ it is easy to verify that

$$
\begin{aligned}
[Z_L, Z_R]_{r_*} &= -2i\omega T_2(\omega), & r_* &\to +\infty, \\
[Z_L, Z_R]_{r_*} &= -2i\omega T_1(\omega), & r_* &\to -\infty,
\end{aligned}
\quad (1.13)
$$

---

[2] $[A, B]_r = A_{,r} \cdot B - A \cdot B_{,r}.$

and since the Wronskian is constant, it follows that

$$T_1(\omega) = T_2(\omega) = T(\omega). \tag{1.14}$$

Similarly

$$\begin{aligned}
[Z_L, Z_L^*]_{r_*} &= 2i\omega(|R_2(\omega)|^2 - 1), & r_* &\to -\infty, \\
[Z_L, Z_L^*]_{r_*} &= -2i\omega|T_2(\omega)|^2, & r_* &\to +\infty,
\end{aligned} \tag{1.15}$$

and consequently

$$|R_2|^2 + |T_2|^2 = 1. \tag{1.16}$$

By a similar procedure applied to Z_R we easily find that

$$|R_1|^2 + |T_1|^2 = 1. \tag{1.17}$$

This means that R_1 and R_2 can differ only by a phase factor and that

$$|R|^2 + |T|^2 = 1 \tag{1.18}$$

holds in general. This equation establishes the symmetry and the unitarity of the S-matrix, and it expresses the conservation of energy because it says that if a wave of unitary amplitude is incident on one side of the potential barrier, it splits into a reflected and a transmitted wave such that the sum of the squares of their amplitudes is still one.

The existence of conservation laws for the scattering of gravitational waves by a black hole raised an interesting question: is it possible to establish a similar conservation law for the polar perturbations of a static, spherically symmetric spacetime generated either by an electromagnetic source or by a nonrotating star? That such a law should exist was known on theoretical grounds: A. Ashtekar, J. Friedmann, R. Sorkin, and R. Wald had told us that the existence of a *conserved symplectic current* can in principle be inferred for any field theory derived from a suitably defined Lagrangian action. However, Chandra wanted to derive the conserved current by using a procedure similar to that used for Schwarzschild black holes. In that case, the central point of the derivation was to show that the Wronskian of two independent solutions of the wave equations is a constant. Conversely, the equations for the polar perturbations of a star are a fourth-order linear differential system: what would be the role played by a Wronskian in this context? The solution of the problem required a considerable amount of hard work on the equations, but at the end the result was rewarding: we found that there exists a vector $\vec{E}$ which satisfies the following equation [11]:

$$\frac{\partial}{\partial x^\alpha} E^\alpha = 0, \quad \alpha = (x^2 = r, x^3 = \cos\theta). \tag{1.19}$$

The vanishing of the ordinary divergence implies that, by Gauss's theorem, the flux of $\vec{E}$ across a closed surface surrounding the star is a constant.

Chapter 1. Gravitational Waves, Stars and Black Holes

In order to write explicitly the components of the vector $\vec{E}$ (I shall omit the details of its derivation) we write the metric of a generic static, spherically symmetric spacetime in the following form:

$$ds^2 = e^{2\nu}(dt)^2 - e^{2\mu_2}(dr)^2 - e^{2\mu_3}d\theta^2 - e^{2\psi}d\varphi^2, \qquad (1.20)$$

where the metric functions depend only on r and θ. We shall restrict to the case when this metric represents the spacetime generated by an unperturbed star composed by a perfect fluid, though in [11] we derived a similar conservation law for charged solutions of Einstein's equations as well. The axisymmetric perturbations of the spacetime (1.20) can be described by the line element

$$ds^2 = e^{2\nu}(dt)^2 - e^{2\psi}(d\varphi - q_2 dr - q_3 d\theta - \omega dt)^2 - e^{2\mu_2}(dr)^2 - e^{2\mu_3}(d\theta)^2. \qquad (1.21)$$

It should be noted that the number of unknown functions in eq. (1.21) is seven, one more than needed. However, this extra degree of freedom disappears when the boundary conditions of the problem are fixed. As a consequence of a generic perturbation, the metric functions will experience small changes with respect to their unperturbed values, which we assume to be known:

$$\begin{aligned}
\nu &\longrightarrow \nu + \delta\nu. & \mu_2 &\longrightarrow \mu_2 + \delta\mu_2, \\
\psi &\longrightarrow \psi + \delta\psi, & \mu_3 &\longrightarrow \mu_3 + \delta\mu_3, \\
\omega &\longrightarrow \delta\omega, & q_2 &\longrightarrow \delta q_2, & q_3 &\longrightarrow \delta q_3.
\end{aligned} \qquad (1.22)$$

Since each element of fluid in the interior of the star undergoes an infinitesimal displacement from its equilibrium position, identified by the Lagrangian displacement $\vec{\xi}$, the energy density and the pressure will change by an infinitesimal amount:

$$\epsilon \longrightarrow \epsilon + \delta\epsilon, \quad p \longrightarrow p + \delta p. \qquad (1.23)$$

Under the assumption of axisymmetric perturbation, all perturbed quantities depend on t, r, and θ. If we now write Einstein's equations supplemented by the hydrodynamical equations and the conservation of baryon number, expand all tensors in tensorial spherical harmonics, and Fourier-expand the time dependent quantities, we find that, as for black holes, the equations decouple into two sets, the *polar* and the *axial*, but with a major difference: the *polar* perturbations involve the same metric variables $(\delta\nu, \delta\mu_2, \delta\psi, \delta\mu_3)$ as for black holes, but now they are coupled to the thermodynamical variables

$$\left\{ \begin{array}{c} \delta\nu \\ \delta\mu_2 \\ \delta\psi \\ \delta\mu_3 \end{array} \right\} \longrightarrow \left\{ \begin{array}{c} \delta\epsilon \\ \delta p \\ \xi_r \\ \xi_\theta \end{array} \right\}. \qquad (1.24)$$

Conversely, the *axial* perturbations $[\delta\omega, \delta q_2, \delta q_3]$ do not induce motion in the fluid except for a stationary rotation. However, we shall see that the fluid plays a role, though different from that played in the polar case. In terms of the perturbed metric and fluid variables the E_2-component of the polar vector $\vec{E}$ is

$$\begin{aligned}E_2 &= r^2 e^{\nu-\mu_2}\sin\theta\big\{[\delta\mu_3,\delta\mu_3^*]_2 + [\delta\psi,\delta\psi^*]_2 \\ &\quad - \big(\delta\nu_{,2}\delta(\psi+\mu_3)^* - c.c\big) + \big(\delta\mu_2\delta(\psi+\mu_3)^*_{,2} - c.c.\big) \\ &\quad + \big[2((\epsilon+p)\delta(\psi+\mu_3-\mu_2)^* - \delta p)e^{\nu+\mu_2}\xi_2 - c.c\big]\big\},\end{aligned}$$
(1.25)

and the E_3-component can be obtained by interchanging 2 with 3.

Equation (1.25) includes, as expected, Wronskians of the polar functions $[\delta\mu_3, \delta\mu_3^*]_2$ and $[\delta\psi, \delta\psi^*]_2$, and it reduces to the Wronskian of the solutions of the Zerilli equation as indicated in section 1.2, when the source terms ϵ and p are zero. We derived a similar expression for $\vec{E}$ when the source is an electromagnetic field. G. Burnett and R. Wald [12] subsequently showed that in the Einstein-Maxwell case our conservation law can be obtained by constructing a symplectic current associated to the perturbed equations derived from a Lagrangian variational principle.

The conserved current $\vec{E}$ represents the flux of gravitational energy which develops through the star and propagates outside. Indeed, it can also be derived from the second variation of the Einstein pseudo-tensor $t_E^{\mu\nu}$ [13, 14]. The reason for choosing the Einstein pseudo-tensor is that among the infinite number of pseudo-tensors that can be defined for the gravitational field, all differing by a divergenceless term, $t_E^{\mu\nu}$ is the only one the second variation of which retains the divergence-free property, provided only the equations governing the static spacetime and its linear perturbations are satisfied. This property derives from the fact that the Einstein pseudo-tensor is a Noether operator for the gravitational field.

In addition, Rafael Sorkin pointed out that the contribution of the source should be introduced not by adding the second variation of the stress-energy tensor of the source $T^{\mu\nu}$, but through a suitably defined Noether operator, the form of which he derived for the electromagnetic case. This operator does not coincide with $T^{\mu\nu}$, but it gives the same conserved quantities. It should be mentioned that the Noether operator to be added to the Einstein pseudo-tensor when the source is a fluid has been derived only very recently by Vivek Iyer [15].

The existence of a conservation law for a spacetime with a perfect fluid source suggested to Chandra that the nonradial oscillations of stars should be reformulated as a problem in scattering theory [8].

> In general relativity, any distribution of matter (or more generally energy of any sort) induces a curvature of the spacetime—a

Chapter 1. Gravitational Waves, Stars and Black Holes

potential well. Matter implies gravity and gravity implies matter. Therefore, instead of picturing the non-radial oscillations of a star as caused by some unspecified external perturbation, we can picture them as caused by incident gravitational radiation. Viewed in this manner, the reflection and absorption of incident gravitational waves by black holes and the non-radial oscillations of stars, become different aspects of the same basic theory. But how different—as we shall see!

After completing the first paper on the flux integral, Chandra and I started to work on the perturbed equations, and we reduced them to an interesting form [16], fairly different from that obtained by Thorne and his collaborators, who first developed the theory of stellar perturbations in general relativity in 1967 [17].

1.3 The polar equations

If one expands the perturbed metric tensor and the stress-energy tensor of the fluid in tensorial spherical harmonics, under the hypothesis of axisymmetric perturbations the polar metric functions and the thermodynamical variables turn out to have the following angular dependence:

$$
\begin{aligned}
\delta\nu &= N_\ell(r) P_l(\cos\theta) e^{i\omega t}, \\
\delta\mu_2 &= L_\ell(r) P_l(\cos\theta) e^{i\omega t}, \\
\delta\mu_3 &= [T_\ell(r) P_l + V_\ell(r) P_{l,\theta,\theta}] e^{i\omega t}, \\
\delta\psi &= [T_\ell(r) P_l + V_\ell(r) P_{l,\theta} \cot\theta] e^{i\omega t}, \\
\delta p &= \Pi_\ell(r) P_l(\cos\theta) e^{i\omega t}, \\
\delta\epsilon &= E_\ell(r) P_l(\cos\theta) e^{i\omega t}, \\
2(\epsilon+p) e^{\nu+\mu_2} \xi_r(r,\theta) e^{i\omega t} &= U_\ell(r) P_l e^{i\omega t}, \\
2(\epsilon+p) e^{\nu+\mu_3} \xi_\theta(r,\theta) e^{i\omega t} &= W_\ell(r) P_{l,\theta} e^{i\omega t},
\end{aligned}
\quad (1.26)
$$

where $P_l(\cos\theta)$ are the Legendre polynomials. After separating the variables the relevant Einstein's equations become

$$(T_\ell - V_\ell + L_\ell) = -W_\ell, \quad (1.27a)$$

$$\left[\frac{d}{dr} + \left(\frac{1}{r} - \nu_{,r}\right)\right](2T_\ell - kV_\ell) - \frac{2}{r}L_\ell = -U_\ell, \quad (1.27b)$$

$$\frac{1}{2}e^{-2\mu_2}\left[\frac{2}{r}N_{\ell,r} + \left(\frac{1}{r} + \nu_{,r}\right)(2T_\ell - kV_\ell)_{,r} - \frac{2}{r}\left(\frac{1}{r} + 2\nu_{,r}\right)L_\ell\right]$$
$$+ \frac{1}{2}\left[-\frac{1}{r^2}(2nT_\ell + kN_\ell + \omega^2 e^{-2\nu}(2T_\ell - kV_\ell)\right] = \Pi_\ell, \quad (1.27c)$$

$$(T_\ell - V_\ell + N_\ell)_{,r} - \left(\frac{1}{r} - \nu_{,r}\right)N_\ell - \left(\frac{1}{r} + \nu_{,r}\right)L_\ell = 0, \quad (1.27d)$$

$$V_{\ell,r,r} + \left(\frac{2}{r} + \nu_{,r} - \mu_{2,r}\right) V_{\ell,r}$$
$$+ \frac{e^{2\mu_2}}{r^2}(N_\ell + L_\ell) + \omega^2 e^{2\mu_2 - 2\nu} V_\ell = 0, \qquad (1.27e)$$

where $k = l(l+1)$ and $2n = (l-1)(l+2)$. After some reduction, the hydrodynamical equations and the conservation of baryon number provide the following expressions for the fluid variables:[3]

$$\Pi_\ell = -\frac{1}{2}\omega^2 e^{-2\nu} W_\ell - (\epsilon + p) N_\ell,$$
$$E_\ell = Q\Pi_\ell + \frac{e^{-2\mu_2}}{2(\epsilon + p)}(\epsilon_{,r} - Qp_{,r}) U_\ell, \qquad (1.28)$$

and

$$U_\ell = \frac{[(\omega^2 e^{-2\nu} W_\ell)_{,r} + (Q+1)\nu_{,r}(\omega^2 e^{-2\nu} W_\ell) + 2(\epsilon_{,r} - Qp_{,r})N_\ell](\epsilon + p)}{[\omega^2 e^{-2\nu}(\epsilon + p) + e^{-2\mu_2}\nu_{,r}(\epsilon_{,r} - Qp_{,r})]}, \qquad (1.29)$$

where

$$Q = \frac{(\epsilon + p)}{\gamma p}, \qquad \gamma = \frac{(\epsilon + p)}{p}\left(\frac{\partial p}{\partial \epsilon}\right)_{\text{entropy=const}} \qquad (1.30)$$

and γ is the adiabatic exponent.

Outside the star, the source vanishes and the polar equations can be reduced to the Zerilli equation (1.2), with the following identification:

$$Z_\ell^+(r) = \frac{r}{nr + 3M}(3MV_\ell(r) - rL_\ell(r)). \qquad (1.31)$$

A remarkable simplification of eqs. (1.27a–1.27e) is possible. Equations (1.27a), (1.28), and (1.29) show that the fluid variables $[W_\ell, U_\ell, E_\ell, \Pi_\ell]$ can be expressed as combinations of the metric perturbations $[T_\ell, V_\ell, L_\ell, N_\ell]$ and their first derivatives. Therefore, after their direct substitution on the right-hand side of the last four eqs. (1.27b–1.27e) a set of new equations which involves exclusively the perturbations of the metric functions $[T_\ell, V_\ell, L_\ell, N_\ell]$ can be derived. The final set is

$$X_{\ell,r,r} + \left(\frac{2}{r} + \nu_{,r} - \mu_{2,r}\right) X_{\ell,r} + \frac{n}{r^2} e^{2\mu_2}(N_\ell + L_\ell) + \omega^2 e^{2(\mu_2 - \nu)} X_\ell = 0,$$

$$(r^2 G_\ell)_{,r} = n\nu_{,r}(N_\ell - L_\ell) + \frac{n}{r}(e^{2\mu_2} - 1)(N_\ell + L_\ell) + r(\nu_{,r} - \mu_{2,r})X_{\ell,r}$$
$$+ \omega^2 e^{2(\mu_2 - \nu)} r X_\ell,$$

---

[3] We restrict our analysis to adiabatic perturbations of fluid stars.

Chapter 1. Gravitational Waves, Stars and Black Holes

$$
\begin{aligned}
-\nu_{,r} N_{\ell,r} &= -G_\ell + \nu_{,r}[X_{\ell,r} + \nu_{,r}(N_\ell - L_\ell)] \\
&\quad + \frac{1}{r^2}(e^{2\mu_2} - 1)(N_\ell - rX_{\ell,r} - r^2 G_\ell) - e^{2\mu_2}(\epsilon + p)N_\ell \\
&\quad + \frac{1}{2}\omega^2 e^{2(\mu_2-\nu)}\left\{ N_\ell + L_\ell + \frac{r^2}{n}G_\ell \right. \\
&\quad \left. + \frac{1}{n}[rX_{\ell,r} + (2n+1)X_\ell] \right\},
\end{aligned}
\quad (1.32)
$$

$$
L_{\ell,r}(1-D) + L_\ell\left[\left(\frac{2}{r} - \nu_{,r}\right) - \left(\frac{1}{r} + \nu_{,r}\right)D\right] + X_{\ell,r} + X_\ell\left(\frac{1}{r} - \nu_{,r}\right)
$$
$$
+ DN_{\ell,r} + N_\ell\left(D\nu_{,r} - \frac{D}{r} - F\right)
$$
$$
+ \left(\frac{1}{r} + E\nu_{,r}\right)\left[N_\ell - L_\ell + \frac{r^2}{n}G_\ell + \frac{1}{n}(rX_{\ell,r} + X_\ell)\right] = 0,
$$

where

$$
\begin{aligned}
A &= \frac{1}{2}\omega^2 e^{-2\nu}, \\
B &= \frac{e^{-2\mu_2}\nu_{,r}}{2(\epsilon+p)}(\epsilon_{,r} - Qp_{,r}), \\
D &= 1 - \frac{A}{2(A+B)} \\
&= 1 - \frac{\omega^2 e^{-2\nu}(\epsilon+p)}{\omega^2 e^{-2\nu}(\epsilon+p) + e^{-2\mu_2}\nu_{,r}(\epsilon_{,r} - Qp_{,r})}, \\
E &= D(Q-1) - Q, \\
F &= \frac{\epsilon_{,r} - Qp_{,r}}{2(A+B)} = \frac{2[\epsilon_{,r} - Qp_{,r}](\epsilon+p)}{2\omega^2 e^{-2\nu}(\epsilon+p) + e^{-2\mu_2}\nu_{,r}(\epsilon_{,r} - Qp_{,r})},
\end{aligned}
\quad (1.33)
$$

and V_ℓ and T_ℓ have been replaced by X_ℓ and G_ℓ defined as

$$
\begin{aligned}
X_\ell &= nV_\ell, \\
G_\ell &= \nu_{,r}[\frac{n+1}{n}X_\ell - T_\ell]_{,r} + \frac{1}{r^2}(e^{2\mu_2} - 1)[n(N_\ell + T_\ell) + N_\ell] \\
&\quad + \frac{\nu_{,r}}{r}(N_\ell + L_\ell) - e^{2\mu_2}(\epsilon + p)N_\ell \\
&\quad + \frac{1}{2}\omega^2 e^{2(\mu_2-\nu)}[L_\ell - T_\ell + \frac{2n+1}{n}X_\ell].
\end{aligned}
\quad (1.34)
$$

Equations (1.32) describe the perturbations of the gravitational field in the interior of the star, with no reference to the motion of the fluid.

Once these equations have been solved, the fluid variables can be obtained in terms of the metric functions from eqs. (1.27a), (1.28), and (1.29).

This fact is remarkable: it shows that all the information on the dynamical evolution of a perturbed star is encoded in the gravitational field, a result which expresses the physical content of Einstein's theory of gravity. Moreover, it should be stressed that the decoupling of the equations governing the metric perturbations from those governing the hydrodynamical variables is possible in general and requires no assumptions about the equation of state of the fluid. Thus, if one is interested exclusively in the study of the emitted gravitational radiation, one can solve the system (1.32) and disregard the fluid behaviour.

Equations (1.32) have to be integrated for each value of the frequency from $r = 0$, where all functions must be regular, up to the boundary of the star. There, the spacetime becomes vacuum and spherically symmetric, and the perturbed metric functions and their first derivatives must be matched continuously with the functions that describe the polar perturbations of a Schwarzschild black hole (for a detailed discussion of the boundary conditions see [16] and [18]).

It was subsequently shown by J. R. Ipser and R. H. Price [19] that the equations describing the polar gravitational perturbations decoupled from the fluid variables can be reduced to a fourth-order system.

1.4 A Schroedinger equation for the axial perturbations

The equations for the axial perturbations are much simpler than the polar ones. Their radial behaviour is completely described by a function $Z_\ell(r)$, which satisfies the following Schroedinger-like equation:

$$\frac{d^2 Z_\ell}{dr_*^2} + [\omega^2 - V_\ell(r)] Z_\ell = 0, \tag{1.35}$$

where $r_* = \int_0^r e^{-\nu + \mu_2} dr$, and

$$V_\ell(r) = \frac{e^{2\nu}}{r^3} [l(l+1)r + r^3(\epsilon - p) - 6m(r)], \quad \nu_{,r} = -\frac{p_{,r}}{\epsilon + p}. \tag{1.36}$$

Outside the star ϵ and p vanish and eq. (1.36) reduces to the Regge-Wheeler potential barrier (1.1). It should be stressed that the potential depends on how the energy density and the pressure are distributed inside the star in its equilibrium configuration.

Since an axial gravitational wave incident on a star does not induce fluid motion, for a long time these perturbations have been considered trivial. But this is not true if we adopt the scattering approach: the absence of fluid motion simply means that an incident axial wave experiences a potential scattering as it does in the case of a Schwarzschild black hole. There is however an important difference. The Schwarzschild potential vanishes at

Chapter 1. Gravitational Waves, Stars and Black Holes   15

the black hole horizon, and it has a maximum at $r_{\max} \sim 3M$. Conversely, due to the centrifugal contribution $\ell(\ell+1)/r^2$ the potential barrier of a perturbed star tends to infinity at $r = 0$. In addition, for a Schwarzschild black hole the Schroedinger-like equation describes a problem of scattering by a one-dimensional potential barrier, whereas in the case of a star it describes the scattering by a central potential.

Since the axial perturbations are described by a Schroedinger equation, the axial component of the energy flux can be derived from the Wronskians of independent solutions, as in the black hole case. However, due to the different boundary conditions, the evaluation of this flux requires the application of the Regge theory of potential scattering in a central field. This theory can be generalized to be applicable to the polar perturbations as well, and to explicitly compute the energy flux associated to the vector $\vec{\mathbf{E}}$ [20].

1.5 The quasi-normal modes of a star

In our approach the nonradial oscillations of stars are thought to be induced by the incidence of polar or axial gravitational waves on the spacetime curvature generated by the star. In this view, a resonant scattering occurs when the star is in a quasi-stationary state that decays, i.e. when it oscillates in a quasi-normal mode.

The quasi-normal modes are solutions of the axial and polar equations that satisfy the following boundary conditions. As in the black hole case, at radial infinity only pure outgoing waves must prevail, whereas the pure ingoing wave condition at the black hole horizon is replaced by the requirement that all perturbed functions have a regular behaviour at $r = 0$. Furthermore, they must match continuously with the exterior perturbation at the surface of the star. Both the polar and the axial quasi-normal modes satisfy the same boundary conditions, but the underlying scattering problem is very different in the two cases. In fact, since a polar perturbation excites the fluid motion, the amount of radiation which leaks out of the star depends on the exchange of energy between the fluid and the gravitational field, whereas the scattering of axial gravitational waves is a pure scattering by a spherically symmetric, static potential.

In studying the theory of stellar perturbations in the framework of general relativity, one encounters new phenomena that do not have Newtonian counterparts. A first example is the existence of new families of modes of vibration, which are modes of the radiative field. They appear because the spacetime is not simply a medium in which gravitational waves propagate: it has its own dynamics and spectrum, as is clearly shown by the existence of the quasi-normal modes of black holes. Spacetime modes exist for stars also but, because the boundary conditions are different, their spectrum will be much different from that of black holes. One of these new families consists of the highly damped polar and axial w-modes, discovered by K. Kokkotas and B. Schutz [21]. Actually it was later shown that there exist two families of such modes [22], but we shall not go into such detail in the present context. The w-modes are modes of vibration in which the motion of the fluid is barely excited, if at all; it is not excited in the axial case. In an article which appeared in *Physics World* in 1991, Bernard Schutz makes an interesting analogy that vividly illustrates the nature of these modes [23]:

> Consider a violin played in an infinitely large room. The air by itself does not have conventional outgoing-wave modes: any sound waves are coming in from somewhere and going out somewhere else. But put a violin string in the room, and there appears a family of modes with purely outgoing sound waves that exchange a small amount of energy with the string, and die away very fast. These modes are strongly damped, and the weaker the coupling of the string to the air, the faster they damp away, so that in the limit of a vacuum around the string, they go away entirely.

Typical values of the lowest w-mode range between ≈ 8 and $12\,\mathrm{kHz}$ (the frequency of the w-modes increases with the order of the mode), and the corresponding damping times are ≈ 0.02–$0.1\,\mathrm{ms}$.

Chandra and I brought to light a further family of spacetime modes [24]. Unlike the w-modes they are slowly damped, and therefore I shall call them the s-modes. They exist only for the axial perturbations, and their appearance is related to the depth of the potential well inside the star, as the following illustrative example shows. Let us compare the shape of the axial potential barriers generated by homogeneous stars of increasing compactness, i.e. of decreasing ratio R/M. It should be remembered that homogeneous stars can exist only if their radius R exceeds $9R_s/8$, or equivalently, if $R/M > 2.25$. In figure 1.1 it is shown how the potential well inside the star becomes deeper as the value of R/M decreases and the star shrinks. In the exterior the potential coincides with the Regge-Wheeler potential that has a maximum at $r \approx 3M$. When $R/M < 2.6$ the potential well in the interior becomes deep enough to allow the existence of one or more quasi-normal modes. In table 1.2 the characteristic frequencies and damping times of the $\ell = 2$ s-modes of homogeneous stars with $M = 1.35 M_\odot$ and different values of R/M are listed.

It should be stressed that the modes that one finds when the radius of the star approaches the limiting value are not related to the quasi-normal modes of a Schwarzschild black hole, because both the boundary conditions and the underlying scattering process are different. Moreover, the progressive increasing of the damping time for these modes means that they are more effectively trapped by the curvature of the star.

R/M	ν_0 in kHz	τ in s	R/M	ν_0 in kHz	τ in s
2.4	8.6293	1.52×10^{-3}	2.28	4.4333	10.8
-	-	-		6.0168	2.50×10^{-1}
-	-	-		7.5462	1.44×10^{-2}
-	-	-		8.9891	1.83×10^{-3}
2.3	5.6153	0.54	2.26	2.6041	5.38×10^{3}
	7.5566	1.16×10^{-2}		3.5427	1.69×10^{2}
	9.3319	1.02×10^{-3}		4.4802	1.22×10^{1}
	-	-		5.4127	1.37×10^{-1}

Table 1.2: Characteristic frequencies and damping times of the $\ell = 2$ s-modes of homogeneous stars, with $M = 1.35 M_\odot$ and increasing compactness.

The existence of the s-modes was proved by using homogeneous stars as a model, and we have seen that they appear only if R/M is sufficiently close to the limiting value 2.25. It would be interesting to understand whether this constraint on R/M derives from the particular choice of the model we have used, or whether it could be relaxed by the use of a different equation of state.

There is further information that one can derive from the knowledge of the frequencies and damping times of the quasi-normal modes. In Newtonian theory the frequency of the **f**-mode scales with the mean density of the star. In geometric units

$$\omega_f = \sqrt{\frac{2\ell(\ell+1)}{2\ell+1} \left(\frac{M}{R^3}\right)}. \tag{1.37}$$

This relation has been generalized by N. Andersson and K. Kokkotas [25], who have determined both the frequency and the damping time of the **f**-mode for several equations of state proposed in the literature for neutron stars. They find the following relations:

$$\begin{aligned}\omega_f &= 0.39 + 44.45\sqrt{\left(\frac{M}{R^3}\right)}, \\ \tau_f &= 0.1 - \left(\frac{M}{R}\right) + 2.69\left(\frac{M}{R}\right)^2,\end{aligned} \tag{1.38}$$

where M and R are expressed in km, ω_f in kHz, and τ_f in ms. These two relations provide estimates for both M and R, good within 5% if compared with the true values. It should be noted that the frequency of the f-mode ranges in the interval $\approx$ 1–2 kHz, and the damping time is $\approx$ 0.1–0.5 s. A further relation is provided by the damping time of the lowest **w**-mode

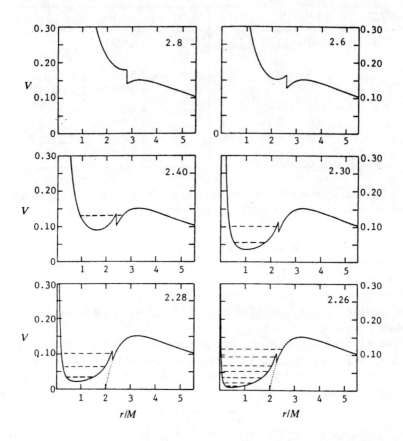

Figure 1.1: The potential barrier of the axial perturbations of homogeneous stars is plotted for different values of the ratio R/M ranging from 2.8 to 2.26.

computed for the same models:

$$\frac{1}{\tau_{w_0}} = 0.104 - 0.063 \left(\frac{M}{R}\right). \tag{1.39}$$

Andersson and Kokkotas have also studied how the axial quasi-normal modes are excited when an initial Gaussian pulse is scattered by the potential barrier of the axial perturbations of homogeneous stars. Their simulation shows that, in principle, the various modes can be excited. However, it is not yet known how the different modes are excited in some realistic situations, for example, during the last stages of the gravitational collapse, when the newborn star wildly oscillates releasing gravitational waves, or when a mass, smaller than the star mass, is scattered or captured by the big one [26, 27].

1.6 Slowly rotating stars

The theory of stellar perturbations developed for static stars can be generalized to the case when the star is rotating so slowly that the distortion from spherical symmetry is quadratic in the angular velocity Ω and may be ignored [28]. The unperturbed configuration is described by the following metric [29, 30]:

$$ds^2 = e^{2\nu}(dt)^2 - e^{2\psi}(d\varphi - \omega dt)^2 - e^{2\mu_2}(dr)^2 - e^{2\mu_3}(d\theta)^2, \qquad (1.40)$$

where ν, ψ, μ_2, μ_3 differ from those of a static star by quantities of order Ω^2, while ω (that is zero in the nonrotating case) is a first-order quantity in Ω. The equations governing ν, ψ, μ_2, μ_3 are given in section 1.3. The equation for ω is

$$\varpi_{,r,r} + \frac{4}{r}\varpi_{,r} - (\mu_2 + \nu)_{,r}\left(\varpi_{,r} + \frac{4}{r}\varpi\right) = 0, \qquad (1.41)$$

where

$$\varpi = \Omega - \omega. \qquad (1.42)$$

In the vacuum outside the star, $\mu_2 + \nu = 0$ and the solution of eq. (1.41) reduces to $\varpi = \Omega - 2Jr^{-3}$, where J is the angular momentum of the star.

In [28] we showed that the axial perturbations of a slowly rotating star couple to the polar perturbations, and vice versa. The way this coupling works for the axial perturbations is illustrated by the following equation:[4]

$$\sum_{l=2}^{\infty}\left\{\frac{d^2 Z_l^1}{dr_*^2} + \omega^2 Z_l^1 - \frac{e^{2\nu}}{r^3}[l(l+1)r + r^3(\epsilon - p) - 6m(r)]Z_l^1\right\} C_{l+2}^{-3/2}(\mu)$$

$$= re^{2\nu - 2\mu_2}(1 - \mu^2)^2 \sum_{l=2}^{\infty} S_l^0(r, \mu), \qquad (1.43)$$

where

$$S_l^0 = \varpi_{,r}[(2W_l^0 + N_l^0 + 5L_l^0 + 2nV_l^0)P_{l,\mu} + 2\mu V_l^0 P_{l,\mu,\mu}]$$
$$+ 2\varpi W_l^0(Q - 1)\nu_{,r}P_{l,\mu}, \qquad (1.44)$$

and Q has been defined in eq. (1.30). $\mu = \cos\theta$, and $C_{l+2}^{-3/2}(\mu)$ and $P_l(\mu)$ are respectively the Gegenbauer and the Legendre polynomials.

Equation (1.43) holds from the center of the star up to radial infinity, provided outside the star ϵ, p, and W are set to zero. As described in previous sections, if the star does not rotate the axial and the polar perturbations are described by two distinct sets of equations: eqs. (1.32) for

---
[4]The equations describing the coupling of the polar with the axial perturbations were subsequently determined by Y. Kojima [31].

the polar variables N_l^0, $L_l^0 V_l^0$, W_l^0, etc., and eqs. (1.35) for the axial function Z_ℓ^0. If the rotation is switched on ($\varpi \neq 0$), the axial function of first order in Ω, Z_ℓ^1, couples as indicated in eq. (1.43) with the polar functions $(W_l^0, N_l^0, L_l^0 V_l^0)$ of zero order in Ω, i.e. evaluated in the case of no rotation.

It should be noted that the coupling function is the quantity ϖ which is responsible for the dragging of inertial frames in the Lense-Thirring effect. Thus, rotating stars exert a dragging not only of the bodies but also of the waves, and consequently an incoming polar gravitational wave can convert, through the fluid oscillations it excites, some of its energy into outgoing axial waves.

I would like to stress that this phenomenon is a purely relativistic effect with no counterpart in Newtonian theory.

Equation (1.43) is not yet separated. When the angular dependence is removed, one finds that the axial and the polar perturbations couple according to the following rules:

- *The Laporte rule*: the polar modes belonging to *even* ℓ can couple only with the axial modes belonging to *odd* ℓ, and conversely.

- *The selection rule*: $l = m + 1$ or $l = m - 1$.

- *The propensity rule* [32]: the transition $l \to l+1$ is strongly favoured over the transition $l \to l-1$. This derives from the manner in which the behaviour of the axial function is affected by the polar source near the origin.

As a consequence of this coupling, new families of modes are likely to emerge. For example, in [28] we studied the axial perturbations of a slowly rotating polytropic star with polytropic index $n = 1.5$, and we showed that if one scatters $\ell = 2$ polar gravitational waves on the potential barrier of eq. (1.43), for some value of the frequency of the incident wave the $m = 3$ axial perturbation induced by the coupling behaves as a pure outgoing wave at radial infinity. These "induced" axial resonances are characterized by damping times considerably longer than those of the polar modes of order zero in Ω (up to a hundred times longer).

1.7 Concluding remarks

The existence of an energy conservation law governing the nonradial oscillations of a spherical star, which was derived in analogy with the conservation law governing the scattering of gravitational waves by a Schwarzschild black hole, provides an additional constraint on the theory and allows us to recast the problem of stellar perturbations as a problem in scattering theory. The scattering approach proves extremely powerful in casting light on some aspects of the theory that were obscure in previous formulations.

Chapter 1. Gravitational Waves, Stars and Black Holes   21

The existence of the slowly damped axial modes in ultra-compact stars, the coupling between the polar and axial perturbations in slowly rotating stars, and the resonances induced by this coupling naturally emerge in this framework, though they could also have been discovered by other approaches.

The scattering approach is also applicable when the star is Newtonian, i.e. when its equilibrium configuration is built in the Newtonian framework and the curvature it generates is very shallow. Indeed, we showed that the frequencies of oscillation of a Newtonian star can be determined by integrating the polar equations in the limit of small curvature, under the condition that no radiation emerges, as in the case of the dipole oscillations [33].

At the end of this article I would like to add to the scientific illustration of my work with Chandra some personal recollection of our collaboration. It developed over twelve years, and it was certainly based on reciprocal respect, esteem, trust, and common scientific interests. But the real engine was Chandra's genuine enthusiasm for science which he was able to communicate to me by making me feel that, no matter how difficult a problem was, together we could solve it. I am grateful to Chandra for his precious gift of sharing with me his patrimony of knowledge, experience, and craftsmanship—fruits of a life entirely dedicated to science.

References

[1] S. Chandrasekhar, *The Mathematical Theory of Black Holes*, Oxford: Clarendon Press (1984).

[2] T. Regge, J. A. Wheeler, *Phys. Rev.* **108**, 1063 (1957).

[3] F. J. Zerilli, *Phys. Rev.* **D2**, 2141 (1970).

[4] C. V. Vishveshwara, *Phys. Rev.* **D1**, 2870 (1970).

[5] W. H. Press, *Ap. J.* **170**, L105 (1971).

[6] S. Chandrasekhar, S. L. Detweiler, *Proc. R. Soc. Lond.* **A344**, 441 (1975).

[7] J. M. Bardeen, W. H. Press, *J. Math. Phys.* **14**, 7 (1973).

[8] S. Chandrasekhar, *The Series Paintings of Claude Monet and the Landscape of General Relativity*, Dedication address, Inter-University Centre for Astronomy and Astrophysics, 28 December 1992.

[9] V. Ferrari, in *Proceedings of the 7th Marcel Grossmann Meeting*, ed. R. Ruffini & M. Kaiser, Singapore: World Scientific (1995).

[10] V. Ferrari, *Phys. Lett.* **A171**, 271 (1992).

[11] S. Chandrasekhar, V. Ferrari, *Proc. R. Soc. Lond.* **A428**, 325 (1990).

[12] G. Burnett, R. Wald, *Proc. R. Soc. Lond.* **A430**, 57 (1990).

[13] R. Sorkin, *Proc. R. Soc. Lond.* **A435**, 635 (1991).

[14] S. Chandrasekhar, V. Ferrari, *Proc. R. Soc. Lond.* **A435**, 645 (1991).
[15] V. Iyer, *Phys. Rev.* **D55**, 3411 (1997).
[16] S. Chandrasekhar, V. Ferrari, *Proc. R. Soc. Lond.* **A432**, 247 (1991).
[17] K. S. Thorne, A. Campolattaro, *Ap. J.* **149**, 591 (1967).
[18] V. Ferrari, *Phil. Trans. R. Soc. Lond.* **A340**, 423 (1992).
[19] J. R. Ipser, R. H. Price, *Phys. Rev.* **D43**, 1768 (1991).
[20] S. Chandrasekhar, V. Ferrari, *Proc. R. Soc. Lond.* **A437**, 133 (1992).
[21] K. D. Kokkotas, B. F. Schutz, *Proc. Mon. Not. R. Astron. Soc.* **255**, 119 (1992).
[22] M. Leins, H. P. Noellert, M. H. Soffel, *Phys. Rev.* **D48**, 3467 (1993).
[23] B. F. Schutz *Physics World* **4** (8), 24 (1991).
[24] S. Chandrasekhar, V. Ferrari, *Proc. R. Soc. Lond.* **A434**, 449 (1991).
[25] N. Andersson, K. D. Kokkotas, *Phys. Rev. Letters* **77**, 4134 (1996).
[26] V. Ferrari, L. Gualtieri, *Intern. J. of Mod. Phys.* **D36**, 323 (1997).
[27] A. Borrelli, V. Ferrari, L. Gualtieri, in preparation (1996)
[28] S. Chandrasekhar, V. Ferrari, *Proc. R. Soc. Lond.* **A433**, 423 (1991).
[29] J. B. Hartle *Ap. J.* **150**, 1005 (1967).
[30] S. Chandrasekhar, J. C. Miller, *Mon. Not. R. Astron. Soc.* **167**, 63 (1974).
[31] Y. Kojima, *Phys. Rev.* **D46**, 4289 (1992).
[32] U. Fano, *Phys. Rev.* **A32**, 617 (1985).
[33] S. Chandrasekhar, V. Ferrari, *Proc. R. Soc. Lond.* **A450**, 463 (1995).

2

Rotating Relativistic Stars

John L. Friedman

Abstract

A survey is presented of some of the recent work on the structure and stability of rapidly rotating relativistic stars. The discussion includes limits set by causality on the mass and rotation of relativistic stars, recent computations of relativistic instability points, and an instability of axial modes.

2.1 Introduction

Chandrasekhar's work on white dwarfs and relativistic stars underlies a substantial part of relativistic astrophysics. This article broadly summarizes work on the structure and stability of rapidly rotating stars, emphasizing the radial instability and the nonaxisymmetric gravitational-wave-driven instability that Chandra studied and associated limits on the mass and rotation of neutron stars. Early reviews of rotating stars are given by Bardeen (1973) and Thorne (1971), a more recent one by Friedman and Ipser (1993), and a recent review of stability theory by Friedman (1996).

Section 2.2 outlines the equations governing the equilibrium of a rotating star modeled as a perfect fluid. Section 2.3 is a brief guide to work on numerical models; and section 2.4 outlines upper limits set by causality on the mass and rotation of relativistic stars. Finally, relativistic stability is discussed in section 2.5. Instability points along sequences of rapidly rotating relativistic stars have recently been computed for the first time; this work and new results on instability of axial modes (r-modes) are mentioned in this section.

2.2 Spacetime of a rotating star

The metric $g_{\alpha\beta}$ of a stationary axisymmetric rotating fluid has two commuting Killing vectors, ϕ^α, t^α, generating rotations and asymptotic time translations. The fluid's four-velocity has the form

$$u^\alpha = N(t^\alpha + \Omega\phi^\alpha),\tag{2.1}$$

implying the existence of a family of two-surfaces perpendicular to the Killing vectors. The metric can then be expressed in terms of its components in the Killing subspace,

$$e^{2\psi} = \phi^\alpha\phi_\alpha, \quad \omega = -\frac{t^\alpha\phi_\alpha}{\phi^\beta\phi_\beta}, \quad e^{2\nu} = -t^\alpha t_\alpha + \frac{(t^\alpha\phi_\alpha)^2}{\phi^\beta\phi_\beta},\tag{2.2}$$

and a conformal factor $e^{2\mu}$ describing the geometry of the orthogonal two-space:

$$ds^2 = -e^{2\nu}dt^2 + e^{2\psi}(d\phi - \omega dt)^2 + e^{2\mu}(dr^2 + r^2 d\theta^2).\tag{2.3}$$

The matter and geometry are determined by the equation of hydrostatic equilibrium,

$$\frac{\nabla_\alpha p}{\epsilon + p} = -u^\beta \nabla_\beta u^\alpha,\tag{2.4}$$

used in the integral form

$$\exp\int^p \frac{dp}{\epsilon + p} = \beta^{1/2} u^t,\tag{2.5}$$

with $\beta^{1/2}$ the constant injection energy per unit baryon mass; by four independent components of the field equations,

$$G^{\alpha\beta} = 8\pi T^{\alpha\beta},\tag{2.6}$$

for the four potentials ν, ω, ψ, and μ; and by a one-parameter equation of state,

$$\epsilon = \epsilon(p).\tag{2.7}$$

The equilibrium configuration of a neutron star is accurately modeled as a perfect fluid, despite a 1 km crust and a superfluid interior that confines vorticity to microscopic tubes. The small error due to anisotropy in the normal part of the crust is estimated in Baym and Pines (1971; see also Shapiro and Teukolsky 1983). The coarse graining required to approximate a neutron star's velocity field as uniform rotation appears to lead to an even smaller error in the star's structure. In a neutron-star interior neutrons and protons are expected to form superfluids, and in a superfluid, a velocity field is curl-free except in vortex lines. For a uniformly rotating fluid with angular velocity Ω, one has

$$\nabla \times \vec{v} = 2\Omega\hat{z}.\tag{2.8}$$

The star's interior approximates this field by a set of vortices, each of which carries vorticity quantized in units

$$\frac{\pi \hbar}{m_n}. \tag{2.9}$$

The number of vortex lines per unit area is then

$$\frac{2\Omega}{\pi\hbar/m_n} = 1.0 \times 10^3 \, \text{cm}^{-2} \, \Omega[\text{rad/s}]. \tag{2.10}$$

Although the approximation of uniform rotation is consequently invalid on scales shorter than 1 cm, the error in computing the structure of the star on larger scales is negligible. In particular, with T^{ab} approximated by a value, $\langle T^{ab} \rangle$, averaged over several cm, the error in computing the metric is of order

$$\delta g_{ab} \sim \left(\frac{1 \, \text{cm}}{R}\right)^2 \sim 10^{-11}. \tag{2.11}$$

Finally, a one-parameter equation of state is accurate, because within days of formation, neutrino emission cools neutron stars to 10^{10} K = 1 MeV, below the 60 MeV Fermi energy of the interior. The equation of state is thus approximately that of zero-temperature matter.

2.3 Models of rapidly rotating stars

2.3.1 Numerical codes

Models of rotating relativistic stars have been computed by several different authors. Early codes were obtained by Bonazzola and Schneider (1974) and Wilson (1972). Butterworth and Ipser (1976) incorporated more precise asymptotic conditions, obtaining an accurate code that was used to construct polytropes and uniform-density configurations, the relativistic analogs of the Newtonian Maclaurin spheroids. The code was modified by Friedman, Ipser, and Parker (1986) to accommodate a set of proposed equations of state (EOSs) for matter above nuclear density, and several hundred models were constructed to find the characteristics of rotating relativistic stars for a wide range of EOS. Additional models, including some based on an EOS for quark matter, were constructed in the wake of the spurious observation of a 0.5 ms pulsar in SN1987 by these authors (Friedman et al. 1989) and by Lattimer et al. (1990), whose code was similarly based on the Butterworth-Ipser algorithm. Neugebauer and Herlt (1984), Wu et al. (1991), Neugebauer and Herold (1992), and Herold et al. (1993) have obtained models of rapidly rotating neutron stars using a code based on a finite-element technique and using a Newton-Raphson method to solve the field equations.

A new generation of accurate codes, accompanying a new generation of computers, began with the work of Komatsu, Eriguchi, and Hachisu (1989a,1989b; hereinafter KEH), who used a somewhat different algorithm: In the Butterworth-Ipser approach, the equations for each potential are discretized to give a matrix acting on the vector of values of that potential, and the matrix is inverted to solve the equation. Komatsu et al. separate off Laplacian operators with constant coefficients which they invert by numerical integration of an analytic Green's function. The initial KEH work was restricted to differentially rotating polytropes, but subsequent papers report results on several proposed neutron-star EOSs (Eriguchi et al. 1994; Hashimoto et al. 1994). Cook et al. report another implementation of the KEH method, which they used first to examine polytropes and then to construct models based on an updated set of candidates for the EOS of neutron-star matter (Cook it et al. 1992, 1994a, 1994b). A similar code was constructed by Stergioulas and compared with results of the original KEH code (Stergioulas and Friedman 1995). And a comparably accurate code by Bonazzola et al. 1993, based on a pseudo-spectral method, has been used to compute a series of models (Salgado et al. 1994a,1994b).

The Stergioulas code, which has been modified by S. Koranda, is now in the public domain and can be obtained, with documentation, from alpha2.csd.uwm.edu/pub/rns/.

A detailed comparison of recent codes is given in Nozawa et al. (1997). For most equations of state, these recent codes agree to one part in 10^3 for all quantities. They are slightly worse for stiff equations of state, with ϵ discontinuous or nearly discontinuous at the surface. The comparison includes tests of virial identities (Gourgoulhon and Bonazzola 1994; Bonazzola and Gourgoulhon 1994) that hold for asymptotically flat spacetimes.

Also worth mentioning are codes by Bocquet et al. (1995) modeling rapidly rotating neutron stars with an axisymmetric poloidal magnetic field; and by Uryu and Eriguchi (1994) using a coordinate system fitted to the boundary of the star, to obtain a collection of stationary, nonaxisymmetric Newtonian models.

It is not yet clear whether even the fastest pulsars are rapidly rotating, in the sense of having angular velocity close to the Kepler frequency Ω_K, the frequency of a satellite in circular orbit at a star's equator. Most neutron stars can be accurately approximated by Hartle's (1967) slow-rotation approximation (see Glendenning, 1997, and Datta, 1988 for reviews), but a proliferation of accurate codes for rapidly rotating stars, including the public-domain code mentioned above, mean that rapidly rotating models can be easily computed.

2.3.2 The two-dimensional family of equilibria stable against collapse

For a given equation of state, the family of equilibrium configurations stable against collapse is a two-dimensional surface. Its boundary is determined by upper and lower limits on rotation and mass: that is, the region is bounded by four curves, $\Omega = 0$, $\Omega = \Omega_K$, $M = M_{\min}$, and $M = M_{\max}$.

$M_{\max}$ is analogous to the Chandrasekhar limit on the mass of a white dwarf. The dwarf of maximum mass is the configuration at the onset of instability to radial oscillations. More generally, a turning-point instability coincides with an extremum of M at fixed J for a one-parameter equation of state (Friedman, Ipser, and Sorkin 1988).[1] Neutron stars above the maximum mass are unstable to collapse. Below the minimum mass they are unstable to explosion (Blinnikov et al. 1984; Colpi et al. 1989, 1991).

The uncertainty in the nuclear EOS leads to sharp differences in possible models of rotating stars. For the softest equations of state, $1.4 M_\odot$ nonrotating models have radii of about 8 km, while models based on the stiffest equations of state (consistent with the fastest observed pulsars) are much less centrally condensed, with nonrotating radii of about 15 km. For the corresponding rotating models, the ratio of the radii is not substantially different, about 12 km for the softest models compared to 20 km for the stiffest. The ratio of the moments of inertia, however, is enhanced by rotation, and this leads to sensitivity of the maximum frequency of rotation on the EOS. A determination of the maximum rate of rotation would thus constrain the equation of state. The two fastest pulsars have periods of rotation within 3% of one another, providing weak evidence for a nearby limit on rotation. If the true limit on rotation *is* near 1.6 ms, the equation of state about nuclear density would be unexpectedly stiff (Lipunov and Postnov 1988; Friedman et al. 1985).

Friedman et al. (1986) noted that among all stable models, the model with maximum mass appeared also to have the maximum Ω, J, and M_0 (baryon mass). As Cook et al. (1992, 1994a, 1994b) have found, there is often a slight difference between the models of maximum M and Ω. Models with maximum values of J and M_0 can also fail to coincide exactly with the maximum Ω model. The models of maximum M, Ω, J, and M_0 do coincide, however, when the *stable* configuration with maximum mass is also the *equilibrium* configuration with maximum mass (Stergioulas and Friedman 1995).

---

[1] These instability points are also extrema of baryon number N at fixed J and of J at fixed N.

2.4 Limits set by causality on the mass and rotation of relativistic stars

Despite the uncertainty in the equation of state of nuclear matter, one can find upper limits on the mass and rotation of relativistic stars that are independent of the EOS. The limit on mass requires a match to known equation of state at low density, and the resulting limit depends on the matching density, ϵ_m, as was emphasized by Hartle and Sabbadini (1977). Above the matching density, the limit is realized by the stiffest equation of state consistent with causality, namely, $\epsilon = p + $ constant, with $v_{\text{sound}} = 1$. The limit was first computed by Friedman and Ipser (1987); a more recent computation by Koranda et al. (1996) is in close agreement, giving a value

$$M_{\text{rot}}^{\max} \simeq 6.1 \left(\frac{2 \times 10^{14}\,\text{g/cm}^3}{\epsilon_m} \right)^{1/2} M_\odot. \qquad (2.12)$$

(The corresponding value for spherical stars, substitutes 4.8 for 6.1, while the more familiar Rhoades-Ruffini, 1974, value arises from matching to a low-density equation of state at a particular value above nuclear density.)

A limit on rotation is set by causality together with the requirement that the EOS allow stars with masses as large as the largest observed neutron-star mass. Glendenning (1992) first estimated this causally limited period for gravitationally bound stars, with the additional assumption that the EOS matched the known EOS below nuclear density. Recent work by Koranda et al. (1996) points out that dropping the low-density match entirely lowers the minimum frequency by only about 3%.

The EOS yielding a minimum period again has a simple physical interpretation; it allows the most centrally condensed star while still supporting a mass $M_{\text{sph}}^{\max}$. If $M_{\text{sph}}^{\max} = 1.442 M_\odot$ (currently the largest accurately measured mass of neutron stars; Taylor and Weisberg 1989; Arzoumanian et al. 1995), the minimum period consistent with causality is 0.28 ms (0.29 ms with a match at nuclear density to a low-density EOS). The minimum period set by causality exactly scales with $M_{\text{sph}}^{\max}$; with a match at nuclear density to a low-density equation of state, the scaling is still nearly precise:

$$P[\text{ms}] > 0.295 + 0.203 \left(M_{\text{sph}}^{\max}/M_\odot - 1.442 \right). \qquad (2.13)$$

2.5 Stability

2.5.1 Axisymmetric stability

From a relativistic standpoint, the axisymmetric instability to collapse can be regarded as a generalization to relativistic gravity of the Chandrasekhar limit on the mass of a white dwarf, because the upper mass limit coincides

Chapter 2. Rotating Relativistic Stars

with the point of instability to collapse. Nonrotating white dwarfs form a one-parameter sequence of increasing density. At the configuration with maximum mass along this sequence, the fundamental radial mode has zero frequency, because the change from one equilibrium configuration to another with the same mass and larger density is a time-independent radial perturbation. At densities above the maximum mass, the star is unstable.

This connection between maximum mass and instability point holds only for stars whose pulsations and equilibrium are governed by the same effective equation of state. The exact stability criterion that Chandra obtained (1964a, 1964b), however, was more general. It locates an instability point where the second order change in a star's mass vanishes for an adiabatic radial perturbation of a star with arbitrary equation of state.

A Newtonian star is unstable to such perturbations when its average adiabatic index γ is less than $4/3$. General relativity's stronger gravity leads to instability at larger values of γ, and this fact implies the instability of supermassive stars, for which the dominance of radiation pressure yields $\gamma \approx 4/3$.

This near equality also holds for dense dwarfs, and it implies that general relativity can render stars dynamically unstable, when they are nearly Newtonian, with radius

$$R = K \frac{2M}{\gamma - 4/3}, \qquad (2.14)$$

where K is a constant of order unity (Chandrasekhar 1964b; Chandrasekhar and Tooper 1964).

Led to the problem by an earlier, heuristic paper by Iben (1963), Chandra saw that "it would be quite straightforward to develop the analog of Eddington's pulsation theory in the exact framework of general relativity." The first paper (Chandrasekhar 1964a) was completed and corrected "just in time: Misner, Zapolsky and Fowler were already on the trail" (Chandrasekhar 1970a). The previous year, at a lecture by W. Fowler on a supermassive-star model for quasars, Feynman suggested that general relativity might imply instability, because the general-relativistic binding energy increases more rapidly with density than does its Newtonian counterparts, and Fowler followed up on the comment with a post-Newtonian calculation that confirmed Feynman's intuition (Fowler 1964; this account of Feynman and Fowler is taken from Thorne 1990).

Oppenheimer and Volkoff (1939) and Harrison, Wakano, and Wheeler (1958) had already considered the stability of neutron stars using a turning-point criterion; shortly after Chandra's paper, Misner and Zapolsky (1964) found numerically that the onset of dynamical instability occurred at configurations of extremal mass. It was soon understood that this coincidence reflected the fact that, in modeling the pulsations, one was using the same effective equation of state as that used to model the equilibrium stars.

Thus, as Thorne (1967) subsequently emphasized, the turning-point instability at the maximum mass is technically not dynamical: for masses slightly above the maximum, collapse apparently occurs on a timescale set by the nuclear reactions and energy loss needed to keep the contracting matter in its zero-temperature thermodynamic equilibrium state (see Thorne, 1967, 1978, for later references and a review of the turning-point method applied to spherical stars; a somewhat different treatment is given by Zel'dovich and Novikov, 1971).

The turning-point argument is valid for uniformly rotating stars as well (Friedman, Ipser, and Sorkin 1988) and can be stated heuristically as follows. When the mass has a maximum along a curve of constant J, the total baryon number turns over as well, because of the relation (Bardeen 1973)

$$dM = \Omega\, dJ + \mu\, dN. \tag{2.15}$$

At the turning point, nearby models have (to first order in the path parameter ϵ) the same angular momentum, baryon number, and mass. The corresponding perturbation relating two such equilibria is then a time-independent solution to the linearized equations of a perfect fluid in general relativity, but a solution for which the angular momentum of each fluid element changes.

Models on the *high*-density side of the instability point are unstable because the injection energy is a *decreasing* function of central density. The relation can be understood from eq. (2.15) if one considers again a sequence of stars with fixed angular momenta. The turning point is a star with maximum mass and baryon number, and on opposite sides of the turning point are corresponding models with the same baryon number. Because $\mu = \partial M/\partial N$ is a decreasing function of central density, the model on the high-density side of the turning point has greater mass than the corresponding model with smaller central density. For spherical neutron stars, the low-density endpoint of the equilibrium sequence is again an extremum of the mass, in this case a minimum. For rapidly rotating stars, however, only the high-density endpoint of a constant J sequence is an extremum of the mass. As the density is lowered at fixed J, the binding energy decreases and the sequence terminates by mass shedding: the equator rotates with angular velocity equal to that of a particle in Keplerian orbit.

The result has a precise phrasing that reflects the J-N symmetry of eq. (2.15):

Theorem. Consider a two-dimensional family of uniformly rotating stellar models based on an equation of state of the form $p = p(\epsilon)$. Suppose that along a continuous sequence of models labeled by a parameter λ, there is a point λ_0 at which both $\dot{N} = dN/d\lambda$ and $\dot{J}$ vanish and where $(\dot{\Omega}\dot{J}+\dot{\mu}\dot{N})\neq 0$. Then the part of the sequence satisfying $\dot{\Omega}\dot{J} + \dot{\mu}\dot{N} > 0$ is unstable for λ near λ_0 (Friedman et al. 1988).

Cook, Shapiro, and Teukolsky (1994a, 1994b) emphasize the implication that the instability points are extrema of J at constant N, as well as extrema of N and M at constant J.

The stabilizing effect of rotation is intuitively clear. A star supported by both rotation and pressure is less dense and has smaller gravity than the corresponding spherical star. A direct study of rotation on the fundamental mode of relativistic stars was considered by Chandrasekhar and Friedman (1972a, 1972b) following a quasi-static analysis for slowly rotating stars by Hartle, Thorne, and Chitre (1972). But this approach relies on nondegenerate perturbation theory, and for isentropic stars at the instability point, the radial mode is degenerate with the set of zero-frequency convective modes. A revised static-stability criterion (Hartle 1975) overcomes the difficulty; it requires the iterative construction of a "comparison sequence" of differentially rotating equilibria, and explicit calculations have been done only for $n = 3/2$ polytropes (Hartle and Munn 1975).

Fortunately, the turning-point method locates the relevant stability points of neutron stars. As in the case of spherical stars, the onset of axisymmetric instability located by the method is initially secular; for rotating stars, its timescale is long enough to accommodate not only heat transfer, but the viscous transfer of angular momentum needed to keep the rotation uniform.

2.5.2 Nonaxisymmetric stability

Work by Detweiler and Ipser (1973) shows that spherical stars are stable if they are stable against collapse and locally stable against convection. But rotating stars are subject to an additional set of nonaxisymmetric instabilities. Newtonian stars that rotate sufficiently rapidly are unstable to a bar mode, a perturbation having angular dependence $\cos m\phi$, for $m = 2$. Models with viscosity are unstable sooner (at slower rotation) than are perfect-fluid models. By breaking conservation of circulation and allowing a transfer of angular momentum between fluid rings, viscosity makes lower energy states with the same total mass and angular momentum accessible to perturbations that are forbidden for a perfect fluid.

For the uniform-density, uniformly rotating Maclaurin spheroids, the instability occurs at a bifurcation point, where the Jacobi family of triaxial ellipsoids branches off (see Chandrasekhar 1969). These ellipsoids are static in a rotating frame and so is the mode that becomes unstable: at the point of bifurcation it takes the Maclaurin spheroid to a nearby Jacobi ellipsoid. In a conversation with Chandra in 1969, Ostriker raised the question "Does the dissipation of energy by gravitational radiation induce, in the manner of viscosity, a secular instability of the Maclaurin spheroid at the point of bifurcation with the Jacobi sequence" (Chandrasekhar 1970a). Chandra found that it does not (1970b, 1970c). The mode made unstable by viscosity remains stable when one includes radiation reaction, but there are surprises that reverse the meaning of this result. Two weeks after completing a paper that reported stability of the Jacobi mode, Chandra rewrote it; he showed that the sequence *is* unstable after the bifurcation point: The instability sets in, not by a mode that is static in the rotating frame, but by one that is stationary in the inertial frame. Where viscosity conserves total angular momentum but not circulation, gravitational radiation conserves circulation but not total angular momentum. By allowing the total angular momentum to decrease, radiation permits a perfect-fluid star to spin down.

Bernard Schutz and I subsequently found a second surprise. A nonaxisymmetric instability driven by gravitational radiation is a generic feature of rotating perfect-fluid stars in general relativity (Friedman and Schutz 1978; Friedman 1978; Comins 1979a, 1979b): Every rotating, self-gravitating perfect fluid is unstable to nonaxisymmetric perturbations which radiate away its angular momentum. And, for polar perturbations, the instability first sets in, not through the $m = 2$ mode, but through modes of large m. The instability is less dramatic than it sounds. As discussed below, viscosity eliminates the instability in ordinary stars and sharply limits its role even in neutron stars.

A final surprise, recently pointed out by Andersson (Andersson 1997; Friedman and Morsink 1997) is that axial modes for all values of m will be unstable for perfect-fluid models with arbitrarily slow rotation. In a spherical star, axial perturbations are time-independent convective currents that do not change the density and pressure of the star and do not couple to gravitational waves. To see how they become unstable for rotating perfect-fluid stars, it is helpful to recall how the nonaxisymmetric instability arises for any mode.

In spherical stars, gravitational radiation removes positive angular momentum L_z from a mode moving in the positive ϕ direction and negative angular momentum from a backward-moving mode; and it therefore damps all time-dependent nonaxisymmetric modes. Once the angular velocity of the star is sufficiently large, however, a mode that moves backward relative to the star is dragged forward relative to an inertial observer. Gravitational radiation will then remove positive angular momentum from the mode. But a mode that moves backward relative to the fluid has negative angular momentum, because the perturbed fluid does not rotate as fast it did without the perturbation. The radiation thus removes positive angular momentum from a mode whose angular momentum is negative. By making the angular momentum of the perturbation increasingly negative, gravitational radiation drives the mode.

Because the geometry of a spherical star is invariant under parity, perturbations of spherical stars are a sum of parts that are eigenfunctions of parity. A perturbation with angular dependence Y_l^m is said to be polar if it behaves under parity like the scalar Y_l^m ($Y_l^m \to (-1)^l Y_l^m$), axial if its behavior is opposite, changing sign for even l, invariant for odd l. An axial perturbation of a spherical star is a time-independent change in the fluid's four-velocity of the form

$$\delta u^\alpha = \zeta(r)\, \epsilon^{\alpha\beta\gamma\delta} \nabla_\beta t \nabla_\gamma r \nabla_\delta Y_l^m. \qquad (2.16)$$

For a rotating star, the frequency σ of these modes is no longer zero. Conservation of circulation (the curl of the equation of hydrostatic equilibrium) implies in the Newtonian limit (Papaloizou and Pringle 1978) the relation

$$\sigma + m\Omega = \frac{2m\Omega}{l(l+1)}. \qquad (2.17)$$

With this frequency, modes that travel backward relative to the star are pulled forward relative to an inertial frame. They are therefore formally unstable for arbitrarily small values of the star's rotation. Andersson numerically computes the real part of the frequency of these modes in a slow-rotation approximation, while Morsink and I show analytically that that their canonical energy can be made negative; the fact that they are unstable (or marginally stable) follows analytically from this and numerically from Andersson's frequency.

Their growth time τ, however, is proportional to a high inverse power ($\tau \propto \Omega^{-4-2l}$). As in the case of the polar modes discussed below, viscosity will presumably enforce stability except for hot, rapidly rotating neutron stars.

The first papers on the generic nonaxisymmetric instability mentioned only in passing its possible damping by viscosity. Results of a study of Detweiler and Lindblom (1977) suggested that viscosity would stabilize any mode whose growth time was longer than the viscous damping time, and this was confirmed by Lindblom and Hiscock (1983). Our present understanding of the viscosity of neutron stars is summarized by Lindblom and Mendell (1995) and in a more detailed review of the structure and stability of rotating relativistic stars by Friedman and Ipser (1993; see also Lai and Shapiro, 1995 for a reevaluation of bulk viscosity, first discussed by Sawyer, 1989a, 1989b.)

At present, it appears that the gravitational-wave-driven instability can limit neutron-star rotation only in hot stars, with temperatures above the superfluid transition point. For substantially higher temperatures, bulk viscosity may also damp the instability. If neutron stars with weak magnetic fields are formed in the accretion-induced collapse of white dwarfs, their rotation may be limited by this instability. If our understanding of viscosity is roughly correct, however, old neutron stars spun up by accretion will never be hot enough to be unstable to gravitational-wave-driven modes. In particular, Wagoner (1984) had suggested that neutron stars spun up by accretion past the onset of nonaxisymmetric instability would hover near the instability point, radiating in gravitational waves a flux of angular momentum balancing that acquired by accretion. Viscosity apparently rules out the Wagoner mechanism, but Schutz (1997) has suggested that a variant of it, nonaxisymmetric instability driven by accretion on the neutron core of a Thorne-Zytkow object, might be hot enough to be unstable.

The location of the instability points in the exact theory has been found numerically for the first time (Stergioulas 1996; Stergioulas and Friedman 1996), using a variant of an Eulerian method developed by Friedman and Ipser (1993) and Ipser and Lindblom (1991a, 1991b; see also Yoshida and Eriguchi 1995; Lindblom 1995). Independently, Yoshida and Eriguchi (1997) have used a relativistic Cowling approximation to estimate the location of these same instability points. The Stergioulas-Friedman work uses a giant code written by Stergioulas (1996) to find the location of these neutral modes for the $m = 2$, 3, and 4 modes of polytropes. They occur at values of the angular velocity that are substantially smaller than expected from the Newtonian studies of Managan (1985) and of Imamura et al. (1985).

A striking feature of this work is a discovery that the $m = 2$ (bar) mode can become unstable for much softer polytropes than is the case in the Newtonian theory. Uniformly rotating Newtonian polytropes are stable to the $m = 2$ mode unless $n < 0.8$ ($\gamma = 1 + 1/n > 2.2$). For models near the maximum mass, however, relativistic polytropes can exhibit an $m = 2$ instability for n as large as 1.5 ($\gamma > 1.7$).

The destabilizing effect of general relativity that Chandra found for the fundamental radial mode is present for the nonaxisymmetric modes as well. For an $n = 1$ polytrope, modes with $m = 3$–5 are unstable for dimensionless values of Ω that are substantially smaller than in the Newtonian limit. More striking is the fact that the $m = 2$ mode, which in uniformly rotating stars is unstable only for very stiff EOS ($n < 0.8$), is unstable in the most relativistic stars for n as large as 1.5.

Chandra thought that the nonaxisymmetric instability would play a significant part in gravitational collapse. Because of the high numerical viscosity of codes that model the evolution of fluids, one cannot yet use them to investigate the gravitational-wave-driven instability. A recent Lai-Shapiro study (1995) avoids the 3+1 numerical problem by examining the instability in the collapse of classical ellipsoids, within in a post-Newtonian framework, following the earlier work of Miller (1974; done while she was a graduate student with Chandra) and of Detweiler and Lindblom (1977; see also Imamura et al., 1995, for a study of an analogous secular instability in which angular momentum is removed from the star not by gravitational waves but by the coupling of the star to an accretion disk). In these studies, gravitational radiation makes the bar mode ($m = 2$) unstable, and it is important in the collapse. Lai and Shapiro emphasize that enough energy is radiated in gravitational waves that, if neutron stars are formed by the accretion-induced collapse of white dwarfs, the collapsing stars may be candidates for gravitational wave interferometers. Incompressible ellipsoids, however, allow much more rapid rotation, measured by the dimensionless quantity $T/|W|$, than do uniformly rotating neutron stars. A uniformly rotating neutron star reaches its maximum rotation well before the large values of $T/|W|$ considered by Lai and Shapiro. But they argue that with a rotation law that is only slightly differential, $T/|W|$ can be large enough for the bar mode to be important.

The viscosity-driven and dynamical bar instabilities that are a key part of Chandra's work on ellipsoids *are* accessible to numerical evolution, and recent studies (Smith et al. 1996; Houser, Centrella, and Smith 1994; Bonazzola et al. 1996; Pickett et al. 1996; Williams and Tohline 1988; Durisen and Tohline 1985; Tohline et al. 1985) show what can be done. For $T/|W| > 0.27$ (much larger than the maximum value for uniformly rotating neutron stars), Newtonian stars are dynamically unstable to a bar mode. The work on dynamical instability by Houser-Centrella-Smith and the earlier authors examines collapse with substantial differential rotation, and the evolution they trace shows a bar that grows into spiral arms; as it loses angular momentum and drops below the critical rotation needed for dynamical instability, the spiral arms wrap themselves around a symmetric core, and the configuration returns to axisymmetry. Because the axisymmetric form is still rotating rapidly enough to be secularly unstable ($T/|W| > 0.14$), Lai and Shapiro suggest two stages of nonaxisymmetric instability—a dynamical bar instability during the collapse and a postcollapse instability in which the core evolves to a nonaxisymmetric configuration on a secular timescale.

To summarize: A dynamical bar instability is a feature of accretion-induced collapse of white dwarfs in which rotation is rapid enough that $T/|W|$ exceeds 0.27. Whether a postcollapse secular instability arises is less secure. In particular, although the possible formation of rapidly rotating neutron stars from the accretion-induced collapse of white dwarfs was suggested more than 25 years ago, we do not know whether neutron stars form in this way; and we do not know whether the magnetic field of a neutron star formed in this way can be small enough to allow rapid rotation. If the magnetic field does not limit the star's initial rotation, it is likely that gravitational radiation will drive a nonaxisymmetric instability. But our understanding of the dominant mechanisms for effective viscosity, of the equation of state of neutron-star matter, and of the likely amount of differential rotation in the newly formed star is too primitive for us to know which mode will dominate or to be certain that the instability driven by gravitational radiation will not be damped by viscosity.

Finally, for neutron stars spun up by accretion, there is some chance for a nonaxisymmetric instability driven by viscosity. (As mentioned earlier, the viscosity in old, cold stars is almost certainly large enough to damp out the radiation-driven modes.) In constrast to the radiation-driven $m = 2$ mode, which sets in sooner in general relativity than in the Newtonian theory, Bonazzola et al. find, within their approximation scheme, that in relativity the mode driven by viscosity sets in at larger values of rotation (measured by the dimensionless quantity $T/|W|$ or Ω/Ω_K). An exceptionally stiff equation of state is therefore needed to allow a viscosity-driven instability of a $1.4 M_\odot$ star, but stars near their upper mass limits may be unstable for midrange equations of state.

References

Andersson, N. 1997, A new class of unstable modes of rotating relativistic stars, *Ap. J.*, submitted; gr-qc/9706075.

Arzoumanian, Z., Thorsett, S.E., and Taylor, J.H. 1995, in *Millisecond Pulsars: A Decade of Surprise*, ASP Conference Series, Vol. 72, ed. A.S. Fruchter, M. Tavani, and D.C. Backer (San Francisco: Astron. Soc. of the Pacific).

Bardeen J.M. 1973, in *Black Holes—Les Houches 1972*, ed. C. DeWitt and B.S. DeWitt (New York: Gordon and Breach).

Baym, G., and Pines, D. 1971, *Ann. Phys.*, **66**, 816.

Blinnikov, S.I., Novikov, I.D., Perevodchikova, T.V., and Polnarev, A.G. 1984, *Soviet Astron. Lett.*, **10**, 177.

Bocquet, M., Bonazzola, S., and Gourgoulhon, E. 1995, *Astron. Astrophys.*, **301**, 757.

Bonazzola, S., Frieben, J., and Gourgoulhon, E. 1996, *Ap. J.*, **460**, 379.

Bonazzola, S., and Gourgoulhon, E. 1994, *Class. Quantum Grav.*, **11**, 1775.

Bonazzola, S., Gourgoulhon, E., Salgado, M., and Marck, J.A. 1993, *Astron. and Astrophys.*, **278**, 421.

Bonazzola, S., and Schneider, S. 1974, *Ap. J.*, **191**, 273.

Butterworth, E.M., and Ipser, J.R. 1976, *Ap. J.*, **204**, 200.

Chandrasekhar, S. 1964a, *Phys. Rev. Lett.*, **12**, 114.

Chandrasekhar, S. 1964b, *Ap. J.*, **140**, 417.

Chandrasekhar, S. 1969, *Ellipsoidal Figures of Equilibrium* (New Haven: Yale University Press).

Chandrasekhar, S. 1970a, General relativity, unpublished.

Chandrasekhar, S. 1970b, *Phys. Rev. Lett.*, **24**, 611.

Chandrasekhar, S. 1970c, *Ap. J.*, **161**, 561.

Chandrasekhar, S., and Friedman, J.L. 1972a, *Ap. J.*, **175**, 379.

Chandrasekhar, S., and Friedman, J.L. 1972b, *Ap. J.*, **176**, 745.

Chandrasekhar, S., and Tooper, R.F. 1964, *Ap. J.*, **139**, 1396.

Colpi, M., Shapiro, S.L., and Teukolsky, S.A. 1989, *Ap. J.*, **339**, 318,

Colpi, M., Shapiro, S.L., and Teukolsky, S.A. 1991, *Ap. J.*, **369**, 422.

Comins, N. 1979a, *MNRAS*, **189**, 233.

Comins, N. 1979b, *MNRAS*, **189**, 255.

Cook, G.B., Shapiro, S.L., and Teukolsky, S.A. 1992, *Ap. J.*, **398**, 203.

Cook, G.B., Shapiro, S.L., and Teukolsky, S.A. 1994a, *Ap. J.*, **422**, 227.

Cook, G.B., Shapiro, S.L., and Teukolsky, S.A. 1994b, *Ap. J.*, **424**, 823.

Datta, B. 1988, *Found. Cosmic Phys.*, **12**, 151.

Detweiler, S.L., and Ipser, J.R. 1973, *Ap. J.*, **185**, 685.

Detweiler, S.L., and Lindblom, L. 1977, *Ap. J.*, **213**, 193.

Durisen, R.H., and Tohline, J.E. 1985, in *Protostars and Planets II*, ed. D. Black and M. Matthews (Tucson: University of Arizona Press).

Eriguchi, Y., Hachisu, I., and Nomoto, K. 1994, *MNRAS*, **266**, 179.

Fowler, W.A. 1964, *Rev. Mod. Phys.*, **36**, 545 and 1104.

Friedman, J.L. 1978, *Comm. Math. Phys.*, **62**, 247.

Friedman, J.L. 1996, *J. Astrophys. and Astronomy*, **17**, 199.

Friedman, J.L., Imamura, J.N., Durisen, R.H., and Parker, L. 1985, *Ap. J.*, **304**, 115; erratum **351**, 705.

Friedman, J.L., and Ipser, J.R. 1987, *Ap. J.*, **314**, 594.

Friedman, J.L., and Ipser, J.R. 1993, in *Classical General Relativity*, ed. S. Chandrasekhar (Oxford: Oxford University Press); reprinted with corrections from *Phil. Trans. R. Lond. A*, **340**, 391 (1992).

Friedman, J.L., Ipser, J.R., and Parker, L. 1986, *Ap. J.*, **304**, 115; Erratum 1990, *Ap. J.*, **351**, 705.

Friedman, J.L., Ipser, J.R., and Parker, L. 1986, *Phys. Rev Lett.*, **62**, 3015.

Friedman, J.L., Ipser, J.R., and Sorkin, R.D. 1988, *Ap. J.*, **325**, 722.

Friedman, J.L., and Morsink, S. 1997, Axial instability of rotating relativistic stars, *Ap. J.*, submitted; gr-qc/9706073.

Friedman, J.L., and Schutz, B.F. 1978, *Ap. J.*, **222**, 281.

Glendenning, N.K. 1992, *Phys. Rev. D*, **46**, 4161.

Glendenning, N.K. 1997, *Compact Stars: Nuclear Physics, Particle Physics and General Relativity* (New York: Springer).

Gourgoulhon, E., and Bonazzola, S. 1994, *Class. Quantum Grav.*, **11**, 443.

Harrison, B.K., Wakano, M., and Wheeler, J.A. 1958, in *Onzieme Conseil de Physique Solvay, La Structure et l'evolution de l'universe* (Brussels: Stoops).

Hartle, J.B. 1967, *Ap. J.*, **150**, 1005.

Hartle, J.B. 1975, *Ap. J.*, **195**, 203.

Hartle, J.B., and Munn, M.W. 1975, *Ap. J.*, **198**, 467.

Hartle, J.B., and Sabbadini, A.G. 1977, *Ap. J.*, **213**, 831.

Hartle, J.B., Thorne, K.S., and Chitre, S.M. 1972, *Ap. J.*, **176**, 177.

Hashimoto, M., Oyamatsu, K., and Eriguchi, Y. 1994, *Ap. J*, **401**, 618.

Herold, H., Heimberger, J., Ruder, H., and Wu, X. 1993, in *Isolated Pulsars, Proc. Los Alamos Workshop, Taos NM*, ed. K.A. Van Riper, R. Epstein, and C. Ho (Cambridge: Cambridge University Press).

Houser, J.L., Centrella, J.M., and Smith, S.C. 1994, *Phys. Rev. Lett.*, **72**, 1314.

Iben, I. 1963, *Ap. J.*, **138**, 1090.

Imamura, J.N., Friedman, J.L., and Durisen, R.H. 1985, *Ap. J.*, **294**, 474.

Imamura, J.N., Toman, J., Durisen, R.H., Pickett, B.K, and Yang, S. 1995, *Ap. J.*, **444**, 363.

Ipser, J.R., and Lindblom, L. 1991a, *Ap. J.*, **373**, 213.

Ipser, J.R., and Lindblom, L. 1991b, *Ap. J.*, **379**, 285.

Komatsu, H., Eriguchi, Y., and Hachisu, I. 1989a, *MNRAS*, **237**, 355.

Komatsu, H., Eriguchi, Y., and Hachisu, I. 1989b, *MNRAS*, **239**, 153.

Koranda, S., Stergioulas, N., and Friedman, J.L. 1996, Upper limits set by causality on the rotation and mass of uniformly rotating relativisitc stars, *Ap. J.*, in press.

Lai, D., and Shapiro, S.L. 1995, *Ap. J.*, **442**, 259.

Lattimer, J.M., Prakash, M., Masak, D., and Yahil, A. 1990, *Ap. J.*, **355**, 241.

Lindblom, L. 1995, *Ap. J.*, **438**, 265.

Lindblom, L., and Hiscock, W.A. 1983, *Ap. J.*, **267**, 384.

Lindblom, L., and Mendell, G. 1995, *Ap. J.*, **444**, 804.

Lipunov, V.M., and Postnov, K.A. 1988, *Astron. Astrophys.*, **206**, L15.

Managan, R. 1985, *Ap. J.*, **294**, 463.

Miller, B.D. 1974, *Ap. J.*, **187**, 609.

Misner, C.W., and Zapolsky, H.S. 1964, *Phys. Rev. Lett.*, **12**, 635.

Neugebauer, G., and Herlt, H. 1984, *Class. Quantum Grav.*, **1**, 695.

Neugebauer, G., and Herold, H. 1992, in *Relativistic Gravity Research*, Lecture Notes in Physics, No. 410, ed. J. Ehlers and G. Schäfer, (Berlin: Springer), 305, 319.

Nozawa, T., Stergioulas, N., Gourgoulhon, E., and Eriguchi, Y. 1997, to appear.

Oppenheimer, J.R., and Volkoff, G.M. 1939, *Phys. Rev.*, **55**, 374.

Papaloizou, J., and Pringle, J.E. 1978, *MNRAS*, **182**, 423.

Pickett, B.K., Durisen, R.H., and Davis, G.A. 1996, *Ap. J.*, **458**, 714.

Rhoades, C.E., and Ruffini, R. 1974, *Phys. Rev. Lett.*, **32**, 324.

Salgado M., Bonazzola, S., Gourgoulhon, E., and Haensel, P. 1994a, *Astron Astrophys.*, **291**, 155.

Salgado, M., Bonazzola, S., Gourgoulhon, E., and Haensel P. 1994b, *Astron. Astrophys. Suppl.*, **108**, 455.

Sawyer, R.F. 1989a, *Phys. Rev. D*, **39**, 3804.

Sawyer, R.F. 1989b, *Phys. Lett. B*, **233**, 412.

Schutz, B.F. 1997, in *Mathematics of Gravitation*, Vol. 2, *Gravitational Wave Detection*, ed. A. Krolak (Warsaw: Banach Center).

Shapiro, S.L., and Teukolsky, S.A. 1983, *Black Holes, White Dwarfs, and Neutron Stars* (New York: Wiley).

Smith, S.C., Houser, J.L., and Centrella, J.M. 1996, *Ap. J.*, **458**, 236.

Stergioulas, N. 1996, Structure and stability of rotating relativistic stars, Ph.D. thesis, University of Wisconsin-Milwaukee.

Stergioulas, N., and Friedman, J.L. 1995, *Ap. J*, **444**, 306.

Stergioulas, N., and Friedman, J.L. 1996, Nonaxisymmetric neutral modes in rotating relativisitc stars, *Ap. J.*, in press.

Taylor, J.H., and Weisberg, J.M. 1989, *Ap. J.*, **345**, 434.

Thorne, K.S. 1967, in *High Energy Astrophysics*, Vol. 3, ed. C. DeWitt, E. Schatzman, and P. Veron (New York: Gordon and Breach).

Thorne, K.S. 1971, in *Gravitation and Cosmology, Proceedings of the International School of Physics Enrico Fermi*, Course 47, ed. R.K. Sachs (New York: Academic Press).

Thorne, K.S. 1978, in *Theoretical Principles in Astrophysics and Relativity*, ed. N.R. Lebovitz, W.H. Reid, and P.O. Vandervoort (Chicago: University of Chicago Press).

Thorne, K. S. 1990, Foreword in *Selected Papers*, Vol. 5, *Relativistic Astrophysics*, by S. Chandrasekhar (Chicago: University of Chicago Press).

Tohline, J.E., Durisen, R.H., and McCollough, M. 1985, *Ap. J.*, **298**, 220.

Uryu, K., and Eriguchi, Y. 1994, *MNRAS*, **269**, 24; second paper submitted.

Wagoner, R.V. 1984, *Ap. J.*, **278**, 345.

Williams, H.A., and Tohline, J.E. 1988, *Ap. J.*, **334**, 449.

Wilson, J.R. 1972, *Ap. J.*, **176**, 195.

Wu, X., Müther, H., Soffel, M., Herold, H., and Ruder, H. 1991, *Astron. Astrophys.*, **246**, 411.

Yoshida, S., and Eriguchi, Y. 1995, *Ap. J.*, **438**, 830.

Yoshida, S., and Eriguchi, Y. 1997, Neutral points of oscillation modes along equilibrium sequences of rapidly rotating polytropes in general relativity—application of the Cowling approximation, astro-ph/9704111.

Zel'dovich, Ya.B., and Novikov, I.D. 1971, *Relativistic Astrophysics*, Vol. 1 (Chicago: University of Chicago Press).

3

Probing Black Holes and Relativistic Stars with Gravitational Waves

Kip S. Thorne

Abstract

In the coming decade, gravitational waves will convert the study of general relativistic aspects of black holes and stars from a largely theoretical enterprise to a highly interactive, observational/theoretical one. For example, gravitational-wave observations should enable us to observationally map the spacetime geometries around quiescent black holes, study quantitatively the highly nonlinear vibrations of curved spacetime in black-hole collisions, probe the structures of neutron stars and their equation of state, search for exotic types of general relativistic objects such as boson stars, soliton stars, and naked singularities, and probe aspects of general relativity that have never yet been seen, such as the gravitational fields of gravitons and the influence of gravitational-wave tails on radiation reaction.

3.1 Introduction

Subrahmanyan Chandrasekhar and I entered the field of relativistic astrophysics at the same time, in the early 1960s—I as a green graduate student at Princeton; Chandra as an established and famous researcher at the University of Chicago. Over the decades of the '60s, '70s and '80s, and into the '90s, Chandra, I, and our friends and colleagues had the great pleasure of exploring general relativity's predictions about the properties of black holes and relativistic stars. Throughout these explorations Chandra was an inspiration to us all.

When we began, there was no observational evidence that black holes or relativistic stars exist in the universe, much less that they play important roles. However, in parallel with our theoretical studies, astronomers discovered pulsars and quickly deduced that they are spinning neutron stars in which relativistic effects should be strong; astronomers also discovered quasars and gradually, over three decades' time, came to understand that they are powered by supermassive black holes; astronomers discovered compact X-ray sources and quickly deduced that they are binary systems in which gas accretes from a normal star onto a stellar-mass black-hole companion or neutron-star companion; and astronomers discovered gamma ray bursts and, after nearly three decades of puzzlement, have concluded that they are probably produced by the final merger of a neutron-star/neutron-star binary or a neutron-star/black-hole binary.

Despite this growing richness of astrophysical phenomena in which black holes and neutron stars play major roles, those of us who use general relativity to predict the properties of these objects have been frustrated: in the rich astronomical data there as yet is little evidence for the holes' and stars' spacetime warpage, which is so central to our theoretical studies. If we had to rely solely on observations and not at all on theory, we could still argue, in 1997, that a black hole is a flat-spacetime, Newtonian phenomenon and the internal structures of neutron stars are un-influenced by general relativistic effects.

Why this frustration? Perhaps because spacetime warpage cannot, itself, produce the only kinds of radiation that astronomers now have at their disposal: electromagnetic waves, neutrinos, and cosmic rays. To explore spacetime warpage in detail may well require using, instead, the only kind of radiation that such warpage can produce: radiation made of spacetime warpage—gravitational waves.

In this article, I shall describe the prospects for using gravitational waves to probe the warpage of spacetime around black holes and relativistic stars, and to search for new types of general relativistic objects, for which there as yet is no observational evidence. And I shall describe how the challenge of developing data analysis algorithms for gravitational-wave detectors is already driving the theory of black holes and relativistic stars just as hard as the theory is driving the wave-detection efforts. Already, several years before the full-scale detectors go into operation, the challenge of transforming general relativistic astrophysics into an observational science has transformed the nature of our theoretical enterprise. At last, after 35 years of only weak coupling to observation, those of us studying general relativistic aspects of black holes and stars have become tightly coupled to the observational/experimental enterprise.

3.2 Gravitational waves

A gravitational wave is a ripple of warpage (curvature) in the "fabric" of spacetime. According to general relativity, gravitational waves are produced by the dynamical spacetime warpage of distant astrophysical systems, and they travel outward from their sources and through the universe at the speed of light, becoming very weak by the time they reach the Earth.

Einstein discovered gravitational waves as a prediction of his general relativity theory in 1916, but only in the late 1950s did the technology of high-precision measurement become good enough to justify an effort to construct detectors for the waves.

Gravitational-wave detectors and detection techniques have now been under development for nearly 40 years, building on foundations laid by Joseph Weber [1], Rainer Weiss [2], and others. These efforts have led to promising sensitivities in four frequency bands, and theoretical studies have identified plausible sources in each band:

- The Extremely Low Frequency (ELF) Band, 10^{-15} to 10^{-18} Hz, in which the measured anisotropy of the cosmic microwave background radiation places strong limits on gravitational-wave strengths—and may, in fact, have detected waves [3, 4]. The only waves expected in this band are relics of the big bang, a subject beyond the scope of this article. (For some details and references see [3, 4, 5] and references cited therein.)

- The Very Low Frequency (VLF) Band, 10^{-7} to 10^{-9} Hz, in which Joseph Taylor and others have achieved remarkable gravity-wave sensitivities by the timing of millisecond pulsars [6]. The only expected strong sources in this band are processes in the very early universe—the big bang, phase transitions of the vacuum states of quantum fields, and vibrating or colliding defects in the structure of spacetime, such as monopoles, cosmic strings, domain walls, textures, and combinations thereof [7, 8, 9, 10]. These sources are also beyond the scope of this article.

- The Low-Frequency (LF) Band, 10^{-4} to 1 Hz, in which will operate the Laser Interferometer Space Antenna, LISA; see sections 3.3.4 below. This is the band of massive black holes ($M \sim 1000$–$10^8 M_\odot$) in the distant universe, and of other hypothetical massive exotic objects (naked singularities, soliton stars), as well as of binary stars (ordinary, white dwarf, neutron star, and black hole) in our galaxy. Early universe processes should also have produced waves at these frequencies, as in the ELF and VLF bands.

- The High-Frequency (HF) Band, 1 to 10^4 Hz, in which operate earth-based gravitational-wave detectors such as LIGO; see sections 3.3.1–3.3.3 below. This is the band of stellar-mass black holes ($M \sim 1$–$1000 M_\odot$) and of other conceivable stellar-mass exotic objects (naked singularities and boson stars) in the distant universe, as well as of supernovae, pulsars, and coalescing and colliding neutron stars. Early universe processes should also have produced waves at these frequencies, as in the ELF, VLF, and LF bands.

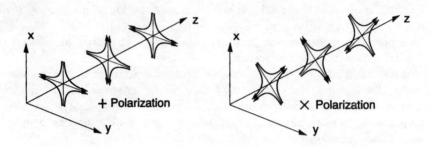

Figure 3.1: The lines of force associated with the two polarizations of a gravitational wave [12].

In this article I shall focus on the HF and LF bands, because these are the ones in which we can expect to study black holes and relativistic stars.

One aspect of a gravitational wave's spacetime warpage—the only aspect relevant to earth-based detectors—is an oscillatory "stretching and squeezing" of space. This stretch and squeeze is described, in general relativity theory, by two dimensionless gravitational-wave fields h_+ and $h_\times$ (the "strains of space") that are associated with the wave's two linear polarizations, conventionally called "plus" (+) and "cross" (×). The fields h_+ and $h_\times$, technically speaking, are the double time integrals of space-time-space-time components of the Riemann curvature tensor; and they propagate through spacetime at the speed of light. The inertia of any small piece of an object tries to keep it at rest in, or moving at constant speed through, the piece of space in which it resides; so as h_+ and $h_\times$ stretch and squeeze space, inertia stretches and squeezes objects that reside in that space. This stretch and squeeze is analogous to the tidal gravitational stretch and squeeze exerted on the Earth by the Moon, and thus the associated gravitational-wave force is referred to as a "tidal" force.

If an object is small compared to the waves' wavelength (as is the case for ground-based detectors), then relative to the object's center, the waves exert tidal forces with the quadrupolar patterns shown in figure 3.1. The names "plus" and "cross" are derived from the orientations of the axes that characterize the force patterns [11].

The strengths of the waves from a gravitational-wave source can be estimated using the "Newtonian/quadrupole" approximation to the Einstein field equations. This approximation says that $h \simeq (G/c^4)\ddot{Q}/r$, where $\ddot{Q}$ is the second time derivative of the source's quadrupole moment and r is the distance of the source from Earth (and G and c are Newton's gravitation constant and the speed of light). The strongest sources will be highly nonspherical and thus will have $Q \simeq ML^2$, where M is their mass and L their size, and correspondingly will have $\ddot{Q} \simeq 2Mv^2 \simeq 4E_{\rm kin}^{\rm ns}$, where v is their internal velocity and $E_{\rm kin}^{\rm ns}$ is the nonspherical part of their internal kinetic energy. This provides us with the estimate

$$h \sim \frac{1}{c^2} \frac{4G(E_{\rm kin}^{\rm ns}/c^2)}{r}; \qquad (3.1)$$

i.e., h is about 4 times the gravitational potential produced at Earth by the mass-equivalent of the source's nonspherical, internal kinetic energy—made dimensionless by dividing by c^2. Thus, in order to radiate strongly, the source must have a very large, nonspherical, internal kinetic energy.

The best known way to achieve a huge internal kinetic energy is via gravity; and by energy conservation (or the virial theorem), any gravitationally induced kinetic energy must be of order the source's gravitational potential energy. A huge potential energy, in turn, requires that the source be very compact, not much larger than its own gravitational radius. Thus, the strongest gravity-wave sources must be highly compact, dynamical concentrations of large amounts of mass (e.g., colliding and coalescing black holes and neutron stars).

Such sources cannot remain highly dynamical for long; their motions will be stopped by energy loss to gravitational waves and/or the formation of an all-encompassing black hole. Thus, the strongest sources should be transient. Moreover, they should be very rare—so rare that to see a reasonable event rate will require reaching out through a substantial fraction of the universe. Thus, just as the strongest radio waves arriving at Earth tend to be extragalactic, so also the strongest gravitational waves are likely to be extragalactic.

For highly compact, dynamical objects that radiate in the high-frequency band, e.g., colliding and coalescing neutron stars and stellar-mass black holes, the internal, nonspherical kinetic energy $E_{\rm kin}^{\rm ns}/c^2$ is of order the mass of the Sun; and, correspondingly, eq. (3.1) gives $h \sim 10^{-22}$ for such sources at the Hubble distance (3000 Mpc, i.e., 10^{10} light-years), $h \sim 10^{-21}$ at 200 Mpc (a best-guess distance for several neutron-star coalescences per year; see section 3.6.2), $h \sim 10^{-20}$ at the Virgo cluster of galaxies (15 Mpc), and $h \sim 10^{-17}$ in the outer reaches of our own Milky Way galaxy (20 kpc). These numbers set the scale of sensitivities that ground-based interferometers seek to achieve: $h \sim 10^{-21}$ to 10^{-22}.

3.3 Gravitational wave detectors in the high- and low-frequency bands

3.3.1 Ground-based laser interferometers

The most promising and versatile type of gravitational-wave detector in the high-frequency band, 1 to 10^4 Hz, is a laser interferometer gravitational-

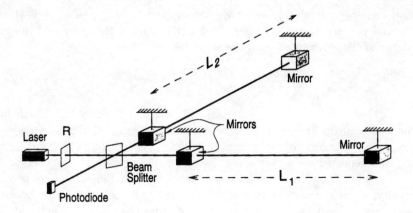

Figure 3.2: Schematic diagram of a laser interferometer gravitational-wave detector [12].

wave detector ("interferometer" for short). Such an interferometer consists of four mirror-endowed masses that hang from vibration-isolated supports, as shown in figure 3.2, and the indicated optical system for monitoring the separations between the masses [11, 12]. Two masses are near each other, at the corner of an "L", and one mass is at the end of each of the L's long arms. The arm lengths are nearly equal, $L_1 \simeq L_2 = L$. When a gravitational wave, with frequencies high compared to the ~ 1 Hz pendulum frequency of the masses, passes through the detector, it pushes the masses back and forth relative to each other as though they were free from their suspension wires, thereby changing the arm-length difference, $\Delta L \equiv L_1 - L_2$. That change is monitored by laser interferometry in such a way that the variations in the output of the photodiode (the interferometer's output) are directly proportional to $\Delta L(t)$.

If the waves are coming from overhead or underfoot and the axes of the + polarization coincide with the arms' directions, then it is the + polarization of the waves that drives the masses, and the detector's strain $\Delta L(t)/L$ is equal to the waves' strain of space $h_+(t)$. More generally, the interferometer's output is a linear combination of the two wave fields:

$$\frac{\Delta L(t)}{L} = F_+ h_+(t) + F_\times h_\times(t) \equiv h(t). \tag{3.2}$$

The coefficients F_+ and $F_\times$ are of order unity and depend in a quadrupolar manner on the direction to the source and the orientation of the detector [11]. The combination $h(t)$ of the two h's is called the gravitational-wave strain that acts on the detector; and the time evolutions of $h(t)$, $h_+(t)$, and $h_\times(t)$ are sometimes called *waveforms*.

Chapter 3. Probing Black Holes ... 47

When one examines the technology of laser interferometry, one sees good prospects to achieve measurement accuracies $\Delta L \sim 10^{-16}$ cm (1/1000 the diameter of the nucleus of an atom)—and $\Delta L = 8 \times 10^{-16}$ cm has actually been achieved in a prototype interferometer at Caltech [13]. With $\Delta L \sim 10^{-16}$ cm, an interferometer must have an arm length $L = \Delta L/h \sim 1$ to 10 km in order to achieve the desired wave sensitivities, 10^{-21} to 10^{-22}. This sets the scale of the interferometers that are now under construction.

3.3.2 LIGO, VIRGO, and the international network of gravitational-wave detectors

Interferometers are plagued by non-Gaussian noise, e.g., due to sudden strain releases in the wires that suspend the masses. This noise prevents a single interferometer, by itself, from detecting with confidence short-duration gravitational-wave bursts (though it may be possible for a single interferometer to search for the periodic waves from known pulsars). The non-Gaussian noise can be removed by cross-correlating two, or preferably three or more, interferometers that are networked together at widely separated sites.

The technology and techniques for such interferometers have been under development for 25 years, and plans for km-scale interferometers have been developed over the past 15 years. An international network consisting of three km-scale interferometers at three widely separated sites is now under construction. It includes two sites of the American Laser Interferometer Gravitational Wave Observatory (LIGO) Project [12] and one site of the French/Italian VIRGO Project (named after the Virgo cluster of galaxies) [14].

LIGO will consist of two vacuum facilities with 4-km-long arms, one in Hanford, Washington (in the northwestern United States), and the other in Livingston, Louisiana (in the southeastern United States). These facilities are designed to house many successive generations of interferometers without the necessity of any major facilities upgrade; and after a planned future expansion, they will be able to house several interferometers at once, each with a different optical configuration optimized for a different type of wave (e.g., broadband burst, or narrowband periodic wave, or stochastic wave).

The LIGO facilities are being constructed by a team of about 80 physicists and engineers at Caltech and MIT, led by Barry Barish (the PI), Gary Sanders (the Project Manager), Albert Lazzarini, Rai Weiss, Stan Whitcomb, and Robbie Vogt (who directed the project during the preconstruction phase). This Caltech/MIT team, together with researchers from several other universities, is developing LIGO's first interferometers and their data analysis system. Other research groups from many universities are contributing to R&D for *enhancements* of the first interferometers, or

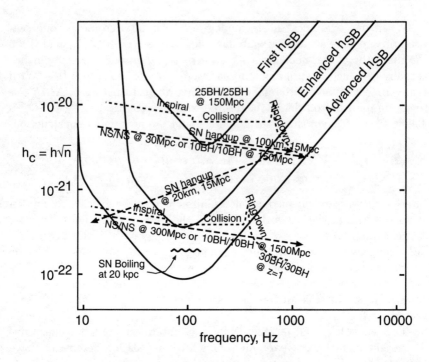

Figure 3.3: LIGO's projected broadband noise sensitivity to bursts $h_{\rm SB}$ [12, 15] compared with the characteristic amplitudes h_c of the waves from several hypothesized sources. The signal-to-noise ratios are $\sqrt{2}$ higher than in [12] because of a factor 2 error in eq. (29) of [11].

are computing theoretical waveforms for use in data analysis, or are developing data analysis techniques for future interferometers. These groups are linked together in a *LIGO Scientific Collaboration* and by an organization called the *LIGO Research Community*. For further details, see the LIGO World Wide Web Site, http://www.ligo.caltech.edu/.

The VIRGO Project is building one vacuum facility in Pisa, Italy, with 3-km-long arms. This facility and its first interferometers are a collaboration of more than a hundred physicists and engineers at the INFN (Frascati, Napoli, Perugia, Pisa), LAL (Orsay), LAPP (Annecy), LOA (Palaiseau), IPN (Lyon), ESPCI (Paris), and the University of Illinois (Urbana), under the leadership of Alain Brillet and Adalberto Giazotto.

The LIGO and VIRGO facilities are scheduled for completion at the end of the 1990s, and their first gravitational-wave searches will be performed in 2001 or 2002. Figure 3.3 shows the design sensitivities for LIGO's *first interferometers* (ca. 2002) [12] and for *enhanced versions* of those interferometers (which are expected to be operating five years or so later) [15], along with a benchmark sensitivity goal for subsequent, more *advanced interferometers* [12, 15].

For each type of interferometer, the quantity shown is the "sensitivity to bursts" that come from a random direction, $h_{\rm SB}(f)$ [12]. This $h_{\rm SB}$ is about 5 times higher than the rms noise level in a bandwidth $\Delta f \simeq f$ for waves with a random direction and polarization, and about $5\sqrt{5} \simeq 11$ times worse than the the rms noise level $h_{\rm rms}$ for optimally directed and polarized waves. (In much of the literature, the quantity plotted is $h_{\rm rms} \simeq h_{\rm SB}/11$.) Along the right-hand branch of each sensitivity curve (above 100 or 200 Hz), the interferometer's dominant noise is due to photon counting statistics ("shot noise"); along the middle branch (10 or 30 Hz to 100 to 200 Hz), the dominant noise is random fluctuations of thermal energy in the test masses and their suspensions; along the steep left-hand branch, the dominant noise is seismic vibrations creeping through the interferometers' seismic isolation system.

The interferometer sensitivity $h_{\rm SB}$ is to be compared with the "characteristic amplitude" $h_c(f) \simeq h\sqrt{n}$ of the waves from a source; here h is the waves' amplitude when they have frequency f, and n is the number of cycles the waves spend in a bandwidth $\Delta f \simeq f$ near frequency f [11, 12]. Any source with $h_c > h_{\rm SB}$ should be detectable with high confidence, even if it arrives only once per year.

Figure 3.3 shows the estimated or computed characteristic amplitudes h_c for several sources that will be discussed in detail later in this article. Among these sources are binary systems made of $1.4 M_\odot$ neutron stars (NS) and binaries made of $10 M_\odot$, $25 M_\odot$, and $30 M_\odot$ black holes (BH), which spiral together and collide under the driving force of gravitational radiation reaction. As the bodies spiral inward, their waves sweep upward in frequency (rightward across the figure along the dashed lines). From the figure we see that LIGO's first interferometers should be able to detect waves from the inspiral of a NS/NS binary out to a distance of 30 Mpc (90 million light-years) and from the final collision and merger of a $25 M_\odot/25 M_\odot$ BH/BH binary out to about 300 Mpc. Comparison with estimated event rates (sections 3.6.2 and 3.7.2 below) suggests, with considerable confidence, that the first wave detections will be achieved by the time the enhanced sensitivity is reached and possibly as soon as the searches by the first interferometers.

LIGO alone, with its two sites which have parallel arms, will be able to detect an incoming gravitational wave, measure one of its two waveforms, and (from the time delay between the two sites) locate its source to within a $\sim 1°$ wide annulus on the sky. LIGO and VIRGO together, operating as a *coordinated international network*, will be able to locate the source (via time delays plus the interferometers' beam patterns) to within a 2-dimensional error box with size between several tens of arcminutes and several degrees, depending on the source direction and on the amount of high-frequency structure in the waveforms. They will also be able to monitor both waveforms $h_+(t)$ and $h_\times(t)$ (except for frequency components above about 1 kHz and below about 10 Hz, where the interferometers' noise becomes severe).

A British/German group is constructing a 600-meter interferometer called GEO600 near Hanover, Germany [16], and Japanese groups, a 300-meter interferometer called TAMA near Tokyo [17]. GEO600 may be a significant player in the interferometric network in its early years (by virtue of cleverness and speed of construction), but because of its short arms it cannot compete in the long run. GEO600 and TAMA will both be important development centers and test beds for interferometer techniques and technology, and in due course they may give rise to km-scale interferometers like LIGO and VIRGO, which could significantly enhance the network's all-sky coverage and ability to extract information from the waves.

3.3.3 Narrowband, high-frequency detectors: interferometers and resonant-mass antennas

At frequencies $f \gtrsim 500$ Hz, the interferometers' photon shot noise becomes a serious obstacle to wave detection. However, narrowband detectors specially optimized for kHz frequencies show considerable promise. These include interferometers with specialized optical configurations ("signal recycled interferometers" [18] and "resonant sideband extraction interferometers" [19]) and large spherical or truncated icosahedral resonant-mass detectors (e.g., the American TIGA [20], Dutch GRAIL [21], and Brazilian OMNI-1 Projects) that are future variants of Joseph Weber's original "bar" detector [1] and of currently operating bars in Italy (AURIGA, Explorer, and Nautilus), Australia (NIOBE), and America (ALLEGRO) [22]. Developmental work for these narrowband detectors is underway at a number of centers around the world.

3.3.4 Low-frequency detectors: the Laser Interferometer Space Antenna (LISA)

The Laser Interferometer Space Antenna (LISA) [23] is the most promising detector for gravitational waves in the low-frequency band, 10^{-4} to 1 Hz (10,000 times lower than the LIGO/VIRGO high-frequency band).

LISA was originally conceived (under a different name) by Peter Bender of the University of Colorado and is currently being developed by an international team led by Karsten Danzmann of the University of Hanover (Germany) and James Hough of Glasgow University (UK). The European Space Agency tentatively plans to fly it sometime in the 2014–2018 time frame as part of ESA's Horizon 2000+ Program of large space missions.

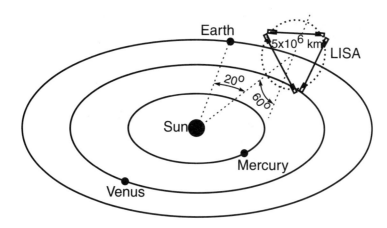

Figure 3.4: LISA's orbital configuration, with LISA magnified in arm length by a factor ~ 10 relative to the solar system.

With NASA participation (which is under study), the flight could be much sooner.

As presently conceived [23], LISA will consist of six compact, drag-free spacecraft (i.e., spacecraft that are shielded from buffeting by solar wind and radiation pressure, and that thus move very nearly on geodesics of spacetime). All six spacecraft would be launched simultaneously in a single Ariane rocket. They would be placed into the same heliocentric orbit as the Earth occupies but would follow 20° behind the Earth; cf. figure 3.4. The spacecraft would fly in pairs, with each pair at the vertex of an equilateral triangle that is inclined at an angle of 60° to the Earth's orbital plane. The triangle's arm length would be 5 million km (10^6 times longer than LIGO's arms!). The six spacecraft would track each other optically, using one-watt YAG laser beams. Because of diffraction losses over the 5×10^6 km arm length, it is not feasible to reflect the beams back and forth between mirrors as is done with LIGO. Instead, each spacecraft would have its own laser; and the lasers would be phase-locked to each other, thereby achieving the same kind of phase-coherent out-and-back light travel as LIGO achieves with mirrors. The six-laser, six-spacecraft configuration thereby would function as three, partially independent and partially redundant, gravitational-wave interferometers.

Figure 3.5 depicts the expected sensitivity of LISA in the same language as we have used for LIGO (fig. 3.3): $h_{\rm SB} = 5\sqrt{5}h_{\rm rms}$ is the sensitivity for high-confidence detection ($S/N = 5$) of a signal coming from a random direction, assuming Gaussian noise.

At frequencies $f \gtrsim 10^{-3}$ Hz, LISA's noise is due to photon counting statistics (shot noise). The sensitivity curve steepens at $f \sim 3 \times 10^{-2}$ Hz

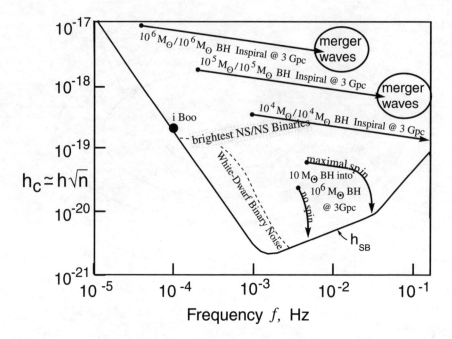

Figure 3.5: LISA's projected sensitivity to bursts h_{SB} compared with the strengths of the waves from several low-frequency sources [23].

because at larger f than that, the waves' period is shorter than the round-trip light travel time in one of LISA's arms. Below 10^{-3} Hz, the noise is due to buffeting-induced random motions of the spacecraft that are not being properly removed by the drag-compensation system. Notice that, in terms of dimensionless amplitude, LISA's sensitivity is roughly the same as that of LIGO's first interferometers (fig. 3.3), but at 100,000 times lower frequency. Since the waves' energy flux scales as $f^2 h^2$, this corresponds to 10^{10} better energy sensitivity than LIGO.

LISA can detect and study, simultaneously, a wide variety of different sources scattered over all directions on the sky. The key to distinguishing the different sources is the different time evolution of their waveforms. The key to determining each source's direction, and confirming that it is real and not just noise, is the manner in which its waves' amplitude and frequency are modulated by LISA's complicated orbital motion—a motion in which the interferometer triangle rotates around its center once per year and the interferometer plane precesses around the normal to the Earth's orbit once per year. Most sources will be observed for a year or longer, thereby making full use of these modulations.

3.4 Stellar core collapse: the births of neutron stars and black holes

In the remainder of this article, I shall describe the techniques and prospects for observationally studying black holes and relativistic stars via the gravitational waves they emit. I begin with the births of stellar-mass neutron stars and black holes.

When the core of a massive star has exhausted its supply of nuclear fuel, it collapses to form a neutron star or a black hole. In some cases, the collapse triggers and powers a subsequent explosion of the star's mantle—a supernova explosion. Despite extensive theoretical efforts for more than 30 years, and despite wonderful observational data from Supernova 1987A, theorists are still far from a definitive understanding of the details of the collapse and explosion. The details are highly complex and may differ greatly from one core collapse to another [24].

Several features of the collapse and the core's subsequent evolution can produce significant gravitational radiation in the high-frequency band. We shall consider these features in turn, the most weakly radiating first, and we shall focus primarily on collapses that produce neutron stars rather than black holes.

3.4.1 Boiling of a newborn neutron star

Even if the collapse is spherical, so it cannot radiate any gravitational waves at all, it should produce a convectively unstable neutron star that "boils" vigorously (and nonspherically) for the first ~ 1 sec of its life [25]. The boiling dredges up high-temperature nuclear matter ($T \sim 10^{12}$K) from the neutron star's central regions, bringing it to the surface (to the "neutrinosphere"), where it cools by neutrino emission before being swept back downward and reheated. Burrows [26] has pointed out that the boiling should generate $n \sim 100$ cycles of gravitational waves with frequency $f \sim 100$ Hz and amplitude large enough to be detectable by LIGO/VIRGO throughout our galaxy and its satellites. Neutrino detectors have a similar range, and there could be a high scientific payoff from correlated observations of the gravitational waves emitted by the boiling's mass motions and neutrinos emitted from the boiling neutrino-sphere. With neutrinos to trigger on, the sensitivities of LIGO detectors should be about twice as good as shown in figure 3.3.

Recent 3+1-dimensional simulations by Müller and Janka [27] suggest an rms amplitude $h \sim 2 \times 10^{-23}(20\,\mathrm{kpc}/r)$ (where r is the distance to the source), corresponding to a characteristic amplitude $h_c \simeq h\sqrt{n} \sim 2 \times 10^{-22}(20\,\mathrm{kpc}/r)$; cf. Fig. 3.3. (The older 2+1-dimensional simulations gave h_c about 6 times larger than this [27] but presumably were less reliable.) LIGO should be able to detect such waves throughout our galaxy with an amplitude signal-to-noise ratio of about $S/N = 2.5$ in each of its two enhanced 4 km interferometers, and its advanced interferometers should do the same out to 80 kpc distance. (Recall that the $h_{\rm SB}$ curves in fig. 3.3 are drawn at a signal-to-noise ratio of about 5). Although the estimated event rate is only about one every 40 years in our galaxy and not much larger out to 80 kpc, if just one such supernova is detected the correlated neutrino and gravitational-wave observations could bring very interesting insights into the boiling of a newborn neutron star.

3.4.2 Axisymmetric collapse, bounce, and oscillations

Rotation will centrifugally flatten the collapsing core, enabling it to radiate as it implodes. If the core's angular momentum is small enough that centrifugal forces do not halt or strongly slow the collapse before it reaches nuclear densities, then the core's collapse, bounce, and subsequent oscillations are likely to be axially symmetric. Numerical simulations [28, 29] show that in this case the waves from collapse, bounce, and oscillation will be quite weak: the total energy radiated as gravitational waves is not likely to exceed $\sim 10^{-7}$ solar masses (about 1 part in a million of the collapse energy) and might often be much less than this; and correspondingly, the waves' characteristic amplitude will be $h_c \lesssim 3 \times 10^{-21} (30\,{\rm kpc}/r)$. These collapse-and-bounce waves will come off at frequencies ~ 200 Hz to ~ 1000 Hz and will precede the boiling waves by a fraction of a second. Though a little stronger than the boiling waves, they probably cannot be seen by LIGO/VIRGO beyond the local group of galaxies and thus will be a very rare occurrence.

3.4.3 Rotation-induced bars and breakup

If the core's rotation is large enough to strongly flatten the core before or as it reaches nuclear density, then a dynamical or secular instability is likely to break the core's axisymmetry. The core will be transformed into a barlike configuration that spins end-over-end like an American football, and that might even break up into two or more massive pieces. As we shall see below, the radiation from the spinning bar or orbiting pieces *could* be almost as strong as that from a coalescing neutron-star binary (section 3.6.2) and thus could be seen by the LIGO/VIRGO first interferometers out to the distance of the Virgo cluster (where the supernova rate is several per year), by enhanced interferometers out to ~ 100 Mpc (supernova rate several thousand per year), and by advanced interferometers out to several hundred Mpc (supernova rate $\sim$ (a few) $\times 10^4$ per year); cf. figure 3.3. It is far from clear what fraction of collapsing cores will have enough angular momentum to break their axisymmetry, and what fraction of those will actually radiate at this high rate; but even if only $\sim 1/1000$ or $1/10^4$ do so, this could ultimately be a very interesting source for LIGO/VIRGO.

Several specific scenarios for such nonaxisymmetry have been identified:

Centrifugal hangup at ~ 100 km radius: If the precollapse core is rapidly spinning (e.g., if it is a white dwarf that has been spun up by accretion from a companion), then the collapse may produce a highly flattened, centrifugally supported disk with most of its mass at radii $R \sim 100$ km, which then (via instability) may transform itself into a bar or may bifurcate. The bar or bifurcated lumps will radiate gravitational waves at twice their rotation frequency, $f \sim 100$ Hz—the optimal frequency for LIGO/VIRGO interferometers. To shrink on down to ~ 10 km size, this configuration must shed most of its angular momentum. *If* a substantial fraction of the angular momentum goes into gravitational waves, then independently of the strength of the bar, the waves will be nearly as strong as those from a coalescing binary. The reason is this: The waves' amplitude h is proportional to the bar's ellipticity e, the number of cycles n of wave emission is proportional to $1/e^2$, and the characteristic amplitude $h_c \simeq h\sqrt{n}$ is thus independent of the ellipticity and is about the same whether the configuration is a bar or is two lumps [30]. The resulting waves will thus have h_c roughly half as large, at $f \sim 100$ Hz, as the h_c from a NS/NS binary (half as large because each lump might be half as massive as a NS), and the waves will chirp upward in frequency in a manner similar to those from a binary (section 3.6.2).

It may very well be, however, that most of the core's excess angular momentum does *not* go into gravitational waves, but instead goes largely into hydrodynamic waves as the bar or lumps, acting like a propeller, stir up the surrounding stellar mantle. In this case, the radiation will be correspondingly weaker.

Centrifugal hangup at ~ 20 km radius: Lai and Shapiro [31] have explored the case of centrifugal hangup at radii not much larger than the final neutron star, say $R \sim 20$ km. Using compressible ellipsoidal models, they have deduced that, after a brief period of dynamical bar-mode instability with wave emission at $f \sim 1000$ Hz (explored by Houser, Centrella, and Smith [32]), the star switches to a secular instability in which the bar's angular velocity gradually slows while the material of which it is made retains its high rotation speed and circulates through the slowing bar. The slowing bar emits waves that sweep *downward* in frequency through the LIGO/VIRGO optimal band $f \sim 100$ Hz, toward ~ 10 Hz. The characteristic amplitude (fig. 3.3) is only modestly smaller than for the upward-sweeping waves from hangup at $R \sim 100$ km, and thus such waves should be detectable near the Virgo cluster by the first LIGO/VIRGO interferometers, near 100 Mpc by enhanced interferometers, and at distances of a few 100 Mpc by advanced interferometers.

Successive fragmentations of an accreting, newborn neutron star: Bonnell and Pringle [33] have focused on the evolution of the rapidly spinning, newborn neutron star as it quickly accretes more and more mass from the pre-supernova star's inner mantle. If the accreting material carries high angular momentum, it may trigger a renewed bar formation, lump formation, wave emission, and coalescence, followed by more accretion, bar and lump formation, wave emission, and coalescence. Bonnell and Pringle speculate that hydrodynamics, not wave emission, will drive this evolution, but that the total energy going into gravitational waves might be as large as $\sim 10^{-3} M_\odot$. This corresponds to $h_c \sim 10^{-21}(10\,\text{Mpc}/r)$.

3.5 Pulsars: spinning neutron stars

As the neutron star settles down into its final state, its crust begins to solidify (crystalize). The solid crust will assume nearly the oblate axisymmetric shape that centrifugal forces are trying to maintain, with poloidal ellipticity ϵ_p proportional to the square of the angular velocity of rotation. However, the principal axis of the star's moment of inertia tensor may deviate from its spin axis by some small "wobble angle" θ_w, and the star may deviate slightly from axisymmetry about its principal axis; i.e., it may have a slight ellipticity $\epsilon_e \ll \epsilon_p$ in its equatorial plane.

As this slightly imperfect crust spins, it will radiate gravitational waves [34]: ϵ_e radiates at twice the rotation frequency, $f = 2f_{\text{rot}}$ with $h \propto \epsilon_e$, and the wobble angle couples to ϵ_p to produce waves at $f = f_{\text{rot}} + f_{\text{prec}}$ (the precessional sideband of the rotation frequency) with amplitude $h \propto \theta_w \epsilon_p$. For typical neutron-star masses and moments of inertia, the wave amplitudes are

$$h \sim 6 \times 10^{-25} \left(\frac{f_{\text{rot}}}{500\,\text{Hz}}\right)^2 \left(\frac{1\,\text{kpc}}{r}\right) \left(\frac{\epsilon_e \text{ or } \theta_w \epsilon_p}{10^{-6}}\right). \quad (3.3)$$

The neutron star gradually spins down, due in part to gravitational-wave emission but perhaps more strongly due to electromagnetic torques associated with its spinning magnetic field and pulsar emission. This spin-down reduces the strength of centrifugal forces and thereby causes the star's poloidal ellipticity ϵ_p to decrease, with an accompanying breakage and resolidification of its crust's crystal structure (a "starquake"; [35], sec. 10.10 and references therein). In each starquake, θ_w, ϵ_e, and ϵ_p will all change suddenly, thereby changing the amplitudes and frequencies of the star's two gravitational "spectral lines" $f = 2f_{\text{rot}}$ and $f = f_{\text{rot}} + f_{\text{prec}}$. After each quake, there should be a healing period in which the star's fluid core and solid crust, now rotating at different speeds, gradually regain synchronism. By monitoring the amplitudes, frequencies, and phases of the two gravitational-wave spectral lines, and by comparing with timing of the electromagnetic pulsar emission, one might learn much about the physics of the neutron-star interior.

How large will be the quantities ϵ_e and $\theta_w \epsilon_p$? Rough estimates of the crustal shear moduli and breaking strengths suggest an upper limit in the range $\epsilon_{\max} \sim 10^{-4}$ to 10^{-6}, and it might be that typical values are far below this. We are extremely ignorant, and correspondingly there is much to be learned from searches for gravitational waves from spinning neutron stars.

One can estimate the sensitivity of LIGO/VIRGO (or any other broadband detector) to the periodic waves from such a source by multiplying the waves' amplitude h by the square root of the number of cycles over which one might integrate to find the signal, $n = f\hat{\tau}$, where $\hat{\tau}$ is the integration time. The resulting effective signal strength, $h\sqrt{n}$, is larger than h by

$$\sqrt{n} = \sqrt{f\hat{\tau}} = 10^5 \left(\frac{f}{1000\,\text{Hz}}\right)^{1/2} \left(\frac{\hat{\tau}}{4\,\text{months}}\right)^{1/2}. \quad (3.4)$$

Four months of integration is not unreasonable in targeted searches; but for an all-sky, all-frequency search, a coherent integration might not last longer than a few days because of computational limitations associated with having to apply huge numbers of trial neutron-star spin-down corrections and earth-motion doppler corrections [36].

Equations (3.3) and (3.4) for $h\sqrt{n}$ should be compared (i) to the detector's rms broadband noise level for sources in a random direction, $\sqrt{5}h_{\text{rms}}$, to deduce a signal-to-noise ratio, or (ii) to h_{SB} to deduce a sensitivity for high-confidence detection when one does not know the waves' frequency in advance [11]. Such a comparison suggests that the first interferometers in LIGO/VIRGO might possibly see waves from nearby spinning neutron stars, but the odds of success are very unclear.

The deepest searches for these nearly periodic waves will be performed by narrowband detectors, whose sensitivities are enhanced near some chosen frequency at the price of sensitivity loss elsewhere—signal-recycled interferometers [18], resonant-sideband-extraction interferometers [19], or resonant-mass antennas [20, 21] (section 3.3.3). With "advanced-detector technology" and targeted searches, such detectors might be able to find with confidence spinning neutron stars that have [11]

$$(\epsilon_e \text{ or } \theta_w \epsilon_p) \gtrsim 3 \times 10^{-10} \left(\frac{500\,\text{Hz}}{f_{\text{rot}}}\right)^2 \left(\frac{r}{1000\,\text{pc}}\right)^2. \quad (3.5)$$

There may well be a large number of such neutron stars in our galaxy; but it is also conceivable that there are none. We are extremely ignorant.

Some cause for optimism arises from several physical mechanisms that might generate radiating ellipticities large compared to 3×10^{-10}:

- It may be that, inside the superconducting cores of many neutron stars, there are trapped magnetic fields with mean strength $B_{\rm core} \sim 10^{13}$G or even 10^{15}G. Because such a field is actually concentrated in flux tubes with $B = B_{\rm crit} \sim 6 \times 10^{14}$G surrounded by field-free superconductor, its mean pressure is $p_B = B_{\rm core} B_{\rm crit}/8\pi$. This pressure could produce a radiating ellipticity $\epsilon_e \sim \theta_w \epsilon_p \sim p_B/p \sim 10^{-8} B_{\rm core}/10^{13}$G (where p is the core's material pressure).

- Accretion onto a spinning neutron star can drive precession (keeping θ_w substantially nonzero) and thereby might produce measurably strong waves [37].

- If a neutron star is born rotating very rapidly, then it may experience a gravitational-radiation-reaction-driven instability first discovered by Chandrasekhar [38] and elucidated in greater detail by Friedman and Schutz [39]. In this "CFS instability", density waves travel around the star in the opposite direction to its rotation but are dragged forward by the rotation. These density waves produce gravitational waves that carry positive energy as seen by observers far from the star, but negative energy from the star's viewpoint; and because the star thinks it is losing negative energy, its density waves get amplified. This intriguing mechanism is similar to that by which spiral density waves are produced in galaxies. Although the CFS instability was once thought ubiquitous for spinning stars [39, 40], we now know that neutron-star viscosity will kill it, stabilizing the star and turning off the waves, when the star's temperature is above some limit $\sim 10^{10}$K [41] and below some limit $\sim 10^9$K [42]; and correspondingly, the instability may operate only during the first few years of a neutron star's life, when $10^9 {\rm K} \lesssim T \lesssim 10^{10}$K.

3.6 Neutron-star binaries and their coalescence

3.6.1 NS/NS and other compact binaries in our galaxy

The best understood of all gravitational-wave sources are binaries made of two neutron stars. The famous Hulse-Taylor [43, 44] binary pulsar, PSR 1913+16, is an example. At present PSR 1913+16 has an orbital frequency of about 1/(8 hours) and emits its waves predominantly at twice this frequency, roughly 10^{-4} Hz, which is in LISA's low-frequency band (fig. 3.5); but it is too weak for LISA to detect. LISA will be able to search for brighter NS/NS binaries in our galaxy with periods shorter than this.

If conservative estimates [45, 46, 47] based on the statistics of binary pulsar observations are correct, there should be many NS/NS binaries in our galaxy that are brighter in gravitational waves than PSR 1913+16. Those estimates suggest that one compact NS/NS binary is born every 10^5

years in our galaxy and that the brightest NS/NS binaries will fall in the indicated region in figure 3.5, extending out to a high-frequency limit of $\simeq 3 \times 10^{-3}$ Hz (corresponding to a remaining time to coalescence of 10^5 years). The birthrate might be much higher than one per 10^5 years, according to progenitor evolutionary arguments [46, 48, 49, 50, 51, 52], in which case LISA would see brighter and higher-frequency binaries than shown in figure 3.5. LISA's observations should easily reveal the true compact NS/NS birthrate and also the birthrates of NS/BH and BH/BH binaries—classes of objects that have not yet been discovered electromagnetically. For further details see [53, 23]; for estimates of LISA's angular resolution when observing such binaries see [54].

3.6.2 *The final inspiral of a NS/NS binary*

As a result of their loss of orbital energy to gravitational waves, the PSR 1913+16 NSs are gradually spiraling inward at a rate that agrees with general relativity's prediction to within the measurement accuracy (a fraction of a percent) [44]—a remarkable but indirect confirmation that gravitational waves do exist and are correctly described by general relativity. If we wait roughly 10^8 years, this inspiral will bring the waves into the LIGO/VIRGO high-frequency band. As the NSs continue their inspiral, over a time of about 15 minutes the waves will sweep through the LIGO/VIRGO band, from ~ 10 Hz to $\sim 10^3$ Hz, at which point the NSs will collide and merge. It is this last 15 minutes of inspiral, with $\sim 16,000$ cycles of waveform oscillation, and the final merger, that the LIGO/VIRGO network seeks to monitor.

To what distance must LIGO/VIRGO look in order to see such inspirals several times per year? Beginning with our galaxy's conservative, pulsar-observation-based NS/NS event rate of one every 100,000 years (section 3.6.1) and extrapolating out through the universe, one infers an event rate of several per year at 200 Mpc [45, 46, 47]. If arguments based on simulations of binary evolution are correct [46, 48, 49, 50, 51, 52] (section 3.6.1), the distance for several per year could be as small as 23 Mpc—though such a small distance entails stretching all the numbers to near the breaking point of plausibility [46]. If one stretches all numbers to the opposite, most pessimistic extreme, one infers several per year at 1000 Mpc [46]. Whatever may be the true distance for several per year, once LIGO/VIRGO reaches that distance, each further improvement of sensitivity by a factor of 2 will increase the observed event rate by $2^3 \simeq 10$.

Figure 3.3 compares the projected LIGO sensitivities [12] with the wave strengths from NS/NS inspirals at various distances from Earth. From that comparison we see that LIGO's first interferometers can reach 30 Mpc, where the most extremely optimistic estimates predict several per year; the enhanced interferometers can reach 300 Mpc where the binary-pulsar-

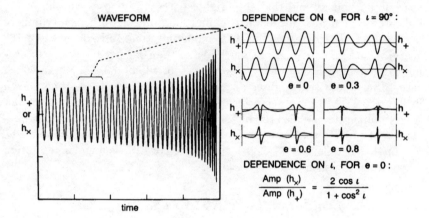

Figure 3.6: Waveforms from the inspiral of a compact binary (NS/NS, NS/BH, or BH/BH), computed using Newtonian gravity for the orbital evolution and the quadrupole-moment approximation for the wave generation [12].

based, conservative estimates predict ~ 10 per year; the advanced interferometers can reach 1000 Mpc, where even the most extremely pessimistic of estimates predict several per year.

3.6.3 Inspiral waveforms and the information they carry

Neutron stars have such intense self-gravity that it is exceedingly difficult to deform them. Correspondingly, as they spiral inward in a compact binary, they do not gravitationally deform each other significantly until several orbits before their final coalescence [55, 56]. This means that the inspiral waveforms are determined to high accuracy by only a few, clean parameters: the masses and spin angular momenta of the stars, and the initial orbital elements (i.e., the elements when the waves enter the LIGO/VIRGO band). The same is true for NS/BH and BH/BH binaries. The following description of inspiral waveforms is independent of whether the binary's bodies are NSs or BHs.

Though tidal deformations are negligible during inspiral, relativistic effects can be very important. If, for the moment, we ignore the relativistic effects—i.e., if we approximate gravity as Newtonian and the wave generation as due to the binary's oscillating quadrupole moment [11]—then the shapes of the inspiral waveforms $h_+(t)$ and $h_\times(t)$ are as shown in figure 3.6.

The left-hand graph in figure 3.6 shows the waveform increasing in amplitude and sweeping upward in frequency (i.e., undergoing a "chirp") as the binary's bodies spiral closer and closer together. The ratio of the amplitudes of the two polarizations is determined by the inclination ι of the orbit to our line of sight (lower right in fig. 3.6). The shapes of the individual waves, i.e., the waves' harmonic content, are determined by the orbital eccentricity (upper right). (Binaries produced by normal stellar evolution should be highly circular due to past radiation reaction forces, but compact binaries that form by capture events, in dense star clusters that might reside in galactic nuclei [57], could be quite eccentric.) If, for simplicity, the orbit is circular, then the rate at which the frequency sweeps or "chirps," df/dt (or equivalently the number of cycles spent near a given frequency, $n = f^2(df/dt)^{-1}$) is determined solely, in the Newtonian/quadrupole approximation, by the binary's so-called *chirp mass*, $M_c \equiv (M_1 M_2)^{3/5}/(M_1 + M_2)^{1/5}$ (where M_1 and M_2 are the two bodies' masses). The amplitudes of the two waveforms are determined by the chirp mass, the distance to the source, and the orbital inclination. Thus (in the Newtonian/quadrupole approximation), by measuring the two amplitudes, the frequency sweep, and the harmonic content of the inspiral waves, one can determine as direct, resulting observables, the source's distance, chirp mass, inclination, and eccentricity [58, 30]. (For binaries at cosmological distances, the observables are the "luminosity distance," "redshifted" chirp mass $(1+z)M_c$, inclination, and eccentricity; cf. section 3.7.2.)

As in binary pulsar observations [44], so also here, relativistic effects add further information: they influence the rate of frequency sweep and produce waveform modulations in ways that depend on the binary's dimensionless ratio $\eta = \mu/M$ of reduced mass $\mu = M_1 M_2/(M_1 + M_2)$ to total mass $M = M_1 + M_2$ and on the spins of the binary's two bodies. These relativistic effects are reviewed and discussed at length in [59, 60]. Two deserve special mention: (i) As the waves emerge from the binary, some of them get backscattered one or more times off the binary's spacetime curvature, producing wave *tails*. These tails act back on the binary, modifying its radiation reaction force and thence its inspiral rate in a measurable way. (ii) If the orbital plane is inclined to one or both of the binary's spins, then the spins drag inertial frames in the binary's vicinity (the "Lense-Thirring effect"), this frame dragging causes the orbit to precess, and the precession modulates the waveforms [59, 61, 62].

Remarkably, the relativistic corrections to the frequency sweep—tails, spin-induced precession, and others—will be measurable with rather high accuracy, even though they are typically $\lesssim 10\%$ of the Newtonian contribution, and even though the typical signal-to-noise ratio will be only ~ 9. The reason is as follows [63, 64, 59]:

The frequency sweep will be monitored by the method of "matched filters"; in other words, the incoming, noisy signal will be cross-correlated with theoretical templates. If the signal and the templates gradually get out of phase with each other by more than $\sim 1/10$ cycle as the waves sweep through the LIGO/VIRGO band, their cross-correlation will be significantly reduced. Since the total number of cycles spent in the LIGO/VIRGO band will be $\sim 16,000$ for a NS/NS binary, ~ 3500 for NS/BH, and ~ 600 for BH/BH, this means that LIGO/VIRGO should be able to measure the frequency sweep to a fractional precision $\lesssim 10^{-4}$, compared to which the relativistic effects are very large. (This is essentially the same method as Joseph Taylor and colleagues use for high-accuracy radio-wave measurements of relativistic effects in binary pulsars [44].)

Analyses using the theory of optimal signal processing predict the following typical accuracies for LIGO/VIRGO measurements based solely on the frequency sweep (i.e., ignoring modulational information) [65]: (i) The chirp mass M_c will typically be measured, from the Newtonian part of the frequency sweep, to $\sim 0.04\%$ for a NS/NS binary and $\sim 0.3\%$ for a system containing at least one BH. (ii) *If* we are confident (e.g., on a statistical basis from measurements of many previous binaries) that the spins are a few percent or less of the maximum physically allowed, then the reduced mass μ will be measured to $\sim 1\%$ for NS/NS and NS/BH binaries and $\sim 3\%$ for BH/BH binaries. (Here and below NS means a $\sim 1.4 M_\odot$ neutron star and BH means a $\sim 10 M_\odot$ black hole.) (iii) Because the frequency dependences of the (relativistic) μ effects and spin effects are not sufficiently different to give a clean separation between μ and the spins, if we have no prior knowledge of the spins, then the spin/μ correlation will worsen the typical accuracy of μ by a large factor, to $\sim 30\%$ for NS/NS, $\sim 50\%$ for NS/BH, and a factor ~ 2 for BH/BH. These worsened accuracies might be improved somewhat by waveform modulations caused by the spin-induced precession of the orbit [61, 62], and even without modulational information, a certain combination of μ and the spins will be determined to a few percent. Much additional theoretical work is needed to firm up the measurement accuracies.

To take full advantage of all the information in the inspiral waveforms will require theoretical templates that are accurate, for given masses and spins, to a fraction of a cycle during the entire sweep through the LIGO/VIRGO band. Such templates are being computed by an international consortium of relativity theorists (Blanchet and Damour in France, Iyer in India, Will and Wiseman in the United States) [66], using post-Newtonian expansions of the Einstein field equations, of the sort pioneered by Chandrasekhar [67, 68]. This enterprise is rather like computing the Lamb shift to high order in powers of the fine structure constant, for comparison with experiment and testing of quantum electrodynamics. Cutler and Flanagan [69] have estimated the order to which the computations must be carried in order that systematic errors in the theoretical templates will not significantly impact the information extracted from the LIGO/VIRGO observational data. The answer appears daunting: radiation reaction effects must be computed to three full post-Newtonian orders (six orders in v/c =(orbital velocity)/(speed of light)) beyond Chandra's leading-order radiation reaction, which itself is five orders in v/c beyond the Newtonian theory of gravity, so the required calculations are $O[(v/c)^{6+5}] = O[(v/c)^{11}]$. By clever use of Padé approximants, these requirements might be relaxed [70].

In the late 1960s, when Chandra and I were first embarking on our respective studies of gravitational waves, Chandra set out to compute the first five orders in v/c beyond Newton, i.e., in his own words, "to solve Einstein's equations through the 5/2 post-Newtonian," thereby fully understanding leading-order radiation reaction and all effects leading up to it. Some colleagues thought his project not worth the enormous personal effort that he put into it. But Chandra was prescient. He had faith in the importance of his effort, and history has proved him right. The results of his "5/2 post-Newtonian" [68] calculation have now been verified to accuracy better than 1% by observations of the inspiral of PSR 1913+16; and the needs of LIGO/VIRGO data analysis are now driving the calculations onward from $O[(v/c)^5]$ to $O[(v/c)^{11}]$. This epitomizes a major change in the field of relativity research: At last, 80 years after Einstein formulated general relativity, experiment has become a major driver for theoretical analyses.

Remarkably, the goal of $O[(v/c)^{11}]$ is achievable. The most difficult part of the computation, the radiation reaction, has been evaluated to $O[(v/c)^9]$ beyond Newton by the French/Indian/American consortium [66], and $O[(v/c)^{11}]$ is now being pursued.

These high-accuracy waveforms are needed only for extracting information from the inspiral waves after the waves have been discovered; they are not needed for the discovery itself. The discovery is best achieved using a different family of theoretical waveform templates, one that covers the space of potential waveforms in a manner that minimizes computation time instead of a manner that ties quantitatively into general relativity theory [59, 71]. Such templates are under development.

3.6.4 NS/NS merger waveforms and the information they carry

The final merger of a NS/NS binary should produce waves that are sensitive to the equation of state of nuclear matter, so such mergers have the potential to teach us about the nuclear equation of state [12, 59]. In essence, LIGO/VIRGO will be studying nuclear physics via the collisions of atomic nuclei that have nucleon numbers $A \sim 10^{57}$—somewhat larger than physicists are normally accustomed to. The accelerator used to drive these "nuclei" up to half the speed of light is the binary's self-gravity, and the radiation by which the details of the collisions are probed is gravitational.

Unfortunately, the NS/NS merger will emit its gravitational waves in the kHz frequency band ($600\,\text{Hz} \lesssim f \lesssim 2500\,\text{Hz}$) where photon shot noise will prevent the waves from being studied by the standard, "workhorse," broadband interferometers of figure 3.3. However, it may be possible to measure the waves and extract their equation-of-state information using a "xylophone" of specially configured narrowband detectors (signal-recycled or resonant-sideband-extraction interferometers, and/or spherical or icosahedral resonant-mass detectors; section 3.3.3 and [59, 72]). Such measurements will be very difficult and are likely only when the LIGO/VIRGO network has reached a mature stage.

A number of research groups ([73] and references therein) are engaged in numerical simulations of NS/NS mergers, with the goal not only to predict the emitted gravitational waveforms and their dependence on the equation of state, but also (more immediately) to learn whether such mergers might power the γ-ray bursts that have been a major astronomical puzzle since their discovery in the early 1970s.

NS/NS mergers are a promising explanation for γ-ray bursts because (i) at least one burst is known, from absorption lines in an optical afterglow, to come from cosmological distances [74], (ii) the bursts have a distribution of number versus intensity that suggests most lie at near-cosmological distances, (iii) their event rate is roughly the same as that conservatively estimated for NS/NS mergers ($\sim$ 1000 per year out to cosmological distances; $\sim$ 1 per year at 300 Mpc); and (iv) it is plausible that the final NS/NS merger will create a γ-emitting fireball with enough energy to account for the bursts and optical afterglow [75, 76]. If enhanced LIGO interferometers were now in operation and observing NS/NS inspirals, they could report definitively whether the γ-bursts are produced by NS/NS binaries; and if the answer were yes, then the combination of γ-burst data and gravitational-wave data could bring valuable information that neither could bring by itself. For example, it would reveal when, to within a few msec, the γ-burst is emitted relative to the moment the NSs first begin to touch; and by comparing the γ and gravitational times of arrival, we could test whether gravitational waves propagate with the speed of light to a fractional precision of $\sim (0.01 \sec)/(10^9 \mathrm{lyr}) = 3 \times 10^{-19}$.

3.6.5 NS/BH Mergers

A NS spiraling into a black hole of mass $M \gtrsim 10 M_\odot$ should be swallowed more or less whole. However, if the BH is less massive than roughly $10 M_\odot$, and especially if it is rapidly rotating, then the NS will tidally disrupt before being swallowed. Little is known about the disruption and accompanying waveforms. To model them with any reliability will likely require full numerical relativity, since the circumferences of the BH and NS will be comparable and their physical separation at the moment of disruption will be of order their sizes. As with NS/NS, the merger waves should carry equation-of-state information and will come out in the kHz band, where their detection will require advanced, specialty detectors.

3.7 Black-hole binaries

3.7.1 BH/BH inspiral, merger, and ringdown

We turn, next, to binaries made of two black holes with comparable masses (BH/BH binaries). The LIGO/VIRGO network can detect and study waves from the last few minutes of the life of such a binary if its total mass is $M \lesssim 1000 M_\odot$ (stellar-mass black holes), cf. figure 3.3; and LISA can do the same for the mass range $1000 M_\odot \lesssim M \lesssim 10^8 M_\odot$ (supermassive black holes), cf. figure 3.5.

The timescales for the binary's dynamics and its waveforms are proportional to its total mass M. All other aspects of the dynamics and waveforms, after time scaling, depend solely on quantities that are dimensionless in geometrized units ($G = c = 1$): the ratio of the two BH masses, the BH spins divided by the squares of their masses, etc. Consequently, the black-hole physics to be studied is the same for supermassive holes in LISA's low-frequency band as for stellar-mass holes in LIGO/VIRGO's high-frequency band. LIGO/VIRGO is likely to make moderate-accuracy studies of this physics; and LISA, flying later, can achieve high accuracy.

The binary's dynamics and its emitted waveforms can be divided into three epochs: *inspiral*, *merger*, and *ringdown* [77]. The inspiral epoch terminates when the holes reach their last stable orbit and begin plunging toward each other. The merger epoch lasts from the beginning of plunge until the holes have merged and can be regarded as a single hole undergoing large-amplitude, quasi-normal-mode vibrations. In the ringdown epoch, the hole's vibrations decay due to wave emission, leaving finally a quiescent, spinning black hole.

The inspiral epoch has been well studied theoretically using post-Newtonian expansions (section 3.6.3), except for the last factor ~ 3 of upward frequency sweep, during which the post-Newtonian expansions may fail. The challenge of computing this last piece of the inspiral is called the "intermediate binary black hole problem" (IBBH) and is a subject of current research in my own group and elsewhere. The merger epoch can be studied theoretically only via supercomputer simulations. Techniques for such simulations are being developed by several research groups, including an eight-university American consortium of numerical relativists and computer scientists called the Binary Black Hole Grand Challenge Alliance [78]. Chandrasekhar and Detweiler [79, 80] pioneered the study of the ringdown epoch using the Teukolsky formalism for first-order perturbations of spinning (Kerr) black holes (see Chandra's classic book [81]), and the ringdown is now rather well understood except for the strengths of excitation of the various vibrational modes, which the merger observations and computations should reveal.

The merger epoch, as yet, is very poorly understood. We can expect it to consist of large-amplitude, highly nonlinear vibrations of spacetime curvature—a phenomenon of which we have very little theoretical understanding today. Especially fascinating will be the case of two spinning black holes whose spins are not aligned with each other or with the orbital angular momentum. Each of the three angular momentum vectors (two spins, one orbital) will drag space in its vicinity into a tornado-like swirling motion—the general relativistic "dragging of inertial frames"—so the binary is rather like two tornados with orientations skewed to each other, embedded inside a third, larger tornado with a third orientation. The dynamical evolution of such a complex configuration of coalescing spacetime warpage, as revealed by its emitted waves, might bring us surprising new insights into relativistic gravity [12].

3.7.2 BH/BH signal strengths and detectability

Flanagan and Hughes [77] have recently estimated the signal strengths produced in LIGO and in LISA by the waves from equal-mass BH/BH binaries for each of the three epochs, inspiral, merger, and ringdown; and along with signal strengths, they have estimated the distances to which LIGO and LISA can detect the waves. In their estimates, Flanagan and Hughes make plausible assumptions about the unknown aspects of the waves. The estimated signal strengths are shown in figure 3.7 for the first LIGO interferometers, figure 3.8 for advanced LIGO interferometers, and figure 3.9 for LISA. Because LIGO and LISA can both reach out to cosmological distances, these figures are drawn in a manner that includes cosmological effects: they are valid for any homogeneous, isotropic model of our universe. This is achieved by plotting observables that are extracted from the measured waveforms: the binary's "redshifted" total mass $(1+z)M$ on the horizontal axis (where z is the source's cosmological redshift) and its "luminosity distance" ([82], pp. 40–43) on the right axis. The signal-to-noise ratio (left axis) scales inversely with the luminosity distance.

We have no good observational handle on the coalescence rate of stellar-mass BH/BH binaries. However, for BH/BH binaries with total mass $M \sim 5\text{--}50 M_\odot$ that arise from ordinary main-sequence progenitors, estimates based on the progenitors' birthrates and on simulations of their subsequent evolution suggest a coalescence rate in our galaxy of one per 1–30 million years [51, 48]. These rough estimates imply that to see one coalescence per year with $M \sim 5\text{--}50 M_\odot$, LIGO/VIRGO must reach out to a distance $\sim 300\text{--}900$ Mpc. Other plausible scenarios (e.g., BH/BH binary formation in dense stellar clusters that reside in globular clusters and galactic nuclei [57]) could produce higher event rates and larger masses, but little that is reliable is known about them (cf. sec. I.A.ii of [77]).

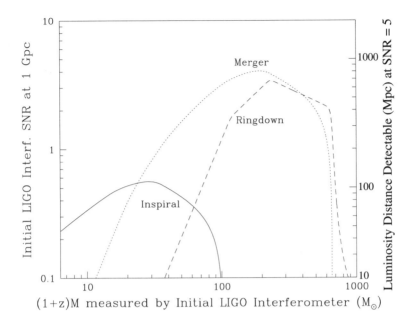

Figure 3.7: The inspiral, merger, and ringdown waves from equal-mass black-hole binaries as observed by LIGO's first interferometers: The luminosity distance to which the waves are detectable (right axis) and the signal-to-noise ratio for a binary at 1 Gpc (left axis), as functions of the binary's redshifted total mass (bottom axis). (Figure adapted from Flanagan and Hughes [77].)

For comparison, the first LIGO interferometers can reach 300 Mpc for $M = 50 M_\odot$ but only 40 Mpc for $M = 5 M_\odot$ (fig. 3.7); enhanced interferometers can reach about 10 times farther, and advanced interferometers about 30 times farther (fig. 3.8). These numbers suggest that (i) if waves from BH/BH coalescences are not detected by the first LIGO/VIRGO interferometers, they are likely to be detected along the way from the first interferometers to the enhanced; and (ii) BH/BH coalescences might be detected sooner than NS/NS coalescences (cf. section 3.6.2).

For binaries with $(1 + z)M \gtrsim 40 M_\odot$, the highly interesting merger signal should be stronger than the inspiral signal, and for $M \gtrsim 100 M_\odot$, the ringdown should be stronger than inspiral (fig. 3.7). Thus, it may well be that early in the life of the LIGO/VIRGO network, observers and theorists will be struggling to understand the merger of binary black holes by comparing computed and observed waveforms.

LIGO's advanced interferometers (fig. 3.8) can see the merger waves for $20 M_\odot \lesssim M \lesssim 200 M_\odot$ out to a cosmological redshift $z \simeq 5$; and for binaries at $z = 1$ in this mass range, they can achieve a signal-to-noise ratio (as-

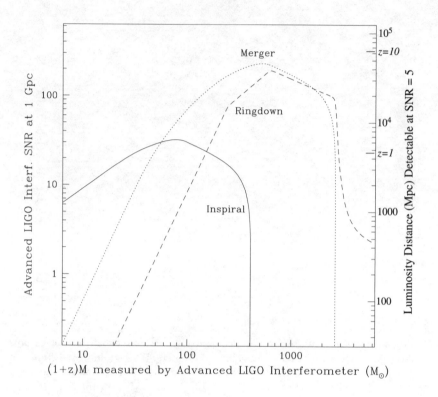

Figure 3.8: The waves from equal-mass black-hole binaries as observed by LIGO's advanced interferometers; cf. the caption for fig. 3.7. On the right side is shown not only the luminosity distance to which the signals can be seen (valid for any homogeneous, isotropic cosmology), but also the corresponding cosmological redshift z, assuming vanishing cosmological constant, a spatially flat universe, and a Hubble constant $H_0 = 75$ km/sec/Mpc. (Figure adapted from Flanagan and Hughes [77].)

suming optimal signal processing [77]) of about 25 in each interferometer.

While these numbers are impressive, they pale by comparison with LISA (fig. 3.9), which can detect the merger waves for $1000 M_\odot \lesssim M \lesssim 10^5 M_\odot$ out to redshifts $z \sim 3000$ (far earlier in the life of the universe than the era when the first supermassive black holes are likely to have formed). Correspondingly, LISA can achieve signal-to-noise ratios of thousands for mergers with $10^5 M_\odot \lesssim M \lesssim 10^8 M_\odot$ at redshifts of order unity, and from the inspiral waves can infer the binary's parameters (redshifted masses, luminosity distance, direction, etc.) with high accuracy [54].

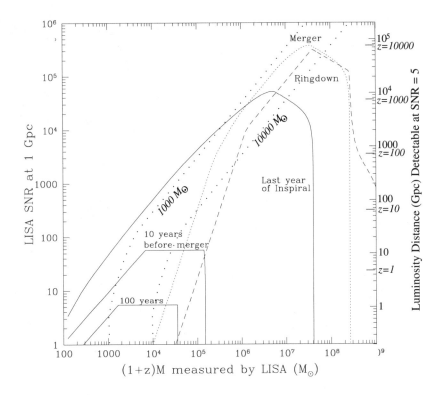

Figure 3.9: The waves from equal-mass, supermassive black-hole binaries as observed by LISA in one year of integration time; cf. the captions for figs. 3.7 and 3.8. The widely spaced dots are curves of constant binary mass M, for use with the right axis, assuming vanishing cosmological constant, a spatially flat universe, and a Hubble constant $H_0 = 75$ km/sec/Mpc. The bottommost curves are the signal strengths after one year of signal integration, for BH/BH binaries 10 years and 100 years before their merger. (Figure adapted from Flanagan and Hughes [77].)

Unfortunately, it is far from obvious whether the event rate for such supermassive BH/BH coalescences will be interestingly high. Conservative estimates suggest a rate of ~ 0.1 per year, while plausible scenarios for aspects of the universe about which we are rather ignorant can give rates as high as 1000 per year [83].

If the coalescence rate is only 0.1 per year, then LISA should still see ~ 3 BH/BH binaries with $3000 M_\odot \lesssim M \lesssim 10^5 M_\odot$ that are ~ 30 years away from their final merger. These slowly inspiraling binaries should be visible, with one year of integration, out to a redshift $z \sim 1$ (bottom part of fig. 3.9).

3.8 Payoffs from binary coalescence observations

Among the scientific payoffs that should come from LIGO/VIRGO's and/or LISA's observations of binary coalescence are the following; others have been discussed above.

3.8.1 Christodoulou memory

As the gravitational waves from a binary's coalescence depart from their source, the waves' energy creates (via the nonlinearity of Einstein's field equations) a secondary wave called the "Christodoulou memory" [84, 85, 86]. This memory, arriving at Earth, can be regarded rigorously as the combined gravitational field of all the gravitons that have been emitted in directions other than toward the Earth [85]. The memory builds up on the timescale of the primary energy emission profile and grows most rapidly when the primary waves are being emitted most strongly: during the end of inspiral and the merger. Unfortunately, the memory is so weak that in LIGO only advanced interferometers have much chance of detecting and studying it—and then, only for BH/BH coalescences and not for NS/NS [87]. LISA, by contrast, should easily be able to measure the memory from supermassive BH/BH coalescences.

3.8.2 Testing general relativity

Corresponding to the very high post-Newtonian order to which a binary's inspiral waveforms must be computed for use in LIGO/VIRGO and LISA data analysis (section 3.6.3), measurements of the inspiral waveforms can be used to test general relativity with very high accuracy. For example, in scalar-tensor theories (some of which are attractive alternatives to general relativity [88]), radiation reaction due to emission of scalar waves places a unique signature on the measured inspiral waveforms—a signature that can be searched for with high precision [89]. Similarly, the inspiral waveforms can be used to measure with high accuracy several fascinating general relativistic phenomena in addition to the Christodoulou memory: the influence of the tails of the emitted waves on radiation reaction in the binary (section 3.6.3), the Lense-Thirring orbital precision induced by the binary's spins (section 3.6.3), and a unique relationship among the multipole moments of a quiescent black hole which is dictated by a hole's "two-hair theorem" (section 3.8.4).

The ultimate test of general relativity will be detailed comparisons of the predicted and observed waveforms from the highly nonlinear spacetime-warpage vibrations of BH/BH mergers (section 3.7.1).

3.8.3 Cosmological measurements

Binary inspiral waves can be used to measure the universe's Hubble constant, deceleration parameter, and cosmological constant [58, 30, 90, 91]. The keys to such measurements are the following: (i) Advanced interferometers in LIGO/VIRGO will be able to see NS/NS inspirals out to cosmological redshifts $z \sim 0.3$ and NS/BH out to $z \sim 2$. (ii) The direct observables that can be extracted from the observed waveforms include a source's luminosity distance (measured to an accuracy $\sim 10\%$ in a large fraction of cases) and its direction on the sky (to accuracy ~ 1 square degree)—accuracies good enough that only one or a few electromagnetically observed clusters of galaxies should fall within the 3-dimensional gravitational error boxes. This should make possible joint gravitational/electromagnetic statistical studies of our universe's magnitude-redshift relation, with gravity giving luminosity distances and electromagnetism giving the redshifts [58, 30]. (iii) Another direct gravitational observable is any redshifted mass $(1+z)M$ in the system. Since the masses of NSs in binaries seem to cluster around $1.4M_\odot$, measurements of $(1+z)M$ can provide a handle on the redshift, even in the absence of electromagnetic aid; so gravitational-wave observations alone may be used, in a statistical way, to measure the magnitude-redshift relation [90, 91].

LISA, with its ability to detect BH/BH binaries with $M \sim 1000$–$100,000 M_\odot$ out to redshifts of thousands, could search for the earliest epochs of supermassive black-hole activity—if the universe is kind enough to grant us a large event rate.

3.8.4 Mapping quiescent black holes; searching for exotic relativistic bodies

Ryan [92] has shown that, when a white dwarf, neutron star, or small black hole spirals into a much more massive, compact central body, the inspiral waves will carry a "map" of the massive body's external spacetime geometry. Since the body's spacetime geometry is uniquely characterized by the values of the body's multiple moments, we can say equivalently that the inspiral waves carry, encoded in themselves, the values of all the body's multipole moments.

By measuring the inspiral waveforms and extracting their map (i.e., measuring the lowest few multipole moments), we can determine whether the massive central body is a black hole or some other kind of exotic compact object [92]; see below.

The inspiraling object's orbital energy E at fixed frequency f (and correspondingly at fixed orbital radius a) scales as $E \propto \mu$, where μ is the object's mass; the gravitational-wave luminosity $\dot E$ scales as $\dot E \propto \mu^2$; and the time to final merger thus scales as $t \sim E/\dot E \propto 1/\mu$. This means that the smaller is μ/M (where M is the central body's mass), the more orbits are spent in the central body's strong-gravity region, $a \lesssim 10GM/c^2$, and thus the more detailed and accurate will be the map of the body's spacetime geometry encoded in the emitted waves.

For holes observed by LIGO/VIRGO, the most extreme mass ratio that we can hope for is $\mu/M \sim 1M_\odot/300M_\odot$, since for $M > 300M_\odot$ the inspiral waves are pushed to frequencies below the LIGO/VIRGO band. This limit on μ/M seriously constrains the accuracy with which LIGO/VIRGO can hope to map the spacetime geometry. A detailed study by Ryan [93] (but one that is rather approximate because we do not know the full details of the waveforms) suggests that LIGO/VIRGO might *not* be able to distinguish cleanly between quiescent black holes and other types of massive central bodies.

By contrast, LISA can observe the final inspiral waves from objects of any mass $\mu \gtrsim 1M_\odot$ spiraling into central bodies of mass $3\times 10^5 M_\odot \lesssim M \lesssim 3\times 10^7 M_\odot$ out to 3 Gpc. Figure 3.5 shows the example of a $10M_\odot$ black hole spiraling into a $10^6 M_\odot$ black hole at 3 Gpc distance. The inspiral orbit and waves are strongly influenced by the hole's spin. Two cases are shown [94]: an inspiraling circular orbit around a nonspinning hole and a prograde, circular, equatorial orbit around a maximally spinning hole. In each case the dot at the upper left end of the arrowed curve is the frequency and characteristic amplitude one year before the final coalescence. In the nonspinning case, the small hole spends its last year spiraling inward from $r \simeq 7.4GM/c^2$ (3.7 Schwarzschild radii) to its last stable circular orbit at $r = 6GM/c^2$ (3 Schwarzschild radii). In the maximal spin case, the last year is spent traveling from $r = 6GM/c^2$ (3 Schwarzschild radii) to the last stable orbit at $r = GM/c^2$ (half a Schwarzschild radius). The $\sim 10^5$ cycles of waves during this last year should carry, encoded in themselves, rather accurate values for the massive hole's lowest few multipole moments [92, 93] (or, equivalently, a rather accurate map of the hole's spacetime geometry).

If the measured moments satisfy the black-hole "two-hair" theorem (usually incorrectly called the "no-hair" theorem), i.e., if they are all determined uniquely by the measured mass and spin in the manner of the Kerr metric, then we can be sure the central body is a black hole. If they violate the two-hair theorem, then (assuming general relativity is correct) either the central body was an exotic object—e.g., a spinning boson star which should have three "hairs" [95], a soliton star [96], or a naked singularity—rather than a black hole, or else an accretion disk or other material was perturbing its orbit [97]. From the evolution of the waves one can hope to determine which is the case, and to explore the properties of the central body and its environment [98].

Models of galactic nuclei, where massive holes (or other massive central bodies) reside, suggest that inspiraling stars and small holes typically will

Chapter 3. Probing Black Holes ...

be in rather eccentric orbits [99, 100]. This is because they get injected into such orbits via gravitational deflections off other stars, and by the time gravitational radiation reaction becomes the dominant orbital driving force, there is not enough inspiral left to strongly circularize their orbits. Such orbital eccentricity will complicate the waveforms and complicate the extraction of information from them. Efforts to understand the emitted waveforms, for central bodies with arbitrary multipole moments, are just now getting underway [92, 101]. Even for central black holes, those efforts are at an early stage; for example, only recently have we learned how to compute the influence of radiation reaction on inspiraling objects in fully relativistic, nonequatorial orbits around a black hole [102, 103].

The event rates for inspiral into supermassive black holes (or other supermassive central bodies) are not well understood. However, since a significant fraction of all galactic nuclei are thought to contain supermassive holes, and since white dwarfs and neutron stars, as well as small black holes, can withstand tidal disruption as they plunge toward a supermassive hole's horizon, and since LISA can see inspiraling bodies as small as $\sim 1 M_\odot$ out to 3 Gpc distance, the event rate is likely to be interestingly large. Sigurdsson and Rees [100] give a "very conservative" estimate of one inspiral event per year within 1 Gpc distance, and 100 to 1000 sources detectable by LISA at lower frequencies "en route" toward their final plunge.

3.9 Conclusion

It is now 37 years since Joseph Weber initiated his pioneering development of gravitational-wave detectors [1], 26 years since Robert Forward [104] and Rainer Weiss [2] initiated work on interferometric detectors, and about 35 years since Chandra and others launched the modern era of theoretical research on relativistic stars and black holes. Since then, hundreds of talented experimental physicists have struggled to improve the sensitivities of gravitational-wave detectors, and hundreds of theorists have explored general relativity's predictions for stars and black holes.

These two parallel efforts are now intimately intertwined and are pushing toward an era in the not distant future when measured gravitational waveforms will be compared with theoretical predictions to learn how many and what kinds of relativistic objects *really* populate our universe, and how these relativistic objects *really* are structured and *really* behave when quiescent, when vibrating, and when colliding.

3.10 Acknowledgments

My group's research on gravitational waves and their relevance to LIGO/ VIRGO and LISA is supported in part by NSF grants AST-9417371 and

PHY-9424337 and by NASA grant NAGW-4268/NAG5-4351. Large portions of this article were adapted and updated from [105].

References

[1] J. Weber. *Phys. Rev.*, **117**, 306 (1960).

[2] R. Weiss. *Quarterly Progress Report of RLE, MIT*, **105**, 54 (1972).

[3] M. S. Turner. *Phys. Rev. D*, **55**, R435 (1997).

[4] L. P. Grishchuk. *Phys. Rev. D*, **53**, 6784 (1996).

[5] M. Gasperini, M. Giovannini, and G. Veneziano. *Phys. Rev. D*, **52**, R6651 (1995).

[6] V. M. Kaspi, J. H. Taylor, and M. F. Ryba. *Astrophys. J.*, **428**, 713 (1994).

[7] Ya. B. Zel'dovich. *Mon. Not. Roy. Astron. Soc.*, **192**, 663 (1980).

[8] A. Vilenkin. *Phys. Rev. D*, **24**, 2082 (1981).

[9] A. Kosowsky and M. S. Turner. *Phys. Rev. D*, **49**, 2837 (1994).

[10] X. Martin and A. Vilenkin. *Phys. Rev. Lett.*, **77**, 2879 (1996).

[11] K. S. Thorne. In S. W. Hawking and W. Israel, editors, *Three Hundred Years of Gravitation*, 330 (Cambridge University Press, 1987).

[12] A. Abramovici et al. *Science*, **256**, 325 (1992).

[13] A. Abramovici et al. *Phys. Lett. A*, **218**, 157 (1996).

[14] C. Bradaschia et al. *Nucl. Instrum. & Methods*, **A289**, 518 (1990).

[15] B. Barish and G. Sanders et al. LIGO advanced R and D program proposal, Caltech/MIT, unpublished (1996).

[16] J. Hough and K. Danzmann et al. GEO600: proposal for a 600 m laser-interferometric gravitational wave antenna, unpublished (1994).

[17] K. Kuroda et al. In I. Ciufolini and F. Fidecaro, editors, *Gravitational Waves: Sources and Detectors*, 100 (World Scientific, 1997).

[18] B. J. Meers. *Phys. Rev. D*, **38**, 2317 (1988).

[19] M. J. Mizuno, K. A. Strain, P. G. Nelson, J. M. Chen, R. Schilling, A. Rudiger, W. Winkler, and K. Danzman. *Phys. Lett. A*, **175**, 273 (1993).

[20] W. W. Johnson and S. M. Merkowitz. *Phys. Rev. Lett.*, **70**, 2367 (1993).

[21] G. Frossati. *J. Low Temp. Phys.*, **101**, 81 (1995).

[22] G. V. Pallottino. In I. Ciufolini and F. Fidecaro, editors, *Gravitational Waves: Sources and Detectors*, 159 (World Scientific, 1997).

[23] P. Bender et al. *LISA, Laser Interferometer Space Antenna for the Detection and Observation of Gravitational Waves: Pre-Phase A Report*. Max-Planck-Institut für Quantenoptik, MPQ 208 (December 1995).

[24] A. G. Petschek, editor. *Supernovae* (Springer Verlag, 1990).

[25] H. A. Bethe. *Rev. Mod. Phys*, **62**, 801 (1990).

Chapter 3. Probing Black Holes ...

[26] A. Burrows. Private communication (1994).
[27] E. Müller and H.-T. Janka. *Astron. Astrophys.*, **317**, 140 (1997).
[28] L. S. Finn. *Ann. N.Y. Acad. Sci.*, **631**, 156 (1991).
[29] R. Mönchmeyer, G. Schäfer, E. Müller, and R. E. Kates. *Astron. Astrophys.*, **256**, 417 (1991).
[30] B. F. Schutz. *Class. Quant. Grav.*, **6**, 1761 (1989).
[31] D. Lai and S. L. Shapiro. *Astrophys. J.*, **442**, 259 (1995).
[32] J. L. Houser, J. M. Centrella, and S. C. Smith. *Phys. Rev. Lett.*, **72**, 1314 (1994).
[33] I. A. Bonnell and J. E. Pringle. *Mon. Not. Roy. Astron. Soc.*, **273**, L12 (1995).
[34] M. Zimmermann and E. Szedenits. *Phys. Rev. D*, **20**, 351 (1979).
[35] S. L. Shapiro and S. A. Teukolsky. *Black Holes, White Dwarfs, and Neutron Stars* (Wiley, 1983).
[36] P. Brady, T. Creighton, C. Cutler, and B. Schutz. *Phys. Rev. D*, submitted (1997); gr-qc/9702050.
[37] B. F. Schutz. Private communication (1995).
[38] S. Chandrasekhar. *Phys. Rev. Lett.*, **24**, 611 (1970).
[39] J. L. Friedman and B. F. Schutz. *Astrophys. J.*, **222**, 281 (1978).
[40] R. V. Wagoner. *Astrophys. J.*, **278**, 345 (1984).
[41] L. Lindblom. *Astrophys. J.*, **438**, 265 (1995).
[42] L. Lindblom and G. Mendell. *Astrophys. J.*, **444**, 804 (1995).
[43] R. A. Hulse and J. H. Taylor. *Astrophys. J.*, **324**, 355 (1975).
[44] J. H. Taylor. *Rev. Mod. Phys.*, **66**, 711 (1994).
[45] R. Narayan, T. Piran, and A. Shemi. *Astrophys. J.*, **379**, L17 (1991).
[46] E. S. Phinney. *Astrophys. J.*, **380**, L17 (1991).
[47] E. P. J. van den Heuvel and D. R. Lorimer. *Mon. Not. Roy. Astron. Soc.*, **283**, L37 (1996).
[48] A. V. Tutukov and L. R. Yungelson. *Mon. Not. Roy. Astron. Soc.*, **260**, 675 (1993).
[49] H. Yamaoka, T. Shigeyama, and K. Nomoto. *Astron. Astrophys.*, **267**, 433 (1993).
[50] V. M. Lipunov, K. A. Postnov, and M. E. Prokhorov. *Astrophys. J.*, **423**, L121 (1994); and related, unpublished work.
[51] V. M. Lipunov, K. A. Postnov, and M. E. Prokhorov. *New Astronomy*, submitted; astro-ph/9610016.
[52] S. F. P. Zwart and H. N. Spreeuw. *Astron. Astrophys.*, **312**, 670 (1996).
[53] D. Hils, P. Bender, and R. F. Webbink. *Astrophys. J.*, **360**, 75 (1990).

[54] C. Cutler. *Phys. Rev. D*, submitted (1997); gr/qc-9703068.

[55] C. Kochanek. *Astrophys. J.*, **398**, 234 (1992).

[56] L. Bildsten and C. Cutler. *Astrophys. J.*, **400**, 175 (1992).

[57] G. Quinlan and S. L. Shapiro. *Astrophys. J.*, **321**, 199 (1987).

[58] B. F. Schutz. *Nature*, **323**, 310 (1986).

[59] C. Cutler, T. A. Apostolatos, L. Bildsten, L. S. Finn, E. E. Flanagan, D. Kennefick, D. M. Markovic, A. Ori, E. Poisson, G. J. Sussman, and K. S. Thorne. *Phys. Rev. Lett.*, **70**, 1984 (1993).

[60] C. M. Will. In M. Sasaki, editor, *Relativistic Cosmology*, 83 (Universal Academy Press, 1994).

[61] T. A. Apostolatos, C. Cutler, G. J. Sussman, and K. S. Thorne. *Phys. Rev. D*, **49**, 6274 (1994).

[62] L. E. Kidder. *Phys. Rev. D*, **52**, 821 (1995).

[63] C. Cutler and E. E. Flanagan. *Phys. Rev. D*, **49**, 2658 (1994).

[64] L. S. Finn and D. F. Chernoff. *Phys. Rev. D*, **47**, 2198 (1993).

[65] E. Poisson and C. M. Will. *Phys. Rev. D*, **52**, 848 (1995).

[66] L. Blanchet, T. Damour, B. R. Iyer, C. M. Will, and A. G. Wiseman. *Phys. Rev. Lett.*, **74**, 3515 (1995).

[67] S. Chandrasekhar. *Selected Papers of S. Chandrasekhar*, Vol. 5, *Relativistic Astrophysics* (U. Chicago Press, 1990).

[68] S. Chandrasekhar and F. P. Esposito. *Astrophys. J.*, **160**, 153 (1970).

[69] C. Cutler and E. E. Flanagan. *Phys. Rev. D*, in preparation.

[70] T. Damour and B.S. Sathyaprakash. *Phys. Rev. D*, submitted (1997); gr/qc-9708034

[71] B. J. Owen. *Phys. Rev. D*, **53**, 6749 (1996).

[72] S. A. Hughes, D. Kennefick, D. Laurence, and K. S. Thorne. *Phys. Rev. D*, in preparation.

[73] Z. G. Xing, J. M. Centrella, and S. L. W. McMillan. *Phys. Rev. D*, **54**, 7261 (1996).

[74] M. R. Metzger et al. *Nature*, **387**, 878 (1997).

[75] P. Meszaros. *Ann. N.Y. Acad. Sci.*, **759**, 440 (1995).

[76] S. E. Woosley. *Ann. N.Y. Acad. Sci.*, **759**, 446 (1995).

[77] E. E. Flanagan and S. A. Hughes. *Phys. Rev. D*, submitted (1997); gr/qc-9701039.

[78] Numerical Relativity Grand Challenge Alliance (1995); references and information on the World Wide Web, http://jean-luc.ncsa.uiuc.edu/GC.

[79] S. Chandrasekhar and S. L. Detweiler. *Proc. Roy. Soc. A*, **344**, 441 (1975).

[80] S. L. Detweiler. *Astrophys. J.*, **239**, 292 (1980).

[81] S. Chandrasekhar. *The Mathematical Theory of Black Holes* (Oxford University Press, 1983).

[82] E. W. Kolb and M. S. Turner. *The Early Universe* (Addison-Wesley, 1990).

[83] M. G. Haehnelt. *Mon. Not. Roy. Astron. Soc.*, **269**, 199 (1994).

[84] D. Christodoulou. *Phys. Rev. Lett.*, **67**, 1486 (1991).

[85] K. S. Thorne. *Phys. Rev. D*, **45**, 520 (1992).

[86] A. G. Wiseman and C. M. Will. *Phys. Rev. D*, **44**, R2945 (1991).

[87] D. Kennefick. *Phys. Rev. D*, **50**, 3587 (1994).

[88] T. Damour and K. Nordtvedt. *Phys. Rev. D*, **48**, 3436 (1993).

[89] C. M. Will. *Phys. Rev. D*, **50**, 6058 (1994).

[90] D. Markovic. *Phys. Rev. D*, **48**, 4738 (1993).

[91] D. F. Chernoff and L. S. Finn. *Astrophys. J.*, **411**, L5 (1993).

[92] F. D. Ryan. *Phys. Rev. D*, **52**, 5707 (1995).

[93] F. D. Ryan. *Phys. Rev. D*, **56**, 1845 (1997).

[94] L. S. Finn and K. S. Thorne. *Phys. Rev. D*, in preparation.

[95] F. D. Ryan. Spinning boson stars with large self-interaction. *Phys. Rev. D*, in press (1997).

[96] T. D. Lee and Y. Pang. *Physics Reports*, **221**, 251 (1992).

[97] S. K. Chakrabarti. *Phys. Rev. D*, **53**, 2901 (1996).

[98] F. D. Ryan, L. S. Finn, and K. S. Thorne. *Phys. Rev. Lett.*, in preparation.

[99] D. Hils and P. Bender. *Astrophys. J.*, **445**, L7 (1995).

[100] S. Sigurdsson and M. J. Rees. *Mon. Not. Roy. Astron. Soc.*, **284**, 318 (1997).

[101] F. D. Ryan. Scalar waves produced by a scalar charge orbiting a massive body with arbitrary multipole moments. *Phys. Rev. D*, submitted (1997).

[102] Y. Mino, M. Sasaki, and T. Tanaka. *Phys. Rev. D*, **55**, 3497 (1997).

[103] T. C. Quinn and R. M. Wald. *Phys. Rev. D*, **56**, 3381 (1997); gr-qc/9610053.

[104] G. E. Moss, L. R. Miller, and R. L. Forward. *Applied Optics*, **10**, 2495 (1971).

[105] K. S. Thorne. In E. W. Kolb and R. Peccei, editors, *Proceedings of the Snowmass 95 Summer Study on Particle and Nuclear Astrophysics and Cosmology*, 398 (World Scientific, 1995).

4

Astrophysical Evidence for Black Holes

Martin J. Rees

Abstract

The case for collapsed objects in some X-ray binary systems continues to strengthen. But there is now even firmer evidence for supermassive black holes in galactic centres. Gravitational collapse seems to have occurred in the centres of most newly forming galaxies, manifesting itself in a phase of quasarlike activity (which may be reactivated later). These phenomena (especially the gas-dynamical aspects) are still a daunting challenge to theorists, but there is 'cleaner' evidence, based on stellar dynamics, for collapsed objects in the centres of most nearby galaxies. The current evidence does not tell us the spin of the collapsed objects—nor, indeed, whether they are described by Kerr geometry, as general relativity theory predicts. There are now, however, several hopeful prospects of discovering observational signatures that will indeed probe the strong-gravity domain.

4.1 Introduction

It's fitting to start with a text from Chandrasekhar (1975): 'In my entire scientific life... the most shattering experience has been the realisation that an exact solution of Einstein's equations of general relativity, discovered by the New Zealand mathematician Roy Kerr, provides the absolutely exact representation of untold numbers of massive black holes that populate the Universe.'

When Chandra wrote this, the evidence was still controversial (see Israel 1996): belief in black holes was at least partly an act of faith (defined by St Paul as 'the substance of things hoped for: the evidence of things not seen'). But observational progress has been remarkable, especially within the last couple of years.

I shall first address the question: Do massive collapsed objects exist—stellar-mass objects in binaries, and supermassive objects in the centres of galaxies? The evidence points insistently towards the presence of dark objects, associated with deep gravitational potential wells; but it does not in itself tell us about the metric in the innermost region where Newtonian approximations break down.

The second part of this paper addresses a separate question: Do these objects have Schwarzschild/Kerr metrics? Several new observational probes of the strong-field domain close to the hole have recently become feasible, offering real prospects of crucially testing our theories of strong-field gravity.

4.2 Stellar-mass black hole candidates

It was recognised back in the 1960s that X-ray sources in binary systems, fuelled by the accretion of gas captured from a companion, would be black hole candidates if they displayed rapid irregular flickering, and if the inferred mass were too high for them to be conventional neutron stars. The likelihood of such stellar-mass remnants was of course prefigured by Chandra's classic early work.

The discovered systems divide into two categories: those where the companion star is of high mass, of which Cyg X1 is the prototype, and the low-mass X-ray binaries (LMXBs), where the companion is typically below a solar mass. The LMXBs are sometimes called 'X-ray novae', because they flare up to high luminosities: they plainly have a different evolutionary history from systems like Cyg X1. The prototype of this class is A 0600-00, discovered in 1975. At least five further LMXBs have been discovered more recently and have had their masses estimated; other Galactic X-ray sources are suspected, on spectroscopic and other grounds, to be in the same category. The strongest current candidates are listed in table 4.1, adapted from Charles (1997). Fuller discussion of these systems, and the evolutionary scenarios that might lead to them, are given by Tanaka and Lewin (1995) and Wijers (1996).

None of these black hole candidates displays the kind of regular period that is associated, in other systems, with a neutron star's spin rate. Indeed it is gratifying that, as discussed in John Friedman's contribution (chapter 2 of this volume), the putative neutron stars all have masses clustering around $1.4 M_\odot$, and there are no regularly pulsing X-ray sources with dynamically inferred masses much higher than this. However, some of the high-mass sources display interesting quasi periodicities which (as discussed later) may offer probes of the metric.

Of course the only black holes that manifest themselves as conspicuous X-ray sources are the tiny and atypical fraction located in close binaries where mass transfer is currently going on. There may be only a few dozen

Chapter 4. Astrophysical Evidence for Black Holes

	$M_h/M_\odot$	$M_*/M_\odot$
High-mass companions		
Cyg X1	11–21	24–42
LMC X3	5.6–7.8	20
Low-mass companions (X-ray transients)		
V404 Cyg	10–15	∼ 0.6
A 0620-00	5–17	0.2–0.7
Nova Muscae	4.2–6.5	0.5–0.8
GS 2000+25	6–14	∼ 0.7
GROJ1655-40	4.5–6.5	∼ 1.2
N. Oph 77	5–9	∼ 0.4
J0422432	6–14	∼ 0.3

Table 4.1: Stellar-mass black hole candidates and their binary companions

such systems in our Galaxy. However, there is every reason to suspect that the total number of stellar-mass holes is at least 10^7. This is based on the rather conservative estimate that only 1 or 2 percent of supernovae leave black holes rather than neutron stars. Still larger numbers of holes could indeed exist (maybe even in the Galactic halo) as relics of early Galactic history.

4.3 Supermassive holes

Even at the first 'Texas' conference, held in 1963 when quasars had just been discovered, some theorists were suggesting that gravitational energy, released by a supermassive object, was responsible for the powerful emitted radiation. There has been a huge (and increasingly systematic) accumulation of data on quasars, and on the other classes of active galaxies which are now recognised as being related: the Seyfert galaxies, already noted as a distinctive category 50 years ago, and the strong radio galaxies (known since the 1950s). But our understanding has developed in fits and starts. Even if a convincing explanation has eventually emerged, it sometimes seems as though this happened only after every other possibility had been exhausted.

Recent progress in the study of active galactic nuclei (AGNs) brings into sharper focus the question of how and when supermassive black holes formed, and how this process relates to galaxy formation. Even more important has been the discovery of ordinary galaxies with equally large redshifts: these were until recently too faint to be detected, but they can now be studied with the combined resources of the Hubble Space Tele-

	$M_h/M_\odot$	Method
M87	2×10^9	Stars + opt.disc
NGC 3115	10^9	Stars
NGC 4486 B	5×10^8	Stars
NGC 4594 (Sombrero)	5×10^8	Stars
NGC 3377	8×10^7	Stars
NGC 3379	5×10^7	Stars
NGC 4258	4×10^7	Masing H_2O disc
M31 (Andromeda)	3×10^7	Stars
M32	3×10^6	Stars
Galactic centre	2.5×10^6	Stars + 3-D motions

Table 4.2: Supermassive holes.

scope (HST) and the Keck 10 metre telescope. But the most clear-cut and quantitative clues have come from studies of relatively nearby galaxies: the centres of most of these display either no activity or a rather low level, but most seem to harbour dark central masses. I shall summarise this evidence and then outline how it fits in with the broad picture of galaxy formation and evolution that is now coming into focus.

4.3.1 Evidence from the stellar cusp—M31 (Andromeda) and others

Central dark masses—dead quasars—have been inferred from studies of the spatial distribution and velocities of stars in several nearby galaxies (for recent reviews see Kormendy and Richstone 1995; Tremaine 1997; or van den Marel 1996). There is, for instance, strong evidence for a mass of about $3 \times 10^7 M_\odot$ in the centre of Andromeda (M31). Even in such a nearby galaxy as this, the hole's gravitational effects on surrounding stars are restricted to the central 2–3 arc seconds of the galaxy's image. Higher-resolution data from the postrefurbishment HST should crucially clarify what is going on in these systems. A list of the candidates (as of January 1997) is given in table 4.2.

4.3.2 M87: low-level activity and a supermassive hole

Although M87, in the Virgo cluster, was the first in which a tightly bound stellar cusp was claimed (Sargent et al. 1978; Young et al. 1978), the stellar dynamics in the core of this giant elliptical galaxy even now remains rather ambiguous. According to Merritt and Oh (1997) the data are consistent with a central mass of $(1-2) \times 10^9 M_\odot$, but the radial dependence of the projected densities and velocities could be accounted for by a dense stellar core alone, provided that the velocities were suitably anisotropic. Separate evidence for a dark central mass comes, however, from a disc of gas, orbiting in a plane perpendicular to the well-known jet (Ford et al. 1994). One of the complications in studying the central stars in M87 is that the nucleus is not quiescent, but that the inner part of the jet emits non-thermal light as well as radio waves.

X-ray data reveal hot gas pervading M87 itself, as well as the surrounding cluster. If there were a huge central hole, then some of this gas would inevitably be swirling into it, at a rate that can be estimated. This accretion would give rise to more conspicuous activity than is observed if the radiative efficiency were as high as 10 percent. Fabian and Canizares (1988) were the first to highlight this apparent problem with the hypothesis that elliptical galaxies harbour massive central holes. Actually, the quiescence is less surprising because, for low accretion rates, the expected luminosity scales as $\dot{M}^2$ rather than as $\dot{M}$: when accretion occurs at a low rate, and the viscosity is high enough to ensure that the gas swirls in quickly (so the densities are low), the radiative efficiency also is low. The gas inflates into a thick dilute torus, where the kinetic temperature of the ions is close to the virial temperature. Only a small fraction of the binding energy gets radiated during the time it takes for each element of gas to swirl inward and be swallowed. Bremsstrahlung, in particular, is inefficient in this situation; the most conspicuous emission may be in the radio band, resulting from synchrotron emission in the strong magnetic field in the inner part of the accretion flow. The radio and X-ray emission is actually fully consistent with accretion at the expected rate, so the observed nonstellar output from M87 actually corroborates the other evidence for a supermassive hole (Fabian and Rees 1995; Narayan and Yi 1995; Reynolds et al. 1996; Mahadevan 1997; and references cited therein).

Weak central radio sources are found in the centres of surprisingly many otherwise quiescent ellipticals (Sadler et al. 1995, and references cited therein). If these galaxies all harbour massive holes, this emission could similarly be attributed to accretion in a slow, inefficient, mode (Fabian and Rees 1995)

4.3.3 The remarkable case of NGC 4258

Much the most compelling case for a central black hole has been supplied by a quite different technique: amazingly precise mapping of gas motions via the 1.3 cm maser-emission line of H_2O in the peculiar spiral galaxy NGC 4258 (Watson and Wallin 1994; Miyoshi et al. 1995) which lies at a distance of about 6.5 Mpc. The spectral resolution in the microwave line is high enough to pin down the velocities with accuracy of 1 km/sec. The Very Long Baseline Array achieves an angular resolution better than 0.5

milliarcseconds (100 times sharper than the HST, as well as far finer spectral resolution of velocities!). These observations have revealed, right in the galaxy's core, a disc with rotational speeds following an exact Keplerian law around a compact dark mass. The inner edge of the observed disc is orbiting at 1080 km/sec. It would be impossible to circumscribe, within its radius, a stable and long-lived star cluster with the inferred mass of $3.6 \times 10^7 M_\odot$. The circumstantial evidence for black holes has been gradually growing for 30 years, but this remarkable discovery clinches the case completely. The central mass must either be a black hole or something even more exotic.

NGC 4258 poses several puzzles. What determines the sharp inner edge of the 'masing' disc? What is the significance of its inferred tilt and warping? How does this thin disc relate to the (thicker) 'molecular tori' that have been postulated in Seyfert galaxies? All these questions deserve study. It would help, of course, if other similar discs could be found. Another Seyfert galaxy, NGC 1068, may show resemblances, but NGC 4258 may prove to be an unusually fortunate example, because its disc is viewed almost edge-on.

4.3.4 Our Galactic centre

Most nearby large galaxies seem to harbour massive central holes, so our own would seem underendowed if it did not have one too. There has been theoretical advocacy of this view for many years (e.g. Lynden-Bell and Rees 1971). Also, an unusual radio source has long been known to exist right at the dynamical centre of our Galaxy, which can be interpreted in terms of accretion onto a massive hole (Rees 1982; Melia 1994; Narayan, Yi, and Mahadevan 1995). But the direct evidence has until recently been ambiguous (see Genzel, Townes, and Hollenbach 1994, and earlier work reviewed therein). This is because intervening gas and dust in the plane of the Milky Way prevent us from getting a clear optical view of the central stars, as we can in, for instance, M31. A great deal is known about gas motions, from radio and infrared measurements, but these are hard to interpret because gas does not move ballistically like stars, being vulnerable to pressure gradients, stellar winds, and other nongravitational influences.

The situation has, however, been transformed by remarkable observations of stars in the near infrared band, where obscuration by intervening material is less of an obstacle (Eckart and Genzel 1996). These observations have been made using an instrument (ESO's New Technology Telescope in Chile) with sharp enough resolution to detect the transverse ('proper') motions of some stars over a three-year period. The radial velocities are also known, from spectroscopy, so one has full three-dimensional information on how the stars are moving within the central 0.1 pc of our Galaxy. The speeds, up to 2000 km/sec, scale as $r^{-1/2}$ with distance from the centre,

Chapter 4. Astrophysical Evidence for Black Holes   85

consistent with a hole of mass $2.5 \times 10^6 M_\odot$.

In my opinion our Galactic centre now provides the most convincing case for a supermassive hole, with the single exception of NGC 4258.

4.3.5 The cumulative evidence

A summary of the current evidence is given in table 4.2. The data here (as in table 4.1) are developing rapidly, and the list may well be longer by the time this paper appears in print.

A feature of the data in this table, emphasised by Kormendy and Richstone (1995) and by Faber et al. (1997), is a crude proportionality between the hole's mass and that of the central bulge or spheroid in the stellar distribution (which is of course the dominant part of an elliptical galaxy, but only a subsidiary component of a disc system like M31 or our own Galaxy.) This conclusion is only tentative, being vulnerable to various selection effects, but it suggests that the hole may form at the same time as the central stellar population. In section 4.5 I shall briefly discuss formation scenarios for the holes, in the context of the remarkable recent progress achieved by optical astronomers in probing the era when galaxies were still forming.

4.4 The fate of stars near a supermassive hole

4.4.1 Tidal disruption

Even if a galaxy's core were swept so clean of gas that no significant emission resulted from steady accretion, there is a separate process that would inevitably, now and again, liberate a large supply of gas whenever a supermassive hole was present: tidal disruption of stars on nearly radial orbits. A rough estimate, based on models for the stellar distribution and velocities, suggests that in M31 a main-sequence star would pass close enough to the putative hole to be tidally disrupted about once every 10^4 years. These estimates, for M31 and other nearby galaxies, should firm up when postrefurbishment HST data are available: it is a stellar-dynamical (rather than gas-dynamical) problem and is therefore relatively 'clean' and tractable.

What happens to a star when it is disrupted? Earlier investigations by, for instance, Lacy, Townes, and Hollenbach (1982), Rees (1988), Evans and Kochanek (1989), and Canizzo, Lee, and Goodman (1990) are now being supplemented by more detailed numerical modelling (e.g. Khokhlov, Novikov, and Pethick 1993; Frohlov et al. 1994). The tidally disrupted star, as it moves away from the hole, develops into an elongated banana-shaped structure, the most tightly bound debris (the first to return to the hole) being at one end (Evans and Kochanek 1989; Laguna et al. 1993; Kochanek 1994; Rees 1994). There would not be a conspicuous 'prompt' flare signalling the disruption event, because the thermal energy liberated is trapped within the debris. Much more radiation emerges when the bound debris (by then more diffuse and transparent) falls back onto the hole a few months later, after completing an eccentric orbit. The dynamics and radiative transfer are then even more complex and uncertain than in the disruption event itself, being affected by relativistic precession, as well as by the effects of viscosity and shocks (see Rees 1994, 1996, and earlier work cited therein).

The radiation from the inward-swirling debris would be predominantly thermal, with a temperature of order 10^5 K; however, the energy dissipated by the shocks that occur during the circularisation would provide an extension into the X-ray band. High luminosities would be attained—the total photon energy radiated (up to 10^{53} ergs) could be several thousand times more than the photon output of a supernova, though the bolometric correction could be much larger too. The flares would, moreover, not be standardised—what is observed would depend on the hole's mass and spin, the type of star, the impact parameter, and the orbital orientation relative to the hole's spin axis and the line of sight; perhaps also on absorption in the galaxy. To compute what happens involves relativistic gas dynamics and radiative transfer, in an unsteady flow with large dynamic range, which possesses no special symmetry and therefore requires full three-dimensional calculations—a worthy computational challenge to those who have many gigaflops at their disposal.

4.4.2 Can the 'flares' be detected?

Supernova-type searches with 10^4 galaxy-years of exposure should either detect flares due to this phenomenon, or else place limits on the mean mass of central black holes in nearby galaxies. This possible bonus should be an added incentive for such searches. It is not clear whether the best strategy involves monitoring nearby galaxies over a large area of sky or larger numbers of more remote galaxies. Large numbers of distant galaxies are, for instance, being routinely monitored by S. Perlmutter and colleagues in programmes aimed at discovering supernovae at redshifts of order 0.5. It would be surprising if such programmes did not detect such flares—a negative result will itself be interesting. However, if a 'flare' (with the expected duration of months) happened in a distant galaxy, one would not be able to check just how quiescent the galaxy had previously been. It would be easier to be sure that a detected flare was actually due to a disrupted star (and not just an upward fluctuation in the gaseous accretion rate) if it were observed in a closer galaxy that was known to have previously been inactive.

There has already been possible serendipitous detection of one transient event in the nucleus of a galaxy (Renzini et al. 1995), though its peak luminosity was far below what might be expected. X-ray surveys may also detect the events if (like AGNs) their spectra, though peaking in the UV, display a high-energy tail (Sembay and West 1993). The predicted flares offer a robust diagnostic of the massive holes in quiescent galaxies.

4.4.3 'Fossil events' in our Galactic centre?

The rate of tidal disruptions in our Galactic centre would be no more than once per 10^5 years. But each such event could generate a luminosity several times 10^{44} erg/sec for about a year. Were this in the UV, the photon output, spread over 10^5 years, could exceed the current ionization rate: the mean luminosity of the Galactic centre might exceed the median value.

The resultant fossil ionization would set a lower limit to the electron density. The radiation emitted from the event might reach us after a delay if it were reflected off surrounding material. Churazov et al. (1994) have already used this argument to set a nontrivial constraint on the history of the Galactic centre's X-ray output over the last few thousand years. Half the debris from a disrupted star would be ejected on hyperbolic orbits in a fan (which may intersect an orbiting disc in a line). The structure in the central 2 pc could be a single spiral feature (Lacy 1994). One speculative possibility (Rees 1987) is that this feature may be a 'vapour trail' created by such an event.

4.5 AGN demography and black hole formation

The quasar population peaks at redshifts between 2 and 3 but genuinely seems to be 'thinning out' at higher redshifts, corresponding to still earlier epochs: the comoving density of quasars falls by at least 3 for each unit increase in z beyond 3 (see Shaver 1995 for a review). The impressive complementary strengths of HST and the Keck telescope have revealed galaxies with the same range of high redshifts as the quasar population itself. Many of the faint smudges visible in the Hubble Deep Field (Williams et al. 1996), the deepest picture of the sky ever obtained, are galaxies with redshifts of order 3, being viewed at (or even before) the era when their spheroids formed.

Considerations of AGN 'demography', by now well known, suggest that the ultraluminous quasar phase may have a characteristic lifetime set by the 'Eddington timescale' of 4×10^7 years, being associated with the formation of a black hole or the immediate aftermath of this process. Straightforward arithmetic based on the observed numbers of quasars then implies (albeit with substantial numerical uncertainty because of the poorly known luminosity function, etc.) that most large galaxies could indeed have gone through a quasar phase; they would, in consequence, by $z = 2$ (2–3 billion years) have developed central holes of $10^6 M_\odot$–$10^9 M_\odot$.

Physical conditions in the central potential wells, when galaxies were young and gas rich, should have been propitious for black hole formation. Infalling primordial gas would gradually condense into stars, forming the central spheroid of such systems. But star formation would be quenched when the gas reached some threshold central concentration: as the gas evolved (through loss of energy and angular momentum) to higher densities and more violent internal dissipation, radiation pressure would inevitably puff it up and inhibit further fragmentation (Rees 1993; Haehnelt and Rees 1993). Much of whatever gas remains at this stage would then agglomerate into a massive hole.

This argument can be quantified, at least in a crudely approximate way. A differentially rotating self-gravitating gas mass can dissipate its energy (via nonaxisymmetric instabilities) on a dynamical timescale. Its internally generated luminosity can then be expressed in terms of its virial velocity $v = (GM/r)^{1/2}$ as

$$L = v^5/G = 10^{59}(v/c)^5 \text{ erg/sec.} \tag{4.1}$$

Note the straightforward analogy to the 'maximal power' c^5/G familiar to relativists and gravitational wave experimenters. This luminosity reaches the Eddington limit when v is high enough: the gas is then puffed up by radiation pressure, and fragmentation is no longer possible. The criterion is

$$v > 300(M/10^6 M_\odot)^{1/5} \text{ km/sec.} \tag{4.2}$$

This criterion may be fulfilled for the entire gas mass, or for the inner part of a self-gravitating disc. Moreover, while sufficient, 4.2 is by no means necessary: fragmentation may be inhibited at a substantially earlier stage by the effects of higher opacity than electron scattering alone provides, or by magnetic stresses. If fragmentation is inhibited, collapse to a supermassive black hole seems almost inevitable. To evade such an outcome, either:

1. Stars must form (before 4.2 is satisfied) with nearly 100 percent efficiency; moreover, they must all have low mass (so that no material is expelled again) *or*

2. Gas must remain in a self-gravitating disc for hundreds of orbital periods, without the onset of any instability that redistributes angular momentum and allows the inner fraction to collapse enough to cross the threshold when 4.2 applies.

Neither of these 'escape routes' seems at all likely—the first would require the stars to have an initial mass function quite different from what is actually observed in the spheroids of galaxies; the second is contrary to well-established arguments that self-gravitating discs are dynamically unstable.

This process involves complex gas dynamics and feedback from stars; we are still a long way from being able to make realistic calculations. At the moment, the most compelling argument that a massive black hole is an expected byproduct comes from the implausibility of the alternatives. The mass of the hole would depend on that of its host galaxy, though not necessarily via an exact proportionality: the angular momentum of the protogalaxy and the depth of its central potential well are relevant factors too. A more quantitative estimate depends on calculating in full detail when, during the progressive concentration towards the centre, star formation ceases (because of radiation pressure, magnetic fields, or whatever) and the remaining gas evolves instead into a supermassive object.

Once a large mass of gas became too condensed to fragment into stars, it would continue to contract and deflate. Some mass would inevitably be shed, carrying away angular momentum, but the remainder could continue contracting until it underwent complete gravitational collapse. This could be a substantial fraction—for example, if 10 percent of the mass had to be shed in order to allow contraction by a factor of 2, about 20 percent could form a black hole.

Firmer and more quantitative conclusions will have to await elaborate numerical simulations. But on one issue I would already bet strongly. This is that a massive black hole forms directly from gas (some, albeit, already processed through stars), perhaps after a transient phase as a supermassive object, rather than from coalescence of stars or mergers of stellar-mass holes.

The energy radiated during further growth of the hole manifests itself as a quasar. The peak in the quasar population (i.e. redshifts in the range 2–3) signifies the era when large galactic spheroids were forming in greatest profusion. It is worth noting, incidentally, that whereas activity in low-z galaxies may be correlated with some unusual disturbance due to a tidal encounter or merger, this may not be the right way to envisage the more common high-z quasars. Any newly formed galaxy is inevitably 'disturbed', in the sense that it has not yet had time to settle down and relax: no external influence is needed to perturb axisymmetry or to trigger a large inflow of gas.

4.6 Do the candidate holes obey the Kerr metric?

4.6.1 Probing the region near the hole

As already discussed in section 4.3, NGC 4258 offers the clearest evidence so far for a central dark mass. But the observed molecular disc lies a long way out: at around 10^5 gravitational radii. We can exclude all conventional alternatives (dense star clusters, etc.); however, the measurements tell us nothing about the central region where gravity is strong, certainly not whether the putative hole actually has properties consistent with the Kerr metric. The stars in the central parts of M31 and our own Galaxy likewise lie so far out that their orbits are essentially Newtonian.

The phenomena of AGNs are due to material closer to the central mass, but nobody could yet claim that any observed features of AGNs offer a clear diagnostic of a Kerr metric. All we can really infer is that 'gravitational pits' exist, which must be deep enough to allow several percent of the rest mass of infalling material to be converted into kinetic energy and then radiated away from a region compact enough to vary on timescales as short as an hour. General relativity has been resoundingly vindicated in the weak-field limit (by high-precision observations in the solar system and of the binary pulsar) but we still lack quantitative probes of the strongly relativistic region.

The tidal disruption events described in section 4.4 depend crucially on distinctive precession effects around a Kerr metric, but the gas dynamics are so complex and messy that even when a flare is detected it will not serve as a useful diagnostic of the metric in the strong-field domain. On the other hand, the stars whose motions reveal a central dark mass in our Galactic centre, in M31, and in other normal galaxies are in orbits $\gtrsim 10^5$ times larger than the putative holes themselves.

Relativists would seize eagerly on any relatively 'clean' probe of the relativistic domain. In most accretion flows, the emission is concentrated towards the centre, where the potential well is deepest and the motions fastest. Such basic features of the phenomenon as the overall efficiency, the minimum variability timescale, and the possible extraction of energy from the hole itself all depend on inherently relativistic features of the metric— on whether the hole is spinning or not, how it is aligned, etc. There are now several encouraging new possibilities.

4.6.2 X-ray spectroscopy of accretion flows

Optical spectroscopy tells us a great deal about the gas in AGNs. However, the optical inferences pertain to gas that is quite remote from the hole itself. This is because the innermost regions would be so hot that their thermal emission would emerge as more energetic quanta: the optical observations sample radiation that is emitted (or at least reprocessed) further out. The X-rays, on the other hand, come predominantly from the relativistic region. Until recently, however, the energy resolution and sensitivity of X-ray detectors was inadequate to permit the study of line shapes. But this is now changing. The ASCA X-ray satellite was the first to offer sufficient spectral resolution to reveal line profiles, and it therefore opened up the possibility of seeking the substantial gravitational redshifts, as well as large doppler shifts, that would be expected (Fabian et al. 1989, and earlier references

Chapter 4. Astrophysical Evidence for Black Holes   91

cited therein). There is already one convincing case (Tanaka et al. 1995) of a broad asymmetric emission line indicative of a relativistic disc viewed at $\sim 60°$ to its plane, and others are now being found. The value of a/m can in principle be constrained too, because the emission is concentrated closer in, and so displays larger shifts, if the hole is rapidly rotating (Iwasawa et al. 1996).

The appearance of a disc around a hole, taking doppler and gravitational shifts into account, along with light bending, was calculated by Cunningham and Bardeen (1973) and by several other authors. The associated swing in the polarization vector of photon trajectories near a hole was also long ago suggested (Connors, Piran, and Stark 1980) as another diagnostic; but this is still not feasible because X-ray polarimeters are far from capable of detecting the few percent polarization expected.

4.6.3 Stars in relativistic orbits?

These X-ray observations are of course of Seyfert galaxies, whose centres, though not emitting as powerfully as quasars, are by no means inactive. But we still need a 'cleaner' and more quantitative probe of the strong-field regime.

A small star orbiting close to a supermassive hole would behave like a test particle, and its precession would probe the metric in the 'strong-field' domain. These interesting relativistic effects have been computed in detail by Karas and Vokrouhlicky (1993, 1994) and Rauch (1997). Would we expect to find a star in such an orbit?

An ordinary star certainly cannot get there by the kind of 'tidal capture' process that can create close binary star systems. This is because the binding energy of the final orbit (a circular orbit with radius $2r_T$, which has the same angular momentum as an initially near-parabolic orbit with pericentre at r_T) is far higher when the companion is a supermassive hole than when it is also of stellar mass—it scales roughly as $M^{2/3}$. This orbital energy would have to be dissipated within the star, and that cannot happen without destroying it: a star whose orbit brings it within (say) $3r_T$ of a massive black hole may not be destroyed on first passage (as described in section 4.4); however, it is then on a bound elliptical orbit, and the cumulative tidal heating on successive pericenter passages would surely disrupt it before the orbit has circularised. (It would then give a 'flare' similar to that discussed in section 4.4, but with a somewhat longer timescale.)

Syer, Clarke, and Rees (1991) pointed out, however, that an orbit can be 'ground down' by successive impacts on a disc (or any other resisting medium) without being destroyed: the orbital energy then goes almost entirely into the material knocked out of the disc, rather than into the star itself. Other constraints on the survival of stars in the hostile environment around massive black holes—tidal dissipation when the orbit is eccentric, irradiation by ambient radiation, etc.—are explored by Podsiadlowski and Rees (1994) and King and Done (1993).

These stars would not be directly observable, except maybe in our own Galactic centre. But they might have indirect effects: such a rapidly orbiting star in an active galactic nucleus could signal its presence by quasi-periodically modulating the AGN emission.

There was a flurry of interest some years ago when X-ray astronomers detected an apparent 3.4 hour periodicity in the Seyfert galaxy NGC 6814. But it turned out that there was a foreground binary star, with just that period, in the telescope's field of view. But theorists shouldn't be downcast. It is more elevated to make predictions than to explain phenomena a posteriori, and that's all we can now do. There is a real chance that someday observers will find evidence that an AGN is being modulated by an orbiting star, which could act as a test particle whose orbital precession would probe the metric in the domain where the distinctive features of the Kerr geometry should show up clearly.

4.6.4 Gravitational wave capture of compact stars

Objects circling close to supermassive black holes could be neutron stars or white dwarfs, rather than ordinary stars. Such compact stars would be impervious to tidal dissipation and would have such a small geometrical cross section that the 'grinding down' process would be ineffective too. On the other hand, because they are small they can get into very tight orbits by straightforward stellar-dynamical processes. For ordinary stars, the 'point mass' approximation breaks down for encounter speeds above 1000 km/sec—physical collisions are then more probable than large-angle deflections: but there is no reason why a 'cusp' of tightly bound *compact* stars should not extend much closer to the hole. Neutron stars or white dwarfs could exchange orbital energy by close encounters with each other until some got close enough that they either fell directly into the hole, or until gravitational radiation became the dominant energy loss. Gravitational radiation losses tend to circularise an elliptical orbit with small pericentre. Most stars in such orbits would be swallowed by the hole before circularisation, because the angular momentum of a highly eccentric orbit 'diffuses' faster than the energy does due to encounters with other stars, but some would get into close circular orbits (Hils and Bender 1995; Sigurdsson and Rees 1997).

A compact star is less likely than an ordinary star in similar orbit to 'modulate' the observed radiation in a detectable way. But the gravitational radiation (almost periodic because the dissipation timescale involves a factor $(M/m*)$) might eventually be detectable (see below).

4.6.5 The Blandford-Znajek process

Blandford and Znajek (1977) showed that a magnetic field threading a hole (maintained by external currents in, for instance, a torus) could extract spin energy, converting it into directed Poynting flux and electron-positron pairs. This is, in effect, an astrophysically realistic example of the Penrose (1969) process whereby the spin of a Kerr hole can be tapped. It would indeed be exciting if we could point to objects where this was happening. The centres of galaxies display a bewildering variety of phenomena, on scales spanning many powers of 10. The giant radio lobes sometimes spread across millions of light-years—10^{10} times larger than the hole itself. If the Blandford-Znajek process is really going on (Rees et al. 1982), these huge structures may be the most direct manifestation of an inherently relativistic effect around a Kerr hole.

Jets in some AGNs definitely have Lorentz factors γ_j exceeding 10. Moreover, some are probably Poynting dominated and contain pair (rather than electron-ion) plasma. But there is still no compelling reason to believe that these jets are energised by the hole itself, rather than by winds and magnetic flux 'spun off' the surrounding torus. The case for the Blandford-Znajek mechanism would be strengthened if baryon-free jets were found with still higher γ_j, or if the spin of the holes could be independently measured, and the properties of jets turned out to depend on a/m.

4.6.6 Scaling laws and 'microquasars'

Two of the galactic X-ray sources that are believed to involve black holes (see table 4.1) generate double radio structures that resemble miniature versions of the classical extragalactic strong radio sources. The jets have been found to display apparent superluminal motions across the sky, indicating that, like the extragalactic radio sources, they contain plasma that is moving relativistically (Mirabel and Rodriguez 1994).

There is no reason to be surprised by this analogy between phenomena on very different scales. Indeed, the physics of flows around black holes is always essentially the same, apart from very simple scaling laws. If we define $l = L/L_{\rm Ed}$ and $\dot{m} = \dot{M}/\dot{M}_{\rm crit}$, where $\dot{M}_{\rm crit} = L_{\rm Ed}/c^2$, then for a given value of $\dot{m}$, the flow pattern may be essentially independent of M. Linear scales and timescales, at a given value of r/r_g, where $r_g = GM/c^2$, are proportional to M, and densities in the flow for a given $\dot{m}$ then scale as M^{-1}. The physics that amplifies and tangles any magnetic field may be scale independent; the field strength B then scales as $M^{-1/2}$. So the bremsstrahlung or synchrotron cooling timescales (proportional to ρ^{-1} and $B^{-1/2}$ respectively) go as M, implying that $t_{\rm cool}/t_{\rm dyn}$ depends primarily on $\dot{m}$. So also do the ratios involving, for instance, coupling of electron and ions in thermal plasma. Therefore, the efficiencies and the value of l are insensitive to M and depend primarily on $\dot{m}$. Moreover, the form of the spectrum depends on M only rather insensitively (and in a manner that is easily calculated).

The kinds of accretion flow inferred in, for instance, M87, giving rise to a compact radio and X-ray source, along with a relativistic jet, would operate just as well if the hole mass were lower by a hundred million, as in the Galactic LMXB sources. So we can actually study the processes involved in AGNs in microquasars close at hand within our own Galaxy. And we may even be able to see the entire evolution of a strong extragalactic radio source, speeded up by a similar factor.

Some X-ray sources in binary systems display large-amplitude variations, caused presumably by changes in the accretion rate $\dot{M}$. If the compact object in these systems were indeed a black hole, then the efficiency may drop when $\dot{M}$ (and consequently the density of infalling matter) is low, because the material may have such a low density that its cooling timescale is longer than the time it takes to swirl inward towards the horizon. (This is a scaled-down version of the effect discussed in section 4.3.2 in connection with M87 and other low-luminosity galactic nuclei.) If the compact object has a hard surface, e.g. a neutron star, the efficiency is independent of $\dot{M}$ (though of course the spectrum of the emergent radiation may not be). According to Narayan, Garcia, and McClintock (1997) the observed X-ray variations tend to have larger amplitude in binaries where the compact object is of high mass than when it is a neutron star. This is just what would be expected if the high-mass objects were indeed black holes; because the efficiency drops when $\dot{M}$ is small, the luminosity can change by a bigger factor than does $\dot{M}$ itself. This evidence, though gratifyingly consistent with the high-mass objects in binaries being black holes, would still not convince an intelligent sceptic, who could postulate a different theory of strong-field gravity or else that the high-mass compact objects were (for instance) self-gravitating clusters of weakly interacting particles. Narayan et al.'s data might well be explicable on such a hypothesis. Other lines of evidence, however, would already have forced such a committed sceptic into a very tight corner!

4.6.7 Discoseismology

Discs or tori that are maintained by steady flow into a black hole can support vibrational modes (Kato and Fukui 1980; Nowak and Wagoner 1992, 1993). The frequencies of these modes can, as in stars, serve as a probe for the structure of the inner disc or torus. The amplitude depends on the importance of pressure, and hence on disc thickness; how they are excited, and the amplitude they may reach, depends, as in the Sun, on interaction with convective cells and other macroscopic motions superimposed on the mean flow. But the *frequencies* of the modes can be calculated more reliably. In particular, the lowest g-mode frequency is close to the maximum value of the radial epicyclic frequency k. This epicyclic frequency is, in the Newtonian domain, equal to the orbital frequency. It drops to zero at the innermost stable orbit. It has a maximum at about $9GM/c^2$ for a Schwarzschild hole; for a Kerr hole, k peaks at a smaller radius (and a higher frequency for a given M). The frequency is 3.5 times higher for $a/m = 1$ than for the Schwarzschild case.

Nowak and Wagoner pointed out that these modes may cause an observable modulation in the X-ray emission from Galactic black hole candidates. Just such effects have been seen in GRS 1915+105 (Morgan et al. 1996). The amplitude is a few percent (and somewhat larger at harder X-ray energies, suggesting that the oscillations involve primarily the hotter inner part of the disc). The fluctuation spectrum shows a peak in Fourier space at around 67 Hz. This frequency does not change even when the X-ray luminosity doubles, suggesting that it relates to a particular radius in the disc. If this is indeed the lowest g-mode, and if the simple disc models are relevant, then the implied mass is $10.2M_\odot$ for a Schwarzschild hole and $35M_\odot$ for a 'maximal Kerr' hole (Nowak et al. 1997). The mass of this system is not well known. However, this technique offers the exciting prospect of inferring a/m for holes whose masses are independently known.

GRS 1915+105 is one of the objects with superluminal radio jets. The simple scaling arguments of section 4.6.6 imply that the AGNs which it resembles might equally well display oscillations with the same cause. However, the periods would be measured in days, rather than fractions of a second.

4.7 Gravitational radiation as a probe

4.7.1 Gravitational waves from newly forming massive holes?

The gravitational radiation from black holes, as Kip Thorne's paper (chapter 3 in this volume) emphasises, offers potentially impressive tests of general relativity, involving no physics other than the dynamics of spacetime itself.

At first sight, the original formation of the holes might seem the most obvious source of strong wave pulses. However, the wave emission would only be efficient if the holes formed on a timescale as short as r_g/c—something that might happen if they built up via coalescence of smaller holes (cf. Quinlan and Shapiro 1990).

If, on the other hand, supermassive black holes formed as suggested in section 4.5—directly from gas (some, albeit, already processed through stars), perhaps after a transient phase as a supermassive object—then the process would be too gradual to yield efficient gravitational radiation. The least pessimistic scenario from the perspective of gravitational wave astronomers, in the context of these latter ideas, would be one in which a supermassive star accumulates, and then collapses into a hole, on a dynamical timescale, via post-Newtonian instability. But even this yields much weaker gravitational radiation than black hole coalescence. That is because post-Newtonian instability is triggered at a radius $r_i \gg r_g$. Supermassive stars are fragile because of the dominance of radiation pressure: this renders the adiabatic index Γ only slightly above $4/3$ (by an amount of order $(M/M_\odot)^{-1/2}$). Since $\Gamma = 4/3$ yields neutral stability in Newtonian theory, even the small post-Newtonian corrections then destabilise such 'superstars'. The characteristic collapse timescale when instability ensues is longer than r_g/c by the 3/2 power of that factor, and the total gravitational wave energy emitted is lower by the cube. Efficiency might be enhanced if the specific angular momentum when the instability occurred were just above the limit that could be accepted by a newly formed Kerr hole. Material falling inward would then accumulate in a disc or pancake structure with dimensions only a few times r_g; if ordinary viscosity were ineffective in expelling the excess angular momentum, the disc might then become sufficiently asymmetric that gravitational waves could do the job.

If the material were initially of uniform density (a 'top hat' distribution) and then fell in freely, it would all reach the centre simultaneously. However, this would not happen if the precollapse density profile were characteristic of a supermassive object. Different shells of material would reach the centre at times spread by roughly the initial free-fall time, larger than r_g/c by a factor $(r_i/r)^{3/2}$. (In the spherical case, r_i/r_g would be $(M/M_\odot)^{1/2}$.) If other mechanisms for angular momentum transfer could be suppressed, the resultant ring of material would swirl inward, owing to loss of angular momentum via gravitational radiation, in $\sim (M_{\rm ring}/M)^{-1}$ orbital periods. A quasi-steady state could therefore be maintained for the overall free-fall timescale $\sim (r_i/r)^{3/2}(r_g/c)$, during which material drains inward so that the amount stored in the ring maintains itself at $(r_i/r)^{-3/4}M$. The gravitational radiation would be 'efficient' in the sense that it carried away a significant fraction of the rest-mass energy, but this would happen over a longer period than r_g/c, so the amplitude would be lower by $(r_i/r)^{-3/4}$. (I should emphasise that this example is merely illustrative and is obviously not very realistic.)

The important point is that the formation of a hole 'in one go' from a supermassive star is an unpromising source of gravitational waves. If the hole grows more gradually, then the prospects are obviously still worse. On the other hand, if the host galaxy had not yet acquired a well-defined single centre, several separate holes could form and yield strong events when they subsequently coalesce.

The gravitational waves associated with supermassive holes would be concentrated in a frequency range around a millihertz—too low to be accessible to ground-based detectors, which lose sensitivity below 100 Hz,

Chapter 4. Astrophysical Evidence for Black Holes 97

owing to seismic and other background noise. Space-based detectors are needed. One such, proposed by the European Space Agency, is the Laser Interferometric Spacecraft Antenna (LISA)—six spacecraft in solar orbit, configured as two triangles, with a baseline of 5 million km whose length is monitored by laser interferometry.

4.7.2 Coalescing supermassive holes

The guaranteed sources of really intense gravitational waves in LISA's frequency range would be coalescing supermassive black holes. Many galaxies have experienced a merger since the epoch $z > 2$ when, according to 'quasar demography' arguments (section 4.3) they acquired central holes. The holes in the two merging galaxies would spiral together, emitting, in their final coalescence, up to 10 percent of their rest mass as a burst of gravitational radiation in a timescale of only a few times r_g/c. These pulses would be so strong that LISA could detect them with high signal-to-noise even from large redshifts. Whether such events happen often enough to be interesting can to some extent be inferred from observations (we see many galaxies in the process of coalescing), and from simulations of the hierarchical clustering process whereby galaxies and other cosmic structures form. Haehnelt (1994) calculated the merger rate of the large galaxies believed to harbour supermassive holes: it is only about one event per century, even out to redshifts $z = 4$. Mergers of small galaxies are more common—indeed big galaxies are probably the outcome of many successive mergers. We have no direct evidence on whether these small galaxies harbour black holes (nor, if they do, of what the hole masses typically are). However it is certainly possible that enough holes of (say) $10^5 M_\odot$ lurk in small early forming galaxies to yield, via subsequent mergers, more than one event per year detectable by LISA.

4.7.3 Effects of recoil

There would be a recoil due to the nonzero net *linear* momentum carried away by gravitational waves in the coalescence. If the holes have unequal masses, a preferred longitude in the orbital plane is determined by the orbital phase at which the final plunge occurs. For spinning holes there may be a rocket effect perpendicular to the orbital plane, since the spins break the mirror symmetry with respect to this plane (Redmount and Rees 1989, and references cited therein).

The recoil is a strong-field gravitational effect which depends essentially on the lack of symmetry in the system. It can therefore only be properly calculated when fully three-dimensional general relativistic calculations are feasible. The velocities arising from these processes would be astrophysically interesting if they were enough to dislodge the resultant hole from the centre of the merged galaxy, or even eject it into intergalactic space.

LISA is potentially so sensitive that it could detect the nearly periodic waves waves from stellar-mass objects orbiting a $10^5 M_\odot$–$10^6 M_\odot$ hole, even at a range of a hundred Mpc, despite the m/M factor whereby the amplitude is reduced compared with the coalescence of two objects of comparable mass M. The stars in the observed 'cusps' around massive central holes in nearby galaxies are of course (unless almost exactly radial) on orbits that are far too large to display relativistic effects. Occasional captures into relativistic orbits can come about by dissipative processes—for instance, interaction with a massive disc (e.g. Canizzo, Lee, and Goodman 1990; Syer, Clarke, and Rees 1991). But unless the hole mass were above $10^8 M_\odot$ (in which case the waves would be at too low a frequency for LISA to detect), solar-type stars would be tidally disrupted before getting into relativistic orbits. Interest therefore focuses on compact stars, for which dissipation due to tidal effects or drag is less effective. As described in section 4.6.3, compact stars may get captured as a result of gravitational radiation, which can gradually 'grind down' an eccentric orbit with close pericenter passage into a nearly circular relativistic orbit (Hils and Bender 1995; Sigurdsson and Rees 1997). The long quasi-periodic wave trains from such objects, modulated by orbital precession (cf. Karas and Vokrouhlicky 1993; Rauch 1997), in principle carry detailed information about the metric.

The attraction of LISA as an 'observatory' is that even conservative assumptions lead to the prediction that a variety of phenomena will be detected. If there were many massive holes not associated with galactic centres (not to mention other speculative options such as cosmic strings), the event rate could be much enhanced. Even without factoring in an 'optimism factor' we can be confident that LISA will harvest a rich stream of data.

LISA is at the moment just a proposal—even if funded, it is unlikely to fly before 2017. (It will cost perhaps 3 times as much as LIGO Phase 1 but may detect infinitely more events.) Is there any way of learning, before that date, something about gravitational radiation? The dynamics (and gravitational radiation) when two holes merge has so far been computed only for cases of special symmetry. The more general problem—coalescence of two Kerr holes with general orientations of their spin axes relative to the orbital angular momentum—is one of the U.S. 'grand challenge' computational projects. When this challenge has been met (and it will almost certainly not take all the time until 2017) we shall find out not only the characteristic wave form of the radiation, but the recoil that arises because there is a net emission of linear momentum.

This recoil could displace the hole from the centre of the merged galaxy (Valtonen 1996, and references therein)—it might therefore be relevant to the low-z quasars that seem to be asymmetrically located in their hosts (and which may have been activated by a recent merger). Even galaxies that do not harbour a central hole may, therefore, once have done so in the past. The core of a galaxy that has experienced such an ejection event may retain some trace of it (perhaps, for instance, an unusual profile), because of the energy transferred to stars via dynamical friction during the merger process (cf. Ebisuzaki, Makino, and Okumura 1991; Faber et al. 1997).

The recoil might even be so violent that the merged hole breaks loose from its galaxy and goes hurtling through intergalactic space. This disconcerting thought should impress us with the reality and 'concreteness' of the entities whose theoretical properties Chandra did so much to illuminate.

References

Blandford, R.D. and Znajek, R.L. 1977, MNRAS **179**, 433.

Canizzo, J.K., Lee, H.M. and Goodman, J. 1990, ApJ **351**, 38.

Chandrasekhar, S. 1975, lecture reprinted in *Truth and Beauty* (University of Chicago Press, 1987), 54.

Charles, P.A. 1997, in *Proc. 18th Texas Conference* (World Scientific, in press).

Churazov, E. et al. 1994, ApJ Supp **92**, 381.

Connors, P.A., Piran, T. and Stark, R.F. 1980, ApJ **235**, 224.

Cunningham, C.T. and Bardeen, J.M. 1973, ApJ **183**, 237.

Ebisuzaki, T., Makino, J. and Okumura, S.K. 1991, Nature **354**, 212.

Eckart, A. and Genzel, R. 1996, Nature **383**, 415.

Evans, C.R. and Kochanek, C.S. 1989, ApJ (Lett) 346, L13.

Faber, S.M. et al. 1997, Astron J in press.

Fabian, A.C. and Canizares, C.R. 1988, Nature **333**, 829.

Fabian, A.C. and Rees, M.J. 1995, MNRAS **277**, L55.

Fabian, A.C., Rees, M.J., Stella, L. and White, N.E. 1989, MNRAS **238**, 729.

Ford, H.C. et al. 1994, ApJ **435**, L27.

Frolov, V.P. et al. 1994, ApJ **432**, 680.

Genzel, R., Townes, C.H. and Hollenbach, D.J. 1994, Rep Prog Phys **57**, 417.

Haehnelt, M. 1994, MNRAS **269**, 199.

Haehnelt, M. and Rees, M.J. 1993, MNRAS **263**, 168.

Hils, D. and Bender, P.L. 1995, ApJ (Lett) **445**, L7.

Israel, W. 1996, Foundations of Physics **26**, 595.

Iwasawa, K. et al. 1996, MNRAS **282**, 1038.

Karas, V. and Vokrouhlicky, D. 1993, MNRAS **265**, 365.

Karas, V. and Vokrouhlicky, D. 1994, ApJ **422**, 208.

Kato, S. and Fukui, J. 1980, PASJ **32**, 377.

Khokhlov, A., Novikov, I.D. and Pethick, C.J. 1993, ApJ **418**, 163.

King, A.R. and Done, C. 1993, MNRAS **264**, 388

Kochanek, C.S. 1994, ApJ **422**, 508

Kormendy, J. and Richstone, D. 1995, Ann Rev Astr Astrophys **33**, 581.

Lacy, J.H. 1994 in *The Nuclei of Normal Galaxies*, ed R. Genzel and A.I. Harris (Kluwer), 165.

Lacy, J.H., Townes, C.H. and Hollenbach, D.J. 1982, ApJ **262**, 120.

Laguna, P., Miller, W.A., Zurek, W.H. and Davies, M.B. 1993, ApJ (Lett) **410**, L83.

Lynden-Bell, D. and Rees, M.J. 1971, MNRAS **152**, 461.

Mahadevan, R. 1997, ApJ, in press.

Melia, F. 1994, ApJ **426**, 577.

Merritt, D. and Oh, S.P. 1997, Astron J **113**, 1279.

Mirabel, I.F. and Rodriguez, L.F. 1994, Nature **371**, 48.

Miyoshi, K. et al. 1995, Nature **373**, 127.

Morgan, E., Remillard, R. and Greiner, J. 1996, IAU Circular No. 6392.

Narayan, R., Garcia, M.R. and McClintock, J.E. 1997, ApJ (Lett) **478**, L79.

Narayan, R. and Yi, I. 1995, ApJ **444**, 231.

Narayan, R., Yi, I. and Mahadevan, R. 1995, Nature **374**, 623.

Nowak, M.A. and Wagoner, R.V. 1992, ApJ **393**, 697.

Nowak, M.A. and Wagoner, R.V. 1993, ApJ **418**, 187.

Nowak, M.A., Wagoner, R.V., Begelman, M.C. and Lehr, D.E. 1997, ApJ (Lett) **477**, L91.

Penrose, R. 1969, Rev Nuovo Cim **1**, 252.

Podsiadlowski, P. and Rees, M.J. 1994, in *Evolution of X-Ray Binaries*, ed. S. Holt and C. Day (AIP), 403.

Quinlan, G.D. and Shapiro, S.L. 1990, ApJ **356**, 483.

Rauch, K.P. 1997, ApJ, in press.

Redmount, I. and Rees, M.J. 1989, Comm Astrophys Sp Phys **14**, 185.

Rees, M.J. 1982, in *The Galactic Center*, ed. G. Riegler and R.D. Blandford (AIP), 166.

Rees, M.J. 1987, in *Galactic Center*, ed. D. Backer and R. Genzel (AIP Conference Proceedings).

Rees, M.J. 1988, Nature **333**, 523.

Rees, M.J. 1993, Proc Nat Acad Sci **90**, 4840.

Rees, M.J. 1994, in *Nuclei of Normal Galaxies*, ed. R. Genzel and A.I. Harris (Kluwer), 453

Rees, M.J. 1996, in *Gravitational Dynamics*, ed. O. Lahav et al. (Cambridge University Press), 103.

Rees, M.J., Begelman, M.C., Blandford, R.D. and Phinney, E.S. 1982, Nature **295**, 17.

Renzini, A. et al. 1995, Nature **378**, 39.

Reynolds, C. et al. 1996, MNRAS **283**, L111.

Sadler, E.M. et al. 1995, MNRAS **276**, 1373.

Sargent, W.L.W. et al. 1978, ApJ **221**, 731

Sembay, S. and West, R.G. 1993, MNRAS **262**, 141.

Shaver, P. 1995, Ann NY Acad Sci **759**, 87.

Sigurdsson, S. and Rees M.J. 1997, MNRAS **284**, 318.

Syer, D., Clarke, C.J. and Rees, M.J. 1991, MNRAS **250**, 505.

Tanaka, Y. and Lewin, W.H.G. 1995, in *X-Ray Binaries*, ed. W.H.G. Lewin et al. (Cambridge University Press), 126.

Tanaka, Y. et al. 1995, Nature **375**, 659.

Tremaine, S. 1997, in *Some Unsolved Problems in Astrophysics*, ed. J. Bahcall and J.P. Ostriker (Princeton University Press, in press).

Valtonen, M. 1996, Comments Astrophys **18**, 191.

van den Marel, R. 1996, in *New Light on Galactic Evolution*, ed. R. Bander and R. Davies (Kluwer, in press).

Watson, W.D. and Wallin, B.K. 1994, ApJ (Lett) **432**, L35.

Wijers, R.A.M.J. 1996, in *Evolutionary Processes in Binary Stars*, ed. R.A.M.J. Wijers et al. (Kluwer), 327

Williams, R. et al. 1996, Astron J **112**, 1335.

Young, P.J. et al. 1978, ApJ **221**, 721.

5

The Question of Cosmic Censorship

Roger Penrose

Abstract

Cosmic censorship is discussed in its various facets. It is concluded that rather little clear-cut progress has been made to date, and that the question is still very much open.

5.1 The role of cosmic censorship in gravitational collapse

Chandra's famous work on the maximum mass of white dwarf stars (Chandrasekhar 1931) pointed the way to our present-day picture (cf. Hawking and Penrose 1970) that for bodies of too large a mass, concentrated in too small a volume, unstoppable collapse will ensue, leading to a *singularity* in the very structure of space-time. The deduction of a strict occurrence of an *actual* singularity in physical space-time would, however, be based on an assumption that no quantum-mechanical principles intervene to change the nature of space-time from that which is classically described by Einstein's general relativity. Indeed, the term 'singularity', in this physical context, really refers to a region where the conventional classical picture of space-time breaks down, to be replaced, presumably, by whatever physics is to go under the name of 'quantum gravity'. It is the normal expectation that this occurs only when classical space-time curvatures diverge—quantum effects taking over when radii of space-time curvature of the order of the Planck length are attained.

In the standard picture of collapse to a black hole (cf. Penrose 1978, for example), these singularities are not visible to observers at a large distance from the hole, being 'shielded' from view by an *absolute event horizon*. Thus, whatever unknown physics takes place at the singularity itself, its effects are not observable by such an observer. The assumption of *cosmic censorship* is that in a generic gravitational collapse the resulting space-time singularity will indeed be shielded from view in this way. Accordingly, it is taken that the alternative of a *naked singularity*—i.e. a *visible* singularity—would not occur (except, conceivably, for some very special initial collapse configurations which could not be expected to take place in an actual astrophysical circumstance).

It is not hard to conceive of physical situations in which one of the standard criteria for 'unstoppable gravitational collapse' is satisfied. All that is required is for sufficient mass to fall into a small enough region. For the central regions of a large galaxy, for example, the required concentration could occur with the stars in the region still being separated from each other, so there is no reason to expect that there could be some overriding physical principle which conspires always to prevent such unstoppable collapse. However, we cannot simply deduce from this that a black hole will be the result. This deduction requires the *crucial* assumption that *cosmic censorship*, in some form, holds true.

Two familiar mathematical criteria for 'unstoppable collapse' are the existence of a *trapped surface* or of a point whose *future light cone begins to reconverge in every direction along the cone*. In either of these situations, in the presence of some other mild and physically reasonable assumptions, like the nonnegativity of energy (plus the sum of pressures), the nonexistence of closed timelike curves, and some condition of genericity (like the assumption that every causal geodesic contains at least one point at which the Riemann curvature is not lined up in a particular way with the geodesic), it follows (by results in Hawking and Penrose 1970) that a space-time singularity of some kind must occur. (Technically: the space-time manifold must be geodesically incomplete in some timelike direction.)

It appears to be a not uncommon impression among workers in the field that as soon as one of these conditions is satisfied—say the existence of a trapped surface—then a black hole will occur; and, conversely, that a naked singularity will be the result if not. However, it should be made clear that *neither* of these deductions is in fact valid. The deduction that a black hole comes about whenever a trapped surface is formed *requires* the assumption of cosmic censorship. Moreover, the deduction that some kind of space-time singularity comes about (in general situations), whether or not it is a naked one, *requires* some such assumption like that of the existence of a trapped surface (cf. e.g. Hawking and Penrose 1970). Thus, the presence of a trapped surface does not imply the absence of naked singularities; still less does the absence of a trapped surface imply the presence of a naked singularity. These points have relevance to various investigations which attempt to address the issue of cosmic censorship by the study of specific examples. Frequently, the absence of a trapped surface appears to be regarded as a criterion for—or at least a strong indication of—cosmic censorship violation (cf., for example, Shapiro and Teukolsky 1991). It should be clear from the above remarks that the issue is by no means as simple as that.

While the question of cosmic censorship remains very much an open

Chapter 5. The Question of Cosmic Censorship   105

one at the present time—possibly the most important unsolved problem in classical general relativity—progress in certain areas has been made, and I shall attempt to address some of these in the following remarks. However, these should not be regarded as in any way a comprehensive survey of progress in these areas, but merely as a personal assessment of the present status of the subject.

5.2 Plausibility of criteria for unstoppable collapse

Before addressing the issue of cosmic censorship directly, it will be appropriate to make a few remarks concerning the question of whether the above criteria—namely, the existence of a trapped surface or of a reconverging light cone—actually do represent conditions that would be realized when too much mass is concentrated in too small a volume. Various researchers (most particularly Shoen and Yau 1993) have presented arguments to show that sufficiently large mass concentrations do indeed lead to the presence of trapped surfaces. I shall not attempt to summarize this area of work here. The arguments are mathematically quite difficult, but the physical implications of these arguments are still far from clear to me. On the other hand, the reconverging light cone condition is easier to see as representing a plausible criterion. I shall have a few comments to make concerning this case. Basically, the argument for the physical realizability of the reconverging light cone condition is that given in Penrose (1969). Imagine a certain amount of massive material, say of total mass M, and allow it to fall to within a roughly defined region whose diameter is of the general order of $4GM$. We consider a space-time point p somewhere in the middle of this region, and examine the future light cone $\mathcal{C}$ of p. Thus, $\mathcal{C}$ is swept out by the future-endless rays (null geodesics) with past endpoint p. The strict condition that $\mathcal{C}$ 'satisfies the reconverging light cone condition' would be that on every ray γ generating $\mathcal{C}$ there is a place where the divergence of the rays changes sign. If it is assumed that such a ray is geodesically complete in the future direction (and that the energy flux across the ray is nonnegative), then it follows that, to the future of p along the ray, there is a point *conjugate* to p (i.e. a point q, distinct from p, with the property that there is a 'neighbouring ray to γ' which intersects γ in p and again at q; more precisely, there is a nontrivial Jacobi field along γ which vanishes both at p and at q). The idea is that as the material falls in across $\mathcal{C}$ it causes focussing of a sufficient degree that such divergence reversal indeed arises. This is merely a feature of there being enough 'focussing power' in the lensing effect of the Ricci tensor component along the ray (namely, $R_{ab}l^a l^b$, where l^a is a null tangent vector to γ), due to the energy density in the matter falling in across $\mathcal{C}$. There are simple integral expressions that can be written down (cf. Clarke 1993, in particular) which provide sufficient conditions for a conjugate point to arise, so it is merely an order-of-magnitude requirement that there is sufficient infall of material to ensure that the focussing condition will be satisfied. The situation could be made to be qualitatively similar to the original Oppenheimer-Snyder (1939) collapsing dust cloud (pressureless fluid), but where there is no symmetry assumed and no particular equation of state employed (like that of Oppenheimer and Snyder's dust).

However, the strict form of the reconverging light cone condition is that *every* ray through p should encounter sufficient material for divergence reversal to occur. This condition might well be considered to be unreasonably strong. It might not be evidently satisfied if the collapsing material is concentrated in a number of separated bodies, say stars, rather that in a continuous medium. Many of the rays through p might then miss the actual collapsing material, so the focussing along such a ray could be much smaller than required (a point specifically raised with me by Robert Wald), being only a secondary effect due to the nonlocal overall focussing produced by Weyl curvature. In fact, this does not make a serious difference, but to see that it does not is not entirely straightforward. The essential point is that for the purposes of the singularity theorem being appealed to here (Hawking and Penrose 1970), it is not necessary to assume that every future ray through p encounter divergence reversal. All that is required, roughly speaking, is that the elements of area of cross section of the intersection $\mathcal{C} \cap \partial I^+(p)$ of $\mathcal{C}$ with the boundary $\partial I^+(p)$ of the (chronological) future $I^+(p)$ of p should eventually decrease in future directions, at every point of the cross section. At those places where a sufficient amount of the matter directly encounters $\mathcal{C}$, this area decrease will be a feature of the reconverging of the generators of $\mathcal{C}$, as they begin to approach their respective conjugate points. As regards the remaining generators of $\mathcal{C}$, they will eventually encounter 'crossing regions' of $\mathcal{C}$, where two parts of this null hypersurface encounter one another and both cross through into the interior of $I^+(p)$ (so that they do not remain on the region of $\mathcal{C}$ that lies on $\partial I^+(p)$). To see that this must be the case when a sufficient concentration of material encounters $\mathcal{C}$, one may appeal to the qualitative similarity between the situations arising when the collapsing material consists of a continuous and fairly uniform medium, and when it consists of a number of discrete bodies (such as stars or, at a different level of description, the constituent particles of the medium) which closely approximate that medium. This is sufficient for establishing that p constitutes a 'future-trapped set', in the sense that is required for Hawking and Penrose (1970), and the deduction of the presence of a singularity follows just as before. I shall not go into the details of this argument here, it being more appropriate to leave this for some later discussion. (It has a bearing on other matters that have been addressed in the literature, concerning the lensing effects of different kinds of mass distribution; cf. Holz and Wald 1997.)

5.3 Causal definition of naked singularities

Let us take it, then, that there will be certain astrophysical situations in which such 'unstoppable' gravitational collapse actually takes place—'unstoppable' in the sense that it leads to some kind of space-time singularity (in accordance with the singularity theorems). When this occurs, and assuming the form of cosmic censorship which asserts that no such singularity is naked in the sense of being visible to some observer at infinity, then the boundary of the past of the 'set of observers at infinity' constitutes the *absolute event horizon*. External to this horizon are the space-time events which can send signals to infinity; cosmic censorship asserts that no singularity can lie within this external region. This presents us with the classical situation of a black hole.

Normally, the cosmic censorship issue is phrased in some such way, i.e. in terms of observers at large distances from the collapse—and, in precise mathematical discussions, in terms of observers at future null infinity $\mathcal{I}^+$. The absolute event horizon is then the boundary $\partial I^-[\mathcal{I}^+]$ of the past of $\mathcal{I}^+$. However, it may be felt that this is not necessarily the appropriate notion of cosmic censorship. For one could imagine a situation in which an observer, near to a gravitationally collapsing body, witnesses a naked singularity arising, this singularity being visible to that particular observer. We might envisage that, in this universe model, the observer and the observed collapse are both within a large region of mass concentration which ultimately—perhaps on a cosmological timescale—collapses to trap both the observer and observed region, thereby preventing signals from reaching 'infinity'. In such a picture, the singularity would not be 'naked' in terms of the definition which uses $\mathcal{I}^+$, since signals reaching the observer are themselves ultimately trapped. But nevertheless the observer would directly witness the singularity, so the singularity would indeed be naked in a more local sense. There being no theory governing what happens as the result of the appearance of such a singularity, this particular observer would not be able to account, in scientific terms, for whatever physical behaviour is seen. This is the kind of 'unpredictability' that one wishes to avoid in a physical situation. One of the reasons for desiring a cosmic censorship principle, after all, is the elimination of such physical uncertainties. A definition of 'naked singularity' which depends upon what is accessible to observers at infinity ($\mathcal{I}^+$) does not achieve this satisfactorily.

Thus, it seems not unreasonable to posit a somewhat stronger form of cosmic censorship than one phrased in terms of observers at infinity. In accordance with this, an appropriate notion of *strong cosmic censorship* has been formulated (Penrose 1978) which has the added advantage that it turns out to be symmetrical in time. Moreover, for a given space-time $\mathcal{M}$, this notion turns out to be equivalent to *global hyperbolicity* for $\mathcal{M}$.

The condition can be phrased (see Geroch, Kronheimer, and Penrose 1972) in terms of the notion of *indecomposable pasts*—**IP**s—and *indecomposable futures*—**IF**s. An **IP** is a *past-set* (i.e. a subset of $\mathcal{M}$ which is the same as its own chronological past) which, in addition, is not the proper union of two other past-sets. An **IF** is defined correspondingly, with 'future' replacing 'past'. It can be shown (Geroch, Kronheimer, and Penrose 1972) that the **IP**s are precisely the pasts of timelike (or causal) curves in $\mathcal{M}$ and the **IF**s are the futures of such curves. Such a timelike (or causal) curve is said to *generate* the **IP** or **IF** in question. An **IP** is called *proper*—a **PIP**—if it consists of the set of points lying to the chronological past of some *point* of $\mathcal{M}$ (which would be the future endpoint of such a generating curve). A **PIF** is, correspondingly, the (chronological) future of some point of $\mathcal{M}$. A *terminal* **IP**—a **TIP**—is an **IP** which is not a **PIP**; a terminal **IF**—a **TIF**—is an **IF** which is not a **PIF**. Sometimes, the **TIP**s and **TIF**s are called *ideal points* for the space-time $\mathcal{M}$.

Let us assume that $\mathcal{M}$ is *strongly causal* (i.e. that every point of $\mathcal{M}$ has an arbitrarily small neighbourhood such that no causal curve leaves and then reenters it; cf. Hawking and Ellis 1973; Penrose 1972). (In fact, it is sufficient, in what follows, to assume that $\mathcal{M}$ is both *future distinguishing*—i.e. that no two points of $\mathcal{M}$ have the same chronological future—and *past distinguishing*—i.e. that no two points of $\mathcal{M}$ have the same chronological past.) Then the points of $\mathcal{M}$ are canonically in one-to-one correspondence with the **PIP**s of $\mathcal{M}$, and also in one-to-one correspondence with the **PIF**s of $\mathcal{M}$. The **TIP**s and **TIF**s of $\mathcal{M}$, namely, $\mathcal{M}$'s ideal points, provide the points of what is called the *causal boundary* $\partial \mathcal{M}$ of $\mathcal{M}$. The *future* causal boundary $\partial^+ \mathcal{M}$ of $\mathcal{M}$ is the set of **TIP**s of $\mathcal{M}$ and the *past* causal boundary $\partial^- \mathcal{M}$ is the set of **TIF**s of $\mathcal{M}$. As thus defined, $\partial \mathcal{M}$ is a disjoint union of $\partial^+ \mathcal{M}$ with $\partial^- \mathcal{M}$. There are circumstances under which it may be felt that certain of the points of $\partial^+ \mathcal{M}$ should be identified with certain of the points of $\partial^- \mathcal{M}$ (see Geroch, Kronheimer, and Penrose 1972; such situations could be considered to represent violations of cosmic censorship). However, for the purposes of this article, I shall prefer to regard the two sets $\partial^+ \mathcal{M}$ and $\partial^- \mathcal{M}$ as being actually disjoint. The entire union $\partial^+ \mathcal{M} \cup \partial^- \mathcal{M} \cup \mathcal{M} = \partial \mathcal{M} \cup \mathcal{M}$ is the (causal) *closure* $\bar{\mathcal{M}}$ of $\mathcal{M}$.

It is convenient to divide these ideal points into two classes, according to whether they are to represent *singular* points of $\mathcal{M}$ or *points at infinity* for $\mathcal{M}$. The simplest way to make such a distinction is to say that a **TIP** represents a point at (future) *infinity*—an ∞-**TIP**—if it is generated by a timelike curve that is of *infinite length* into the future, and a **TIF** represents a point at (past) infinity—an ∞-**TIF**—if it is generated by a timelike curve that is of infinite length into the past. The remaining **TIP**s and **TIF**s—the singular **TIP**s and **TIF**s—then represent the singular points of $\mathcal{M}$. Although this distinction between points at infinity and singular points appears to be the simplest, it is not always regarded as the most appropriate (cf. Clarke 1993, for example). According to the definition given here (taken from Penrose 1978) one can have a 'point at infinity' for which the space-time curvature diverges as that ideal point is approached. It might be reasonable to call such an ideal point a 'singular point at infinity', but it would be given by an ∞-**TIP** or ∞-**TIF** as defined here, nevertheless, and not by what I am referring to as a 'singular **TIP**' or 'singular **TIF**'.

Now, a naked singularity may be described as a singularity which lies to the future of some point of space-time, but which can also be 'seen' by some observer. The reason for the first part of this description is that one would not want some cosmic censorship principle to exclude the big bang; that is to say, the big bang should not count as a 'naked singularity'. Thus, a naked singularity lies both to the future of some point $p \in \mathcal{M}$ and to the past of some other point $q \in \mathcal{M}$. In terms of **TIP**s, a *naked singular* **TIP** would be a singular **TIP** R which contains the point p and which lies to the past of a point q, i.e.

$$p \in R \text{ and } R \subset I^-(q) \text{ for some } p, q \in \mathcal{M}. \qquad (5.1)$$

Here the standard notation $I^-(q)$ is used for the chronological past of a point q; correspondingly, $I^-[Q]$ stands for the chronological past of a *set* Q, and $I^+(p)$ and $I^+[P]$ stand for the chronological futures of a point p and a set P, respectively (notations already employed above). A point at infinity might also be 'naked', in the same sense, so we can define a naked ∞-**TIP** R in just the same way. We can similarly define a naked singular **TIF** or ∞-**TIF** as a **TIF** S for which

$$S \subset I^+(p) \text{ and } q \in S \text{ for some } p, q \in \mathcal{M}. \qquad (5.2)$$

5.4 Strong cosmic censorship

Many of the reasons for wishing to exclude naked singularities apply also to naked points at infinity. It is no better, from the point of view of uniqueness of evolution, that 'uncontrollable information' be allowed to enter the space-time from infinity than from a singularity. (Anti–de Sitter space is an example of a space-time possessing naked points at infinity; there are indeed reasons of this kind for regarding this model as 'unphysical'.) In any case, as was remarked upon above, some ∞-**TIP**s or ∞-**TIF**s might arise from regions of infinite curvature and could be thought of as, in some sense, 'singular', in any case. Thus, it seems to be a reasonable formulation of the requirement that a space-time $\mathcal{M}$ be in accordance with cosmic censorship that there should be no naked **TIP**s in the above sense, whether they be singular **TIP**s or ∞-**TIP**s. As was shown in Penrose (1979), this condition is *equivalent* to the condition that $\mathcal{M}$ be free of naked **TIF**s—by virtue of the fact that it is equivalent, also, to the condition that $\mathcal{M}$ be *globally hyperbolic*.

The assertion that $\mathcal{M}$ is free of naked singularities in this sense can also be phrased in other equivalent ways. It is convenient to introduce causal relations between **TIP**s as follows. We say that the **TIP** P *causally precedes* the **TIP** Q if $P \subseteq Q$, and that the **TIP** P *chronologically precedes* the **TIP** Q if there exists a point $q \in Q$ such that $P \subset I^-(q)$. Then, by the above definition, a **TIP** is naked iff there is another **TIP** to its chronological future. Defining the causal relations between **TIF**s correspondingly—if R and S are **TIF**s, R causally precedes S if $S \subseteq R$ and R chronologically precedes S if there exists $r \in R$ such that $S \subset I^+(r)$—we see that the **TIF** S is naked iff there is another **TIF** to its chronological past. In this sense, naked singularities (or points at infinity) are *timelike* entities.

We can now formulate our principle of *strong cosmic censorship* as the assertion that naked singularities or points at infinity (in the above sense) do not occur in generic space-times, where it is assumed that Einstein's equations hold with some reasonable equations of state for the matter. This is still a little vague because of occurrence of the words 'generic' and 'reasonable' in the definition. In my opinion, it is probably not particularly helpful to try to be more precise at this stage. There is a fairly well defined intuitive meaning for the word 'generic'; no doubt, when the appropriate theorem comes along, then an appropriately relevant definition of 'generic' will become clearer (as was the case with some singularity theorems; cf. Hawking and Penrose 1970). Without some such restriction, however, counterexamples to cosmic censorship can occur, such as in certain very special situations of spherically symmetrical collapse (cf. Christodoulou 1994; Choptuik 1993). As regards 'reasonable equations of state' the essential point is that we should exclude equations that could lead to singularities in generic situations even in special relativity (e.g. leading to infinite density), such as occurs with 'dust' when caustics arise in the flow lines. The equations of state should also be such as to ensure energy positivity (and perhaps stronger restrictions such as the dominant energy condition).

As yet, there is no mathematical theorem asserting the truth of any appropriate form of cosmic censorship in general relativity; yet, as we have seen above, cosmic censorship is an essential ingredient of the standard picture of gravitational collapse to a black hole. As was indicated in section 5.2 above, one can present convincing arguments to show that situations can occur—and, indeed, *will* occur in appropriate circumstances of gravitational collapse, when sufficient matter is being concentrated in too small a region—in which singularities will arise, according to general relativity. But without a cosmic censorship assumption, there is no guarantee that these singularities will not be naked. If the singular region is not naked, even merely in the weaker sense that observers at *infinity* are not able to 'see' the singular region (e.g. so that no ∞-**TIP** representing a point of $\mathcal{I}^+$ contains a naked singular **TIP**), then there will be some region of the

Chapter 5. The Question of Cosmic Censorship   111

space-time, including the singularity region, that cannot be seen from $\mathcal{I}^+$. In other words, the causal past of $\mathcal{I}^+$ (written $J^-[\mathcal{I}^+]$) is not the whole of the space-time $\mathcal{M}$ so the chronological past $I^-[\mathcal{I}^+]$ cannot exhaust $\mathcal{M}$ either. As stated in section 5.3 above, the boundary $\partial I^-[\mathcal{I}^+]$ defines the *horizon* of the resulting black hole.

There is another crucial role that is played by cosmic censorship in the theoretical discussion of gravitational collapse. Without such an assumption, one cannot deduce the well-known *area-increase* theorem (cf. Floyd and Penrose 1971; Hawking 1972). This theorem asserts that the areas of cross section of a black-hole horizon $\partial I^-[\mathcal{I}^+]$ are nondecreasing into the future. There are various different versions of this theorem, depending upon which version of cosmic censorship is adopted (see Penrose 1978 for a discussion of several of these). The area-increase theorem is important, particularly because of its relation to black-hole entropy and thermodynamics (Bekenstein 1973; Hawking 1975).

5.5 Thunderbolts

One remark should be made here concerning cosmic censorship proposals of this nature. They do not, as they stand, eliminate the possibility of what Hawking (1993) refers to as *thunderbolts*, first considered in Penrose (1978). This is the hypothetical situation according to which a gravitational collapse results in a 'wave of singularity' coming out from the collapse region, which destroys the universe as it goes! On this picture, the entire space-time could remain globally hyperbolic since everything beyond the domain of dependence of some initial hypersurface is cut off ('destroyed') by the singular wave. An observer, whether at infinity or in some finite location in the space-time, is destroyed just at the moment that the singularity would have become visible, so that observer cannot actually 'see' the singularity.

One condition which excludes this particular possibility (Penrose 1978, condition CC4) is

no ∞-**TIP** contains a singular **TIP**.

For an asymptotically flat space-time $\mathcal{M}$, we may expect that the future-null conformal boundary $\mathcal{I}^+$ of $\mathcal{M}$ can be identified with its set of ∞-**TIP**s. In any situation where the above condition is violated, we have an ∞-**TIP** which directly 'sees' the singularity (in the sense of being causally to its future). In the situation where a thunderbolt is present, the 'observer at infinity' represented by that ∞-**TIP** would be destroyed by the wave of infinite curvature at that very moment, but we still have an ideal point there, represented by this ∞-**TIP**. However, the conformal boundary $\mathcal{I}^+$ would cease to be smooth at that point.

In section 5.4, one formulation of the condition of strong cosmic censorship was given as an assertion that 'timelike' singularities (or points at infinity) are to be excluded. The above condition for ruling out thunderbolts can be phrased as the condition that a point at infinity cannot lie *causally* to the future of a singular point (defined in terms of **TIP**s). One could imagine formulating an 'extrastrong' version of cosmic censorship in which *all* causally separated (distinct) **TIP**s are excluded (in the sense that no **TIP** shall properly include another **TIP**; cf. section 5.4) and not merely the timelike-separated ones that are excluded by ordinary strong cosmic censorship. However, this condition would be unreasonable because it would rule out all asymptotically flat space-times! Being a null hypersurface, the future of the conformal boundary $\mathcal{I}^+$ of an asymptotically flat $\mathcal{M}$ contains null generators, and any pair of **TIP**s representing two distinct points of the same generator would be causally separated in the above sense.

However, it might well be reasonable to expect that a slightly weaker extrastrong version of cosmic censorship might be appropriate, in which it is asserted that there is no pair of distinct **TIP**s P, Q such that $P \subset Q$, and where P is a singular **TIP** (and where the corresponding statement in terms of **TIF**s could also be appended, if desired). This would incorporate both strong cosmic censorship and the exclusion of thunderbolts, in the above sense. It remains to be seen whether such a formulation might still be too strong (cf. the article by Israel, in chap. 7 of this volume).

5.6 Some arguments against cosmic censorship

Most of the arguments presented to date which are aimed at disproving cosmic censorship have been concerned with the examination of specific examples. However, there is an inherent difficulty in using arguments of this kind, because any specific example that can be studied in detail is liable to be 'special' in some way or other, and unlikely to be considered to be 'generic' in some appropriate sense. At least, this applies to specific examples that can be studied analytically in detail. It may be that with the further development of numerical techniques and computer power, examples might eventually be considered which could indeed be argued to be appropriately 'generic'. On the other hand, there is the compensating difficulty that with numerical solutions there may be some doubt, in any particular case, whether a seeming singularity is actually a genuine singularity, or whether the singularity is indeed naked. As things stand, specific examples can only give *indications* as to whether cosmic censorship is likely to be true, not definitive answers.

The first example of a gravitational collapse leading to a naked singularity was that given by Yodzis, Seifert, and Muller zum Hagen (1973). They pointed out that even in exactly spherically symmetrical collapse, infinite-curvature naked singularities could arise with dust, owing to the presence of caustics in the family of dust world-lines (at which the den-

Chapter 5. The Question of Cosmic Censorship

sity diverges), provided that these caustics occur before an absolute event horizon is reached. As suggested in section 5.4, such circumstances should not be considered as providing violations of cosmic censorship because the infinite densities that arise have nothing to do with general relativity (or, indeed, with gravity at all) because they occur just as readily with the equations for dust in special relativity.

Such regions of infinite density are sometimes referred to as 'shell-crossing' singularities, since they are regions where the different shells of collapsing material begin to cross one another. However, this terminology is not altogether appropriate because the difficulties with infinite density do not occur in the regions where different dust flows actually cross each other, but at the boundary of such a region, where there is a caustic in the flow lines and one flow becomes three.[1] Nevertheless, where there are indeed several superimposed flows, the energy-momentum tensor of 'dust' cannot be used, but instead one has a sum of a number of different terms of this kind, i.e.

$$T_{ab} = \rho u_a u_b + \cdots + \tau w_a w_b, \tag{5.3}$$

where $\rho, \ldots, \tau$ are the respective densities of the different components of the dust (pressureless fluid) and where each of $u^a, \ldots, w^a$ is a unit future-timelike vector giving the direction its flow. For each component of the flow, the flow world-lines are geodesics, and each of $\rho u_a, \ldots, \tau w_a$ is divergence-free. Of course, regions of infinite density can still arise—whenever one of the systems of flow lines encounters caustics.

We can generalize the above finite sum of terms to a situation in which there is a *continuum* of terms. This gives us an instance of the kind of system that is treated by the *Vlasov* equation. More generally, the Vlasov equation covers the cases when there is a continuous superposition of fluids which possess pressure—rather than being just 'dust', as in the cases considered above (the simpler case of a 'collisionless' fluid).

In the collapse situation studied by Shapiro and Teukolsky (1991), referred to in section 5.1, the Vlasov equation is used but (as was pointed out to me by Alan Rendall) the individual fluid components do not possess pressure (the 'collisionless' case), and it is not clear that the situation is free of the problems that occur with the Yodzis, Seifert, and Muller zum Hagen (1973) example. Building upon earlier ideas of Thorne, who suggested that prolate spheroidal collapse might lead to naked singularities (because of a resemblance to *cylindrical* trapped-surface-free collapse; cf. also Thorne 1972; Chrusciel 1990), Shapiro and Teukolsky consider the collapse of a prolate axisymmetrical body composed of collisionless material, and they argue that naked singularities can arise. They indicate the pres-

---

[1] It does not become merely *two* overlapping flows, for topological reasons. One of the three flows 'counts' as negative and the other two as positive, preserving the net count of overlapping flows, as one passes from one side of the caustic to the other.

ence of regions at which the density diverges and argue from the absence of trapped surfaces that these singularities could well be naked. Moreover, they point out that their singular regions do not arise merely from infinite density, because they extend outside the matter region. However, as was pointed out in section 5.1 above, more needs to be established if we are to ascertain whether these singularities are indeed naked. In particular, we would need to examine the regions of the space-time lying to the future of the singularity, but this is not possible within the framework of the computer calculation that they carry out, since the calculation terminates as soon as the singularity is reached.

Moreover, as was pointed out by Iyer and Wald (1991), collapse situations that appear to resemble that of Shapiro and Teukolsky can be constructed where no trapped surfaces appear before *nonnaked* singularities arise. In both the Iyer-Wald example and the Shapiro-Teukolsky example, there is a reasonable-looking family of constant-time spacelike hypersurfaces according to which the time evolution is described. In neither example are there trapped surfaces before a singularity appears. However, the singularity is clearly *not* naked in the Iyer-Wald example, because their example is simply the ordinary extended Schwarzschild solution described according to a nonstandard time coordinate. This sheds considerable doubt on the Shapiro-Teukolsky suggestion that their singularity is actually naked.

A closely related situation was studied by Tod (1992). In this example, there is a collapsing shell of 'null matter' (a delta-function shell of massless dust) which falls into a region of Minkowski space that it surrounds. The mass density can vary arbitrarily with spatial direction, and the (convex, smooth) shape of the shell, at one particular time, can also be chosen arbitrarily. By choosing this shape to be a suitable prolate ellipsoid it is not hard to ensure that caustics in the collapsing shell—and hence singularities—arise before there are any trapped surfaces. The description is given in terms of standard $t = $ const. hypersurfaces in the interior Minkowski space. Nevertheless, the situation is completely consistent with the conventional picture of gravitational collapse to a black hole. Trapped surfaces do in fact occur in the space-time, but not until after the t-value at which singularities arise. This is again similar to the Shapiro-Teukolsky situation, and there is no reason to expect a violation of cosmic censorship.

Other examples have been described (Choptuik 1993; Christodoulou 1994) in which the collapsing matter consists of a massless scalar field. In some of these, there are naked singularities. However, all these examples are extremely special, owing to the fact that spherical symmetry is assumed. Accordingly, it is hard to see that such examples can shed a great deal of light on the general issue of cosmic censorship. The condition of genericity is far from being satisfied. Moreover, according to a recent result of Christodoulou (1997), 'almost all' examples, even within *this* limited class, are free of naked singularities.

Chapter 5. The Question of Cosmic Censorship

It would thus appear that there is, so far, no convincing evidence against cosmic censorship's being a principle with which classical general relativity accords. Quantum general relativity, on the other hand, does raise some serious problems in this regard. It is hard to avoid the conclusion that the endpoint of the Hawking evaporation of a black hole would be a naked singularity—or at least something that on a classical scale would closely resemble a naked singularity. Nevertheless, these considerations are not directly relevant to what is normally referred to as 'cosmic censorship', which is intended to be a principle applying to *classical* general relativity only. When quantum effects are allowed, negative energy densities are possible— needed for the consistency of the Hawking effect, in which the area-increase property for a black-hole horizon is violated. In any case, unless there are mini–black holes in the universe (and the observational evidence seems to be against this), there would seem to be no direct astrophysical or cosmological role for the 'objects' which represent the final stages of Hawking evaporation, owing to the absurdly long timescales needed for this process when it originates with an astrophysical black hole. (For theoretical considerations, on the other hand, these 'objects' could well be important—but that is another story!)

5.7 Some arguments in favour of cosmic censorship

There being no convincing evidence against cosmic censorship, we must ask whether, on the other hand, there is any convincing evidence in favour of it. Indeed, there are no results that I am aware of which give direct and convincing support to the view that there is a mathematical theorem asserting some form of cosmic censorship in classical general relativity. But are there any plausible general lines of argument aimed in this direction? I am not sure. In Penrose (1979, pp. 625–626), I put forward some rather vague suggestions of this nature, but, to my knowledge, these have not been followed up in a serious way. The idea was to try to show, roughly speaking, that Cauchy horizons are unstable, in some appropriate sense— at least for an initial Cauchy hypersurface Σ which is either compact or appropriately asymptotically flat. The idea would be that in the 'generic' case, the Cauchy horizon $H^+(\Sigma)$ would be replaced by a singularity, so that the maximal space-time consistent with evolution from Σ would in fact be the domain of dependence of Σ. This would have to be globally hyperbolic, i.e. satisfy strong cosmic censorship. There is some evidence for such an instability (for asymptotically flat Σ) in work which shows that the 'inner horizon' (Cauchy horizon) of the Reissner-Nordström space-time (and of the Kerr space-time) is unstable (owing to the occurrence of infinitely blueshifted radiation); cf. Simpson and Penrose (1973), McNamara (1978a, 1978b), and Chandrasekhar and Hartle (1982). For further references concerning the issue of the (in)stability of black-hole Cauchy horizons in general relativity, see Ori (1997) and the article by Israel (chap. 7) in this volume.

On the other hand, there is some evidence that when there is a positive cosmological constant in Einstein's equations, stable Cauchy horizons may be possible. This situation comes about when the surface gravity of the cosmological horizon is greater than that of the Cauchy horizon, which can occur with Reissner–Nordström–de Sitter and Kerr–de Sitter space-times (see Chambers and Moss 1994; cf. also Mellor and Moss 1990, 1992, Brady and Poisson 1992). This situation is closely related to that considered below, in which inequalities arise from 'dropping particles into black holes'. As suggested below, it may well be that cosmic censorship requires a zero (or at least a nonpositive) cosmological constant.

Even if such a general result could be proved, it is not clear to me that this would really establish what is required for a suitable cosmic censorship theorem. It would not seem to rule out the thunderbolts discussed in section 5.5. This would really be necessary in order that the standard picture of gravitational collapse to a black hole can be obtained. What is the theoretical evidence that this picture is indeed likely to be always the correct one, according to classical general relativity? There seems to be little direct mathematical evidence. There are, however, certain rigorous mathematical results that give *indirect* support for cosmic censorship in this form. Ironically, these results have come about from a specific attempt to *disprove* cosmic censorship!

In Penrose (1973), I put forward a family of examples of gravitational collapse in which there is a collapsing spherical shell of null dust, the density being an arbitrary function of direction out from the centre, the region inside the shell being Minkowski space. Shortly afterwards, Gibbons (1972) pointed out that there is no need for the shell to be spherical, and he considered this more general case of a smooth convex collapsing shell of null dust which surrounds a region of Minkowski space. (This is the generalization employed by Tod, 1992, referred to in section 5.6, above.) As the shell collapses inwards, the matter density (the coefficient of a delta function) increases inversely as the area of cross section of the shell until it gets to a point where it can reverse the divergence of an intersecting outgoing light flash that originates in a region within the Minkowski space. If it does this all the way around, then the intersection S of that light flash with the infalling matter shell will be a trapped surface in the region just beyond the shell. All the geometry that needs to be considered for this takes place in Minkowski space. It depends only on the shapes of the collapsing shell and outgoing light flash, and on the matter density distribution on the shell.

Suppose that, in some particular shell geometry and matter distribution, it is possible to find an outgoing light flash for which S does provide us with a trapped surface. Then, according to the standard picture of collapse to a black hole—of which cosmic censorship is the most contentious

Chapter 5. The Question of Cosmic Censorship

part—the space-time will settle down to become a Kerr space-time in the future asymptotic limit. If we assume this to be the case, we find that a certain geometrical inequality must hold true. Suppose the area of $\mathcal{S}$ is A_0, the area of the intersection of the absolute event horizon $\partial I^-[\mathcal{I}^+]$ with the infalling matter shell is A_1, the future limit of the area of cross section of the (Kerr) horizon is A_2, and the area of the horizon of a Schwarzschild black hole of the same mass m is $A_3 = 16\pi m^2$ (units such that $G = c = 1$). We then have
$$A_0 \leq A_1 \leq A_2 \leq A_3 \leq 16\pi m_0^2, \tag{5.4}$$
where m_0 is the rest-mass of the total energy-momentum of the incoming null dust shell.[2] The first inequality follows from the fact that the shell is infalling; the second follows from the area-increase theorem (which, as we recall, requires cosmic censorship); the third follows because the area of the Kerr horizon is smaller than that of Schwarzschild for the same mass; the fourth is a consequence of $m \leq m_0$, a relation which expresses the fact that, although there might be a loss of mass due to gravitational radiation, there will not be a gain (because of the Bondi-Sachs mass-loss theorem and the asymptotic Minkowskian triangle inequality), the radiation being assumed to be entirely outgoing. Note that, in addition to cosmic censorship, there are (reasonable, but unproved) assumptions that the black hole actually *settles down* to become a Kerr space-time in the asymptotic limit (the known theorems simply assuming stationarity) and that the usual asymptotic assumptions for asymptotically flat space-times hold good (both at spacelike and null infinity).

The two quantities A_0 and m_0 depend only on the initial Minkowski space setup. If any such example could be found for which $A_0 > 16\pi m_0^2$, then this would provide a counterexample to the standard picture of gravitational collapse—essentially contradicting cosmic censorship. However, no example of this kind has ever been constructed. Moreover, various versions of the inequality $A_0 \leq 16\pi m_0^2$ have been proved by different authors (some of which refer to a somewhat different situation in which the geometry within a spacelike hypersurface is used); see Gibbons (1972, 1984, 1997), Jang and Wald (1977), Geroch (1973), and Huisken and Ilmanen (1997). Although none of these results directly establishes any form of cosmic censorship, they may be regarded as offering it some considerable support. Cosmic censorship could be said to supply a behind-the-scenes 'reason' why these inequalities are true!

There are also other types of inequalities which have been regarded as 'tests' of cosmic censorship. One may ask the question whether it is possible to 'spin up' a Kerr (or Kerr-Newman) black hole to a degree where its angular momentum exceeds the value for which a horizon is possible,

---

[2] That this is the rest-mass of the incoming shell is a point that was glossed over in Penrose (1973).

by allowing particles to drop into it. The mass, angular momentum, and charge of the particles come into the calculation, and various inequalities relate these to the black hole's geometrical parameters, in order that the horizon be preserved. It appears that these inequalities are always satisfied (cf., for example, Wald 1974; Semiz 1990)—except, curiously, if there is a positive cosmological constant (a situation pointed out to me by Gary Horowitz; cf. Brill et al. 1994). I am not sure of the significance of this final proviso. Of course, it might be the case that cosmic censorship requires a zero cosmological constant. We recall from section 5.4 that a negative cosmological constant (in anti–de Sitter space) leads to naked points at infinity. It is not at all inconceivable that a positive cosmological constant might correspondingly lead to naked singular points. This has relation to the issue of the instability of Cauchy horizons, as noted above.

The question of whether a black-hole horizon can be 'destroyed' by perturbing it with infalling matter is really part of the general question of the stability of a black-hole horizon. An unstable horizon could be expected to lead to a naked singularity. As far as I am aware, the arguments that have been given for horizon stability are fairly firm, but not yet fully conclusive. It would be interesting to know whether the presence of a cosmological constant makes a significant difference. My own feelings are left somewhat uncertain by all these considerations.

5.8 Do we need new techniques?

As will be seen from the preceding remarks, we are still a long way from any definitive conclusions concerning cosmic censorship. It is possible that radically new mathematical techniques will be required for any real progress to be made. My particular preferences would be for techniques related to developments in *twistor theory*. At the most immediate level, twistor theory is concerned with the geometry of the space $\mathbb{P}\mathcal{T}$ of rays (null geodesics) in a space-time $\mathcal{M}$. The points of $\mathcal{M}$ would be regarded as secondary structures, and those of $\mathbb{P}\mathcal{T}$ as being somewhat more fundamental. A point of $\mathcal{M}$ is interpreted, in $\mathbb{P}\mathcal{T}$, in terms of the family of rays through that point. This family has the structure of a sphere in $\mathbb{P}\mathcal{T}$—in fact, a *Riemann sphere*, which is a 1-dimensional complex manifold. The central idea of twistor theory is to call upon the power of *complex analysis* (physically, because of its links with quantum mechanics). For this purpose, the 5-dimensional manifold $\mathbb{P}\mathcal{T}$ must be thought of in terms of a larger complex manifold which, in the case when $\mathcal{M}$ is Minkowski space, turns out to be complex projective 3-space.

There are many problems, as yet unsolved, associated with how twistor theory is to be applied to general (vacuum) space-times, and it will be a long time before it has anything serious to say about cosmic censorship. Nevertheless, it has found a large number of applications (see, in particular,

Bailey and Baston 1990; Mason and Woodhouse 1996). So far, it has not been significantly used to treat global questions in general relativity. The closest it has come to this is in the work of Low (1990, 1994), where some progress is made towards the understanding the causal structure of a space-time in terms of linking properties of spheres (or of loops, in the case of a space-time of 2+1 dimensions) in its space of rays.

In relation to this, it may be noted that there is a connection between cosmic censorship and the topology of $\mathbb{P}\mathcal{T}$. If $\mathcal{M}$ satisfies strong cosmic censorship (i.e. is globally hyperbolic), then the space $\mathbb{P}\mathcal{T}$ is *Hausdorff*, whereas it is not Hausdorff in many cases where cosmic censorship fails. For any real progress to be made towards applying twistor theory to questions such as cosmic censorship, however, some major advances in understanding how the Einstein (vacuum) equations relate to twistor theory are needed. There does seem to be a deep connection between twistor theory and the Einstein equations, however—as yet elusive. This link is mediated through the equations for helicity 3/2 massless fields (Penrose 1992). It has been known for some time that the consistency conditions for such fields (in potential form) are the Einstein vacuum equations (Buchdahl 1958; Deser and Zumino 1976; Julia 1982). The other end of the link is the fact that the space of charges for such fields in *Minkowski* space is *twistor space*. Bringing together all the facets of this connection has proved to be a difficult problem (see Penrose 1996).

Although twistor theory remains a long way from addressing any significant issues of cosmic censorship, it does have relevance to various issues connected with general relativity and space-time geometry (see Huggett and Tod 1985; Penrose and Rindler 1986; Penrose 1996). Perhaps it already has something to say about cosmology. The picture of a big bang leading to a Friedmann-type universe with negative spatial curvature and hyperbolic spatial geometry fits in well with the complex-analytic (Riemann sphere) underlying philosophy, while the flat and closed spatial geometries do not do nearly so well (see Penrose 1997). Although negative spatial curvature cannot really be said to be a 'prediction' of the theory, it is perhaps the nearest to one, in general relativity and cosmology, that the theory has yet come up with.

Acknowledgments

I am particularly grateful to Robert Wald for various important remarks and for his help with the references. I am also grateful to the NSF for support under contract PHY93-96246.

References

Bailey, T.N. and Baston, R.J., eds. (1990) *Twistors in Mathematics and Phy-*

sics, London Mathematical Society Lecture Notes Series, No. 156 (Cambridge University Press, Cambridge)

Bekenstein, J. (1973) Phys. Rev. **D7**, 2333.

Brady, P.R. and Poisson, E. (1992) Class. Quant. Grav. **9**, 121.

Brill, D.R., Horowitz, G.T., Kastor, D., and Traschen, J. (1994) Phys. Rev. **D49**, 840.

Buchdahl, H.A. (1958) Nuovo Cim. **10**, 96.

Chambers, C.M. and Moss, I.G. (1994) Class. Quant. Grav. **11**, 1035.

Chandrasekhar, S. (1931) Astrophys. J. **74**, 81.

Chandrasekhar, S. and Hartle, J.B. (1982) Proc. R. Soc. London **A384**, 301.

Christodoulou, D. (1994) Ann. Math. **140**, 607.

Christodoulou, D. (1997) The instabilities of naked singularities in the gravitational collapse of a scalar field, to appear.

Choptuik, M. (1993) Phys. Rev. Lett. **70**, 8.

Chrusciel, P. (1990) Ann. Phys. **202**, 100.

Clarke, C.J.S. (1993) *The Analysis of Space-Time Singularities*, Cambridge Lecture Notes in Physics (Cambridge University Press, Cambridge).

Deser, S. and Zumino, B. (1976) Phys. Lett. **B62**, 335.

Floyd, R.M. and Penrose, R. (1971) Nature, Phys. Sci., **229**, 177.

Geroch, R. (1973) Ann. N.Y. Acad. Sci. **224**, 108.

Geroch, R., Kronheimer E.H., and Penrose, R. (1972) Proc. R. Soc. London **A347**, 545.

Gibbons, G.W. (1972) Commun. Math. Phys. **27**, 87.

Gibbons, G.W. (1984) in *Global Riemannian Geometry*, eds. T. Willmore and N.J. Hitchin (Ellis Horwood, Chichester).

Gibbons, G.W. (1997) Collapsing shells and the isoperimetric inequality for black holes, hep-th/9701049.

Hawking, S.W. (1972) Commun. Math. Phys. **25**, 152.

Hawking, S.W. (1975) Commun. Math. Phys. **43**, 199.

Hawking, S.W. (1993) in *The Renaissance of General Relativity (in honour of D.W. Sciama)*, eds. G. Ellis, A. Lanza, and J. Miller (Cambridge Univerisity Press, Cambridge).

Hawking, S.W. and Ellis, G.F.R. (1973) *The Large-Scale Structure of Space-Time* (Cambridge University Press, Cambridge)

Hawking, S.W. and Penrose, R. (1970) Proc. R. Soc. London **A314**, 529.

Holz, D.E. and Wald, R.M. (1997) A new method for determining cumulative gravitational lensing effects in inhomogeneous universes, Phys. Rev. D, submitted.

Chapter 5. The Question of Cosmic Censorship 121

Huggett, S.A. and Tod, K.P. (1985) *An Introduction to Twistor Theory*, London Mathematical Society Student Texts (L.M.S., London).

Huisken, G. and Ilmanen, T. (1997) Proof of the Penrose inequality, to appear.

Iyer, V. and Wald, R.M. (1991) Phys. Rev. **D44**, 3719.

Jang, P.S. and Wald, R.M. (1977) J. Math. Phys. **18**, 41.

Julia, B. (1982) Comptes Rendus Acad. Sci. Paris **295**, Sér. II, 113.

Low, R. (1990) Class. Quant. Grav. **7**, 177.

Low, R. (1994) Class. Quant. Grav. **11**, 453.

Mason, L.J. and Woodhouse, N.M.J. (1996) *Integrability, Self-Duality, and Twistor Theory* (Oxford University Press, Oxford).

McNamara, J.M. (1978a) Proc. R. Soc. London **A358**, 499.

McNamara, J.M. (1978b) Proc. R. Soc. London **A364**, 121.

Mellor, M. and Moss, I.G. (1990) Phys. Rev. **D41**, 403.

Mellor, M. and Moss, I.G. (1992) Class. Quant. Grav. **9**, L43.

Oppenheimer, J.R. and Snyder, H. (1939) Phys. Rev. **56**, 455.

Ori, A. (1997) Gen. Rel. and Grav. **29**, 881.

Penrose, R. (1969) Rivista del Nuovo Cim. Numero speciale **1**, 252.

Penrose, R. (1972) *Techniques of Differential Topology in Relativity*, CBMS Regional Conf. Ser. in Appl. Math., No. 7 (S.I.A.M., Philadelphia).

Penrose, R. (1973) Ann. N.Y. Acad. Sci. **224**, 125.

Penrose, R. (1978) in *Theoretical Principles in Astrophysics and Relativity*, eds. N.R. Liebowitz, W.H. Reid, and P.O. Vandervoort (University of Chicago Press, Chicago).

Penrose, R. (1979) in *General Relativity: An Einstein Centenary Survey*, eds. S.W. Hawking and W. Israel (Cambridge University Press, Cambridge).

Penrose, R. (1992) in *Gravitation and Modern Cosmology*, eds. A. Zichichi, N. de Sabbata, and N. Sánchez (Plenum Press, New York).

Penrose, R. (1996) in *Quantum Gravity: International School of Cosmology and Gravitation XIV Course*, eds. P.G. Bergmann, V. de Sabbata, and H.-J. Treder (World Scientific, Singapore).

Penrose, R. (1997) in *The Universe Unfolding*, eds. H. Bondi and M. Weston-Smith (Oxford University Press, Oxford).

Penrose, R. and Rindler, W. (1986) *Spinors and Space-Time*, Vol. 2: *Spinor and Twistor Methods in Space-Time Geometry* (Cambridge University Press, Cambridge).

Shapiro, S. and Teukolsky, S.A. (1991) Phys. Rev. Lett. **66**, 994.

Schoen, R. and Yau, S.-T. (1993) Commun. Math. Phys. **90**, 575.

Semiz, I. (1990) Class. Quant. Grav. **7**, 353.

Simpson, M. and Penrose, R. (1973) Int. J. Theor. Phys. **7**, 183.

Thorne, K.S. (1972) in *Magic without Magic*, ed. J.R. Klauder (Freeman, San Francisco).

Tod, K.P. (1992) Class. Quant. Grav. **9**, 1581.

Wald, R.M. (1974) Ann. Phys. **82**, 548.

Yodzis, P., Seifert, H.-J. and Muller zum Hagen, H. (1973) Commun. Math. Phys. **34**, 135.

6

Black Hole Collisions, Toroidal Black Holes, and Numerical Relativity

Saul A. Teukolsky

Abstract

Computer simulations are on the verge of making important breakthroughs in general relativity theory. Recent theoretical advances that will speed this process along are reviewed. Two examples of insights gained from numerical simulations are then discussed. In the first, the geometry of the event horizon formed by the head-on collision of two black holes is analyzed. The second describes the somewhat surprising result that toroidal black holes can exist. Their existence is reconciled with a number of theorems that appeared to make it unlikely that such black holes could form.

6.1 Introduction

When one discussed one's work with Chandra, he was never satisfied with just hearing about the details. He would always ask about how this particular piece of work fit into the big picture, about where it might lead in the future. In this spirit, before discussing the results of some recent computer simulations of black holes, I would like to first share some thoughts about the future of general relativity. I will restrict my comments to classical gravitation, since quantum gravity will be covered by other participants in the symposium.

In trying to predict the future, it is helpful to look at the past. What factors led to significant breakthroughs in general relativity in the past 30 to 40 years? The two most important were the introduction of new analytic techniques and the impact of new technology.

The most far-reaching new analytic techniques brought to bear on general relativity were the global methods introduced by Penrose, Hawking, Geroch, and others. These new methods led in the 1960s and '70s to many important insights, such as the singularity theorems, the black hole area theorem, improved understanding of the structure of infinity, and so on. Other new analytic techniques, such as spinor techniques, played an important though lesser role. For example, perturbation theory for rotating black holes was intractable until it was treated with spinors.

New technology led to new experimental opportunities in gravitation. For example, X-ray satellites led to the discovery of Cygnus X-1 and other black hole candidates. Laser ranging to the moon, VLBI monitoring of quasars to test the Shapiro delay, and various solar system tests of general relativity using artificial satellites all combined to increase our confidence in the experimental underpinnings of the theory. Before this era, we relied on the three classical tests of the theory discussed in Einstein's original paper. But it was already understood that the gravitational redshift experiment did very little to distinguish general relativity from other possible theories. Eddington's famous measurement of light bending was known by aficionados to be suspect; a generous statement is that his error estimates would not satisfy modern experimenters. Even the perihelion precession of Mercury had lost some of its weight because of the Brans-Dicke theory and the possibility that some of the precession was due to rapid rotation of the Sun. The cumulative impact of the new experiments, together with the realization that general relativity predicts many null results that would not be null in alternative theories of gravity, gave the theory fresh experimental respectability within mainstream physics. The most important of these new experiments, made possible by advances in radio astronomy and high precision atomic clocks, was the discovery of the Hulse-Taylor binary pulsar at Arecibo in 1974. This system is a relativist's dream: a highly accurate clock in rapid orbit through the strong gravitational field of its companion. Continued timing of this system by Taylor and his coworkers has confirmed that the system is losing energy by gravitational wave emission at the rate predicted by general relativity to an accuracy of better than 1%.

Can we identify factors at work now that will lead to breakthroughs in the future? Once again, I think one can identify two important influences.

The first is that technology continues to advance. The most likely experimental area to revolutionize general relativity in the near future is the development of gravitational wave detectors such as LIGO (see chapter 3, by Thorne, in this volume). Previous tests of gravitation theory can be thought of in terms of a perturbation expansion away from Newtonian gravitation. Solar system tests typically involve v^2/c^2 corrections to Newtonian gravity. Even the Hulse-Taylor binary pulsar probes "just" v^5/c^5 corrections. Black holes, by contrast, are manifestations of the full nonlinear theory. Astrophysicists have built up a huge edifice of theory that assumes black holes exist and have the properties of Einstein's theory. For example, black holes are assumed to power quasars and AGNs, even though we have no direct experiments confirming their supposed attributes. With detectors like LIGO, we will finally have the prospect not just of detecting gravitational waves directly, but of using them to probe the features of sources like black holes.

The second factor that will in all probability lead to future breakthroughs is the use of computers. By this, I do not mean computers to do symbolic calculations. Computer algebra, while sometimes very convenient, has not yet had a really significant impact on the field, even though it has been used for over 30 years [1]. It is very helpful for checking long calculations. But computer algebra by and large continues to rely on brute force. Human beings are still much better at clever shortcuts and insights. Your eye can scan a page and by some kind of parallel processing see connections that lead to simplifications. No computer could yet discover the Kerr metric. But once you have the metric, a computer can verify that it really is a solution of Einstein's equations.

In the more distant future, computer algebra will benefit from advances in expert systems and artificial intelligence. Already in medical diagnosis there are expert systems that have been programmed by following around a leading diagnostician and recording what he does. "Why did you order that test?" "What will you do if it comes back negative?" All this gets programmed in and combined with the computer's huge database of facts to produce a system that in the end can be better than any single doctor. In the same way, I foresee computer scientists getting volunteer general relativists to do complicated algebra. They'll stop you and say, "Now why did you substitute for the Christoffel symbols there?" And you'll say, "Well, it looked like I could get the second derivative terms to cancel." And they'll program this all into the system. But I think this is still a long way off!

The imminent breakthroughs I foresee with computers involve the numerical solution of Einstein's equations for many interesting cases. There are three reasons why we are on the verge of important advances in the computer solution of Einstein's equations. Two of these reasons are fairly obvious. First, hardware: Massively parallel machines with thousands of processors will for the first time give us not only the horsepower to do all the arithmetic involved, but also enough memory to compute solutions as a time evolution in three spatial dimensions. Second, software: The programming of parallel machines is very difficult. But software packages are being developed that take care of all the bookkeeping involved in dividing one's problem up and distributing it over thousands of processors. Moreover, algorithms for the solution of partial differential equations are now fairly mature and adequate for an attack on Einstein's equations.

The third reason is equally important: There have been recent advances in our theoretical understanding of Einstein's equations that should translate into advances in numerical solutions. I discuss these in the next section.

6.2 Einstein's equations in hyperbolic form

The most fruitful way of tackling Einstein's equations numerically has been to use the ADM 3+1 decomposition. This involves rewriting Einstein's 4-dimensional representation in terms of time and three spatial coordinates. You specify initial data on the $t=0$ time slice. You then use the ADM equations to have the computer advance the data by a succession of small steps Δt until you have mapped out the spacetime sufficiently far into the future.

Black holes present a potential problem, because they have a singularity inside. You cannot allow your time slice to run into the singularity, otherwise the computer will give you nasty messages like "overflow". The traditional way of dealing with this has been to use "singularity-avoiding" time slicing, such as maximal slicing. These slicings take advantage of the freedom in general relativity to choose coordinates by holding back the advance of proper time in the strong field regions near the singularity. Coordinate time advances at the same rate as proper time far from the black hole where gravity is weak. Such methods, when supplemented by a few other tricks, have been successful in spherical symmetry, but barely adequate in axisymmetry. The problem is that the time slicing becomes very distorted once the black hole forms, stretching out near the black hole surface as it is held back inside the black hole. Since you can distribute only a finite number of grid points on each time slice, you eventually lose the ability to represent the gravitational field accurately and the code crashes.

The new methods involve trying to cut the black hole out of the computational domain. This seems like such an obvious idea: Since no information can get out of a black hole, why integrate the equations inside? The problem up till now has been with the boundary conditions. If you cut a region out of the domain, what boundary conditions should be imposed on the boundary of the excised region? A simple way to see the problem and its solution is to consider the simpler case of the 3+1 decomposition of Maxwell's equations (see, e.g., [2]). We write Maxwell's equations in the following form:

$$[\nabla \cdot \mathbf{E} = 4\pi\rho], \tag{6.1}$$

$$\partial_t \mathbf{A} = -\mathbf{E} - \nabla\phi, \tag{6.2}$$

$$\partial_t \mathbf{E} = \nabla \times (\nabla \times \mathbf{A}) - 4\pi\mathbf{J}$$

$$= \nabla(\nabla \cdot \mathbf{A}) - \nabla^2 \mathbf{A} - 4\pi\mathbf{J}. \tag{6.3}$$

Equation (6.1) is bracketed because it is not an *evolution* equation. It is a *constraint* equation that has to be satisfied at $t=0$ but is then preserved by the evolution equations. Equations (6.2) and (6.3) are the analogs of the ADM equations for the evolution of the 3-metric and the extrinsic curvature: $\partial_t g_{ij} = \cdots$, and $\partial_t K_{ij} = \cdots$. The key point is that the term

Chapter 6. Black Hole Collisions ...

$\nabla(\nabla \cdot \mathbf{A})$ spoils *hyperbolicity* of the system of equations. It is easy to see this if we rewrite the equations in second-order form:

$$\partial_t^2 \mathbf{A} - \nabla^2 \mathbf{A} + \nabla(\nabla \cdot \mathbf{A}) = 4\pi \mathbf{J} - \nabla \partial_t \phi. \tag{6.4}$$

The extra second derivatives of $\mathbf{A}$ spoil the nice wave operator that we would otherwise have on the left-hand side of equation (6.4).

In Maxwell's theory, we know how to deal with this problem: We make use of the gauge freedom to eliminate the offending term. Suppose we try imposing the Coulomb gauge, $\nabla \cdot \mathbf{A} = 0$. This condition is enforced by adding ϕ to the system of equations:

$$\nabla^2 \phi = -4\pi \rho, \tag{6.5}$$
$$\partial_t \mathbf{A} = -\mathbf{E} - \nabla \phi, \tag{6.6}$$
$$\partial_t \mathbf{E} = -\nabla^2 \mathbf{A} - 4\pi \mathbf{J}. \tag{6.7}$$

The key point here is that equation (6.5) is *elliptic*. Both ϕ and $\mathbf{A}$ have action-at-a-distance pieces in this gauge. These pieces cancel exactly, leaving causal behavior for the $\mathbf{E}$ and $\mathbf{B}$ fields. But in a numerical simulation, the noncausal pieces may not cancel exactly. Moreover, one cannot cut out a hole in the computational domain. The potential ϕ satisfies an elliptic equation that requires boundary conditions on any excised region. The traditional singularity-avoiding methods for Einstein's equations have analogous problems: There are gauge fields that tend to propagate acausally, and there does not seem to be any mathematically justifiable way to cut holes out because the equations are not in hyperbolic form.

In Maxwell's equations, the standard way of fixing these problems is to use the Lorentz gauge, $\nabla \cdot \mathbf{A} = -\partial_t \phi$. The equations become

$$\partial_t \phi = -\nabla \cdot \mathbf{A}, \tag{6.8}$$
$$\partial_t \mathbf{A} = -\mathbf{E} - \nabla \phi, \tag{6.9}$$
$$\partial_t \mathbf{E} = -\nabla \times (\nabla \times \mathbf{A}) - 4\pi \mathbf{J}. \tag{6.10}$$

In second-order form, this system is

$$\partial_t \phi = -\nabla \cdot \mathbf{A}, \tag{6.11}$$
$$\Box \mathbf{A} = -4\pi \mathbf{J}. \tag{6.12}$$

Equation (6.11) has characteristic speed 0, while equation (6.12) has characteristic speed c.

A possible disadvantage of the analogous gauge choice in general relativity (the harmonic gauge) is that one has used up all the gauge freedom to get a manifestly hyperbolic system. It may be useful to use the coordinate freedom for other purposes, e.g., to keep a black hole surface at a fixed coordinate location on the grid. Is it possible to achieve a hyperbolic formulation, independent of gauge choice?

In Maxwell's equations, the trick is to take another time derivative:

$$\partial_t^2 \mathbf{E} = -\nabla \times (\nabla \times \mathbf{E}) - 4\pi \partial_t \mathbf{J}. \tag{6.13}$$

Expand the curl curl term and replace $\nabla \cdot \mathbf{E}$ by $4\pi\rho$. This gives

$$\Box \mathbf{E} = -4\pi(\nabla\rho + \partial_t \mathbf{J}) \tag{6.14}$$

with characteristic speed c, and equation (6.9) with characteristic speed 0. The gauge field ϕ is completely arbitrary in this formulation.

Recently, Choquet-Bruhat and York [3, 4] have derived an analogous hyperbolic formulation of Einstein's equations by taking another time derivative of the ADM equations. In this formulation one is left with the full freedom to impose gauge conditions on the spatial coordinates. A hyperbolic formulation using harmonic coordinates has been explored by Bona et al. [5]. Hyperbolic formulations using tetrad formalisms have also been derived [6, 7]. Some of these formulations have been compared, and other formulations derived, by Friedrich [8]. The exciting thing about these new formulations is that they offer the possibility of a mathematically well-posed procedure for excising black holes. Since the characteristics are all ingoing at the surface of a black hole, no explicit boundary condition is required for hyperbolic equations there. A number of different groups are experimenting with computer implementations of these new formulations, and one can expect significant results eventually.

What kinds of information might we learn from computer simulations? I will discuss two examples, both based on simulations using the "old" singularity-avoiding techniques.

6.3 Head-on collisions of black holes

Some of the most interesting questions in numerical relativity concern the formation of black holes from gravitational collapse or collisions. A curious paradox, however, is that the appearance and growth of a black hole cannot be determined in a numerical simulation until after the complete spacetime has been constructed. The reason is that there is no instantaneous criterion for deciding whether a particular event is inside or outside a black hole. Rather, it is necessary to determine the fate of all possible light rays emitted from that event. If any light ray can escape to infinity, the event is outside the black hole. If no light ray can reach infinity, the event is inside the black hole.

The *event horizon*, or surface of the black hole, is the boundary between events that can send light rays to infinity and those that cannot. Thus light rays have to be tracked arbitrarily far into the future to identify this

Chapter 6. Black Hole Collisions ...

boundary. As one evolves a spacetime numerically from one time slice to the next, it is thus impossible to locate the surface of the black hole on each slice concurrently. One must go back after one has determined the fate of light rays to mark the event horizon on any given slice.

In some cases, an *apparent horizon* signals the existence of a black hole. An apparent horizon is also determined by the fate of beams of light, but in this case it is the instantaneous fate of the rays that counts. In a weak gravitational field about a central source, an outward beam of light diverges and its cross-sectional area increases. However, a sufficiently strong gravitational field actually focuses the light so that the area of the beam immediately decreases. If such light beams are converging at all points on a closed surface, we have a region of closed trapped surfaces. The apparent horizon is the outer boundary of the region of closed trapped surfaces. If an apparent horizon is present on any given time slice, then there must be a black hole on that slice, and the apparent horizon must lie inside the hole's event horizon (assuming that the cosmic censorship hypothesis is true) [9]. At late times, when the gravitational field has settled down to a stationary state, the event horizon and the apparent horizon coincide.

An apparent horizon is readily identified from the computed metric on each time slice as it is produced in a numerical code. While its appearance guarantees the presence of a black hole, there are two problems with using an apparent horizon as a diagnostic for black holes. First, a black hole may be present on the slice without an apparent horizon. Second, even when the apparent horizon exists, the event horizon does not coincide with it, especially in the early stages of black hole formation.

Several codes have been developed to map out the event horizon given a numerical spacetime and the apparent horizon [10, 11, 12]. These codes take as input numerically generated metric coefficients and probe the geometry using null geodesics. While in principle the event horizon is determined by emitting a large number of test photons and tracking them to see which escape and which are trapped by the black hole, in practice it is numerically more efficient to trace photons backwards in time from the last time slice of the evolution [11, 12]. Since future-directed photons tend to diverge away from the horizon, either to infinity or into the black hole, past-directed photons are "attracted" onto the horizon.

One of the first applications of a horizon-finding code [10] was to study the event horizons that form during the head-on collision of two black holes to form a single black hole. We simulated such a collision for nonrotating black holes of equal mass to form a Schwarzschild black hole [13]. In this simulation, each black hole is formed from the collapse of a spherical ball of collisionless particles. Each ball collapses because the particles are chosen to have no initial motion. To mimic a collision, we begin the simulation with the two clusters well separated but headed toward each other with a velocity of 0.15c. The two clusters then fall toward each other while individually collapsing to form black holes. The individual black holes originate at the center of each cluster and then grow outward. Their event horizons are tidally distorted by their mutual gravitational interaction. The black holes then merge, in the process converting $\approx 3 \times 10^{-4}$ of the system's mass into energy in the form of gravitational waves radiating outward. At the end of the simulation, there is a single spherical black hole encompassing all the matter. Pictures of this process can be found in [10, 14], where further details are given.

We have mapped out the location of the event horizon in this computed spacetime by numerically propagating light rays ([10], especially fig. 3). The resulting spacetime picture represents a precise calculation of the "pair of pants" description of the event horizon as sketched in textbooks over 20 years ago ([9], fig. 60, or [15], fig. 34.6). Our evolution code found a common apparent horizon appearing at $t \simeq 6.5M$. The horizon code found event horizons appearing much earlier than this. In particular, we found disjoint event horizons appearing within the spheres of matter as early as $t \simeq 0.13M$. The disjoint event horizons are tidally distorted, growing toward each other until they coalesce at $t \simeq 2.16M$, forming a single event horizon (fig. 1 of [10]). We terminated the evolution at $t \simeq 11.7M$. At this time the errors in the simulation owing to "grid stretching" have grown to nearly 10%. However, the system has settled down to an almost static Schwarzschild black hole.

The event horizon is generated by light rays. These generators have several important properties: once on the horizon, they can never leave it; and they can never cross, except possibly when getting onto the horizon. These properties raise some interesting questions about the nature of the event horizon for a head-on collision. For example, what happens right along the inside "seam" of the pair of pants? The theorems forbid light rays to travel up each inside seam and meet at the "crotch".

The behavior of these light rays and their singularities can be understood in terms of the classical theory of caustics. In spherical symmetry with no gravity, the beam emerging from a point caustic traces out the usual Minkowski light cone. If gravity is sufficiently strong to focus the rays, they instead trace out the horizon for a single Schwarzschild black hole. The point caustic of spherical symmetry is a special case; the situation is different when the symmetry is broken, since point caustics are not stable structures under perturbations (see, e.g., [16] for a discussion of caustics in classical optics, or [17] for a discussion of their spacetime properties).

When we trace the generators *backwards* in time in our simulation, we see that some of them cross each other and leave the horizon. Going forwards in time, these *crossovers* are the earliest points along the rays which lie on the horizon. Crossovers are ubiquitous features of black hole

Chapter 6. Black Hole Collisions ...   131

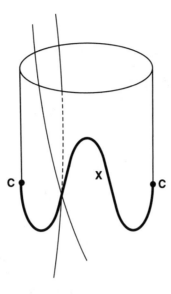

Figure 6.1: Geometry of the event horizon for the head-on collision of two black holes. The horizon is depicted in a spacetime diagram, with time plotted vertically. A representative pair of generators is shown, crossing along the "seam" of the pair of pants. The line of crossovers is shown as a heavy line X. The line of crossovers terminates at the cusp caustics C.

interactions. In our simulation, a line of such crossover points extends from the crotch on the pair of pants down along each inside trouser seam, around each bottom, and a small distance up each outside seam (see fig. 6.1). At the endpoints of the line of crossovers, slightly up the outside of each "leg", are cusp caustics. A line of crossovers terminating in a cusp caustic is a stable feature in axisymmetry. The line of crossovers is produced by the intersection of two separate beams of light, and so it must be *spacelike*. Thus the inside seam of the pair of pants is *not* a generator of the horizon. Somewhat unexpectedly, we find that the crossover line, although spacelike at each point, becomes asymptotically lightlike as it approaches the cusp caustic on the outside seam. It thus joins on smoothly to the generator of the horizon that begins at the caustic and travels up the side of the pair of pants [18].

A similar analysis has been made of the horizon structure of two colliding *vacuum* black holes [14]. The simulations for this case are described in [19, 20, 21]. This situation is different from that described above, where the black holes are "born" and the horizon's origin can be studied. In this case, by contrast, the black holes exist eternally. Thus the trouser legs do not have cusps on the sides but continue into the past. In other respects the horizon structure is similar, with a line of crossovers along the inside seam.

The numerical simulations described in this section allowed one to explore the geometry of the event horizon in detail and spurred analytic understanding of this geometry. This is one prototype of how numerical simulation can be used. In the next section we describe another mode, where numerical simulation leads to the discovery of an unexpected feature of the theory.

6.4 Toroidal black holes

There are several theorems which state that under various conditions a black hole must have spherical topology. It was therefore somewhat surprising when computer simulations of collapsing matter gave rise to black holes that initially form with toroidal topology and only at a later time become spherical. In this section we describe recent work showing that the computational results are in fact consistent with the theorems and can be understood in terms of a simple spacetime model of a toroidal event horizon. This work is more fully described in [22].

The simplest situation arises in the general stationary spacetime, for which it can be shown that any black hole must have the surface topology of a 2-sphere [23]. The first theorem regarding the topology of *nonstationary* black holes is due to Gannon [24]. Assuming a physically reasonable condition of asymptotic flatness, he proved that the surface topology of a smooth black hole must be either a 2-sphere or a torus (provided that the energy-momentum tensor of the matter fields satisfies the dominant energy condition).

Gannon's approach has recently been extended and generalized to yield stronger theorems, under the assumptions of asymptotic flatness, global hyperbolicity, and a suitable energy condition on the matter fields. Under the title of "topological censorship", Friedman, Schleich, and Witt proved a theorem that any two causal curves extending from past to future null infinity may be deformed into each other (in the sense of homotopy) [25]. As Jacobson and Venkataramani have pointed out, a black hole with toroidal surface topology provides a potential mechanism for violating topological censorship by sending a light ray from the infinite past through the hole in the torus and back out to future null infinity [26]. This light ray would not be deformable to a light ray that skirts the horizon altogether. Thus the topological censorship theorem implies that the hole in a toroidal horizon must close up quickly, before a light ray can pass through. Jacobson and Venkataramani have also established a theorem that strengthens a recent result due to Browdy and Galloway that the surface geometry of a black hole at a given time must be a 2-sphere if no new null generators enter the horizon at later times [27]. The theorem of Jacobson and Venkataramani

Chapter 6. Black Hole Collisions ...

limits the time for which a toroidal black hole can persist, albeit in a highly technical way.

In an attempt to make toroidal black holes, we have used a numerical relativity code to simulate the collapse of rotating toroidal configurations of collisionless particles [28]. The code has been subjected to a number of tests to ensure that it is accurate. Among these tests are the propagation of linearized gravitational waves of both polarizations, for which analytic solutions exist; maintaining rotating equilibrium configurations in stable equilibrium for many dynamical timescales; and verifying that the black holes that form from collapse settle down to Kerr holes at late times. In addition, during the course of numerical evolution, we compute a set of physical diagnostics to monitor the code's reliability in the nonlinear regime. We compute the Brill mass and angular momentum of the spatial hypersurfaces and look for trapped regions. The Brill mass, corrected for the loss of gravitational radiation through the outer boundary, is a conserved quantity. Since in axisymmetry gravitational radiation carries no angular momentum, the total angular momentum of the system is another conserved quantity. We locate the apparent horizon during black hole formation and probe its geometry. Once the evolution is completed and the black hole settles down, we map out the event horizon and ergoregion and probe their geometries. Although we cannot prove mathematical theorems by means of a numerical simulation, our tests give us confidence that the results reported in [28] are correct.

We have found an interesting case in which the black hole event horizon initially develops as a toroid [28, 10]. The initial configuration is based on a solution for a rotating toroidal cluster in stable equilibrium [29]. The cluster has an outer circumferential radius of $R/M = 4.5$. To get it to collapse, we reduce the angular momentum of each particle by a factor of 0.5, producing a nonequilibrium cluster with total angular momentum $J/M^2 = 0.70$. Spatial snapshots of the collapsing configuration, together with the location of the apparent and event horizons, are plotted in figures 6 and 7 of [10]. The toroidal horizon first forms entirely within the vacuum, between the origin and the inner edge of the toroidal cluster. It then expands to fill up the doughnut hole, becoming topologically spherical approximately when the outer edge of the horizon reaches the inner edge of the matter toroid.

The event horizon is found after the simulation is complete by propagating light rays backwards in time from the surface of the final equilibrium black hole as described in the previous section. The spacetime reconstruction of the horizon is depicted in figures 4 and 5 of [22]. When we trace the rays backwards in time, some of them leave the horizon at crossovers. For these rays the crossovers are the endpoints of the null generators of the horizon. The line of crossovers forms a spacelike curve.

The computational simulation of a rotating black hole and its horizon structure is thus in complete accord with the various theorems regarding black hole topology. At times when the surface of the rotating black hole has a manifold structure, its surface topology is either toroidal or spherical, in agreement with the work of Gannon [24]. However, Gannon's theorem assumes a smooth black hole, and it is unclear how it could be generalized to apply to the formation stage when new generators are being added. In our computational model, the toroidal black hole has a nonsmooth inner rim where new generators emerge from crossover points (as well as a nonsmooth outer rim at the very early times before the caustic). Also, at the exceptional time at which the toroid pinches off, just prior to becoming spherical, the surface of the black hole is not even a (Hausdorff) manifold.

At late times, when equilibrium has been reached, the topology is spherical, in accord with the results of Hawking [23]. At early times, the topology defined by the Cauchy slicing is temporarily toroidal [30]. However, the spacetime curve traced out by a point on the inner rim of the torus is a line of crossovers and is therefore spacelike. Thus the "hole" in the torus closes up faster than the speed of light. Consequently, no causal signal can link through the torus and escape back to the exterior spacetime region to provide a violation of topological censorship. Finally, at late times when the horizon has its full complement of generators, the rotating black hole has spherical topology, in agreement with the theorems of Browdy and Galloway [27].

Note that the concept of "linking" a torus has some subtleties. Consider a thread linking the eye of needle, where we regard the eye as a 2-dimensional surface in three spatial dimensions. Clearly there is no way to unlink the thread by pulling it sideways through the metal of the eye. Now add the time dimension. Suppose the eye of the needle stops existing at some point in time. Then simply by waiting long enough, one has unlinked the thread. Similarly, in the time-reversed case, when the eye forms at some finite time around the thread, one can move the thread sideways before this happens and end up with an unlinked thread. For a toroidal black hole that forms after a finite time, one can always deform a worldline that would have passed through the torus by moving it "sideways" in spacetime at an early time so that it no longer links the torus. (In topological jargon, the homotopy is trivial.) The real import of the topological censorship results relates to a pair of observers who are initially coincident. One of their worldlines then goes into the doughnut hole while the other tries to go around the black hole and meet up with the first. The theorems restrict the ability to do this if the separation occurs "too late". It is clear that the full implications of the theorem of Jacobson and Venkataramani [26] for this model deserve further study.

Acknowledgments

The simulations described in sections 3 and 4 were carried out together with my long-time collaborator, Stuart Shapiro. I would also like to thank Jeff Winicour for patiently explaining to me some of the subtleties of horizon topology. This research was supported by the NSF Grand Challenge Grant PHY 93-18152/ASC 93-18152 (ARPA supplemented). The work was also supported by NSF grant PHY 94-08378 and NASA grant NAG-2809 to Cornell University. Computing resources were made available by the Cornell Theory Center.

References

[1] For the early history of computer algebra in relativity, see [15], box 14.3.

[2] A. M. Abrahams and J. W. York, in *Astrophysical Sources of Gravitational Radiation*, eds. J.-A. Marck and J.-P. Lasota (Cambridge University Press, Cambridge, in press); preprint gr-qc/9601031.

[3] Y. Choquet-Bruhat and J. W. York, C. R. Acad. Sc. Paris **321**, Série I, 1089 (1995).

[4] A. Abrahams, A. Anderson, Y. Choquet-Bruhat, and J. W. York, Phys. Rev. Lett. **75**, 3377 (1995).

[5] C. Bona, J. Massó, E. Seidel, and J. Stela, Phys. Rev. Lett. **75**, 600 (1995).

[6] M. H. P. M. Van Putten and D. M. Eardley, Phys. Rev. D **53**, 3056 (1996).

[7] F. B. Estabrook, R. S. Robinson, and H. D. Wahlquist, Class. Quantum Gravit. **14**, 1237 (1997).

[8] H. Friedrich, Class. Quantum Gravit. **13**, 1451 (1996).

[9] S. W. Hawking and G. F. R. Ellis, *The Large Scale Structure of Space-Time* (Cambridge University Press, Cambridge, 1973).

[10] S. A. Hughes, C. R. Keeton, P. Walker, K. T. Walsh, S. L. Shapiro, and S. A. Teukolsky, Phys. Rev. D **49**, 4004 (1994).

[11] P. Anninos, D. Bernstein, S. Brandt, J. Libson, J. Massó, E. Seidel, L. Smarr, W.-M. Suen, and P. Walker, Phys. Rev. Lett. **74**, 630 (1995).

[12] J. Libson, J. Massó, E. Seidel, W.-M. Suen, and P. Walker, Phys. Rev. D **53**, 4335 (1996).

[13] S. L. Shapiro and S. A. Teukolsky, Phys. Rev. D **45**, 2739 (1992).

[14] R. A. Matzner, H. E. Seidel, S. L. Shapiro, L. Smarr, W.-M. Suen, S. A. Teukolsky, and J. Winicour, Science **270**, 941 (1995).

[15] C. W. Misner, K. S. Thorne, and J. A. Wheeler, *Gravitation* (W. H. Freeman, San Francisco, 1973).

[16] M. V. Berry and C. Upstill, Prog. Opt. **17**, 256 (1980).

[17] H. Friedrich and J. M. Stewart, Proc. R. Soc. London **A385**, 345 (1983).

[18] Many years ago, Penrose sketched a figure showing the merger of two event horizons. The inside seam is shown without further comment. The behavior at the vertex and up the outside of the pants leg is not included in the figure. Apparently, this work was not noticed by anyone else. The reference is fig. 2 of R. Penrose, in *Gravitational Radiation and Gravitational Collapse*, IAU Symposium No. 64, ed. C. DeWitt-Morette (Reidel, Boston, 1974), 82.

[19] P. Anninos, D. Hobill, E. Seidel, L. Smarr, and W.-M. Suen, Phys. Rev. Lett. **71**, 2851 (1993).

[20] P. Anninos, D. Bernstein, S. R. Brandt, D. Hobill, E. Seidel, and L. Smarr, Phys. Rev. D **50**, 3801 (1994).

[21] P. Anninos, D. Hobill, E. Seidel, L. Smarr, and W.-M. Suen, Phys. Rev. D **52**, 2044 (1995).

[22] S. L. Shapiro, S. A. Teukolsky, and J. Winicour, Phys. Rev. D **52**, 6982 (1995).

[23] S. W. Hawking, Comm. Math. Phys. **25**, 152 (1972).

[24] D. Gannon, Gen. Rel. Grav. **7**, 219 (1976).

[25] J. L. Friedman, K. Schleich, and D. M. Witt, Phys. Rev. Lett. **71**, 1486 (1993).

[26] T. Jacobson and S. Venkataramani, Class. Quantum Gravit. **12**, 1055 (1995).

[27] S. Browdy and G. J. Galloway, J. Math. Phys. **36**, 4952 (1995); G. J. Galloway, in *Differential Geometry and Mathematical Physics*, Contemporary Mathematics Vol. 170, eds. J. Beem and K. L. Duggal (Amer. Math. Soc., Providence, 1994).

[28] A. M. Abrahams, G. B. Cook, S. L. Shapiro, and S. A. Teukolsky, Phys. Rev. D **49**, 5153 (1994).

[29] S. L. Shapiro and S. A. Teukolsky, Astrophys. J. **419**, 636 (1993).

[30] Note that one can choose a time slicing for which the horizon cross sections are never toroidal. Further discussion of horizon topology has been given by M. Siino, gr-qc/9701003 (1997).

7

The Internal Structure of Black Holes

Werner Israel

Abstract

Gravitational collapse to a black hole leaves a decaying tail of gravitational waves. These waves are partially absorbed by the hole and have drastic effects on the geometry near the inner (Cauchy) horizon because of a diverging blueshift. This article reviews recent efforts to analyze these effects, highlighting outstanding problems and gaps.

7.1 Introduction

The interaction of a black hole with its environment, now pinpointed with some confidence as the prime mover in many instances of quasar and AGN activity, involves some of the most difficult analysis in all of mathematical physics. The last two decades have seen remarkable progress in our understanding of this area, and no one has contributed more to the mathematical development than Chandra [1].

But Chandra's interests in what he called "the most perfect macroscopic objects in the universe" [2] were not driven solely by astrophysical priorities. The colliding plane wave solutions he studied with Basilis Xanthopoulos [3] are closely related (via complex coordinate re-identifications) to Kerr-Newman black hole interiors. An elegant product of his collaboration with Jim Hartle [4], entitled "On Crossing the Cauchy Horizon of a Reissner-Nordström Black Hole," is an analysis of the evolution and fate of linear wavelike perturbations inside a spherical charged hole.

It would be in Chandra's royal tradition to keep the historical aspects in view and to recall that the first serious exploration of a (Schwarzschild) black hole's interior was undertaken in 1950 by the Irish mathematical physicist J. L. Synge [5]. Synge, who passed away on March 29, 1995, at the age of 98, was the pioneer who honed the geometrical approach to relativity into a practical tool. In the 1950 paper, too, his motivation was geometrical. It is legal, he argued, to ask what general relativity has to say about the gravitational field of a point mass even if such objects do not exist in nature. He was the first to introduce a pair of lightlike coordinates to surmount the coordinate barrier at the Schwarzschild radius. However, because his coordinates inside that radius were inverse trigonometric functions of Kruskal's, he was led to a latticelike extension of the conventional Kruskal manifold, containing an infinite sequence of $r = 0$ singularities through which an infallen particle oscillates like a shuttlecock. Undoubtedly this picture looked bizarre. And perhaps it did not help that the work appeared in the *Proceedings of the Royal Irish Academy*, a journal of irreproachable reputation, but not delivered with the morning papers. So it went almost completely unnoticed. But to a large degree it anticipated the technique and core results of the 1960 publications of Kruskal and Szekeres, and the latter indeed makes reference to it.

The *New York Times* a few years ago featured a cartoon by Tom Bloom, showing a NASA astronaut plunging head-first into a black hole. Ahead of and below him is a row of mileposts marked with *times*: "this year, next year, sometime, never." This cartoon conveys useful scientific information. It illustrates the most significant property of a black hole's interior: descent into a black hole is basically a progression in time. (Inside a spherical hole, for example, it is well known that the usual radial coordinate r becomes timelike.) Thus the internal structure of a black hole is not really a structural problem at all, but an evolutionary problem, not essentially different from following the motion of a fluid, using Euler's equations, up to onset of turbulence or a shock. This has the advantage that we are protected by causality. We do not need to understand the physics of the innermost regions to make meaningful statements about the outer and *preceding* layers where curvatures are still below Planck levels.

In the wake of a complete gravitational collapse, one expects the geometry outside the event horizon to settle, after some initial agitation, toward a Kerr (or Kerr-Newman) form asymptotically. The exact (stationary) Kerr-Newman geometry can be continued analytically beyond its event horizon. One then encounters (for nonvanishing angular momentum or charge) an inner bifurcate Killing horizon. Penrose [6] in 1968 first drew attention to a pathological feature of this horizon: because its ingoing sheet (called the Cauchy horizon) corresponds to infinite external advanced time, any perturbation must have its time dependence infinitely speeded up there. Subsequent linear perturbative studies on fixed Reissner-Nordström [7, 4] or Kerr-Newman [8] backgrounds confirmed that wavelike disturbances falling into the hole (for instance, the Price wave tail [9] left in the wake of a nonspherical collapse) are blueshifted to infinite energy densities at the Cauchy horizon. The view which then took hold was that Cauchy horizons are unstable and exceptional features, peculiar to exactly stationary Kerr-Newman geometries, with no counterparts in realistic black holes. For the causal structure of a generic black hole, the prototype was

Chapter 7. The Internal Structure of Black Holes 139

taken to be the Schwarzschild geometry, uncomplicated by inner horizons, and terminating in an all-enveloping singularity which (in line with strong cosmic censorship as commonly understood) is spacelike. More recent studies, which it is my purpose to review, point to a picture [10] that conforms to the spirit (and even to the letter, see section 7.6) of Penrose's strong cosmic censorship conjecture [11] but is somewhat richer in structure.

The internal structure of spherical black holes is by now relatively well understood, and this is reviewed in section 7.2. The following three sections discuss the much thornier problem of generic black hole interiors, with emphasis on gaps in understanding and outstanding problems, and including (in section 7.4) a brief exposition of a technique ("covariant double-null dynamics") which has been found useful to analyze these problems. There is a concluding summary in section 7.6.

7.2 Back reaction: spherical models

According to the perturbative analyses [4, 7, 8], divergences at the Cauchy horizon (CH) are rather mild to linear order. Amplitudes of wavelike perturbations typically decay like

$$\delta\Phi \sim (\ln|V|)^{-\frac{1}{2}p+1} \qquad (V \to 0^-), \tag{7.1}$$

where V is a Kruskal advanced time which goes to zero at CH and $\frac{1}{2}p = 2\ell+2$ for a multipole of order ℓ (or $2\ell+3$ if this multipole was absent before the collapse). Thus field amplitudes Φ (in particular, the metric $g_{\mu\nu}$) stay regular on CH to first order, though their derivatives $\partial_V \Phi$ diverge.

Since the Einstein equations are quadratic in first-order gradients $\nabla g_{\mu\nu}$, it was widely expected that taking back reaction into account would make the amplitudes themselves blow up on CH at higher orders of perturbation theory. However, detailed investigations have not borne out these expectations. The basic reason is that the divergences are lightlike: roughly speaking, terms quadratic in $\partial_V \Phi$ are held in check because the associated metric coefficient g^{VV} is zero.

Attempts to include back reaction began with spherical models for a first reconnaissance. Their similar horizon structures suggest replacing the asymptotically Kerr black hole by a spherical charged hole, and the tail of quadrupole gravitational waves by spherical scalar waves. The large blueshift near CH further suggests using an optical approximation which replaces the infalling waves by a stream of lightlike particles ("gravitons").

The earliest models [12,13,14] accordingly considered the effect of radial streams of radiation on the internal geometry of a spherical hole of fixed charge e. I shall briefly summarize the basic equations and the results. Any spherisymmetric geometry can be described by the metric

$$ds^2 = g_{AB}\, dx^A\, dx^B + r^2\, d\Omega^2 \qquad (A, B = 0, 1),$$

where x^A are any pair of coordinates which label the different 2-spheres and the radius $r(x^A)$ is a function of these coordinates. Its gradient defines the mass interior to a given 2-sphere (not counting electrostatic field energy) through the equations

$$(\nabla r)^2 = g^{AB}(\partial_A r)(\partial_B r) = f = 1 - 2m/r + e^2/r^2. \qquad (7.2)$$

The Einstein field equations are then captured in the two-dimensionally covariant equations [13]

$$\partial_A m = 4\pi r^2 T_A^B \partial_B r, \qquad \nabla_A \nabla_B r = -4\pi r T_{AB} + \kappa g_{AB}. \qquad (7.3)$$

These imply that the mass function satisfies the (1+1)-dimensional wave equation

$$\Box m = -16\pi^2 r^3 T_{AB} T^{AB} + 8\pi r f P_\perp, \qquad (7.4)$$

which brings out some of the nonlinearity hidden in the usual form of the Einstein equations. I have defined $\kappa = m - e^2/r$, $P_\perp = T_\theta^\theta$ is the transverse pressure, and, in this way of writing the equations, the stress-energy (T_A^B, $P_\perp$) does not include the Maxwellian contribution due to the electrostatic field of the hole's charge e. In (7.3) and (7.4) it is assumed that $T_A^A = 0$, which holds for the sources of interest here (massless scalar fields, crossflowing lightlike streams).

For an influx of radiation blueshifted near CH according to

$$T_{VV} \sim (\ln|V|)^{-p} V^{-2} \qquad (V \to 0^-) \qquad (7.5)$$

(cf. (7.1)), the right-hand side of (7.4) acts as a diverging source for m provided at least a nominal outflow is present also. (For a pure influx [12], $T_{AB}T^{AB} = 0$ and m suffers little change near CH.) It is found [13] that the mass function diverges at CH like

$$m \sim (\ln|V|)^{-p}|V|^{-1}, \qquad (7.6)$$

an effect sometimes referred to as "mass inflation." More specifically [15], if the outflux is due to backscatter,

$$m \sim |u|^{-p} v^{-(p+2)} e^{\kappa_0(u+v)} \qquad (v \to \infty, \ u \to -\infty), \qquad (7.7)$$

where $v = -\kappa_0^{-1} \ln|V|$ is conventional advanced time ($v = +\infty$ on $\mathcal{I}$ and on CH) and u is internal retarded time, decreasing toward $-\infty$ at the event horizon and the rear of CH (which has no past endpoint, see fig. 7.1). The positive constant κ_0 is the surface gravity of the inner horizon of the static Reissner-Nordström geometry that the external field asymptotically approaches.

Chapter 7. The Internal Structure of Black Holes

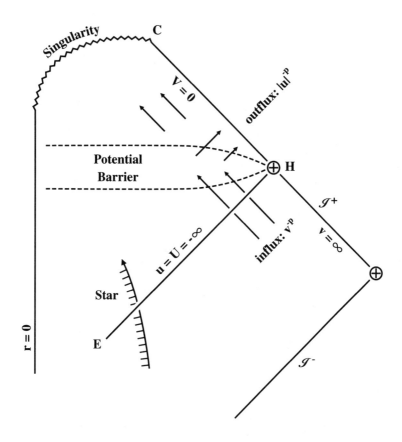

Figure 7.1: Penrose conformal map of a charged spherical hole. EH is the event horizon, CH the Cauchy horizon, and H is a singular point of the mapping: in reality, CH and EH are endless 3-cylinders of different radii which never intersect. The figure depicts an infalling stream of radiation which is partially scattered off a ridge of curvature about midway between the horizons.

No trace of the drastic change (7.6) in the internal geometry is detectable externally, because news of it cannot escape from the hole. Outside observers continue to register a mass close to that of the progenitor star.

Although gross from a global standpoint the singular behavior (7.6) is spread over the surface of CH and hence is pancakelike and locally mild in a sense made precise by Ori [14]. While tidal forces and Weyl curvature ($\sim m/r^3$) become infinite, the cumulative tidal *deformation* of bodies falling toward CH remains bounded (and indeed small) up to the very moment of crossing.

Nor is there any drastic effect on the intrinsic geometry of the lightlike 3-cylinder CH, which is given by the radial function r. Only gradually does CH contract, as the outflow (which is not blueshifted) focuses its generators, finally tapering to a strong spacelike singularity for $r = 0$ (fig. 7.1).

An intuitive understanding of mass inflation can be obtained from a simple mechanical example. Consider a pair of massive thin spherical shells, one moving inwards, the other outwards in a Reissner-Nordström spacetime with central charge e. At the moment when they cross and (let us assume) pass freely through each other, their mutual gravitational potential energy of order $-m_{\text{in}} m_{\text{out}}/r$ (which acts as a debit on the total gravitational mass of the *outer* body) is suddenly transferred from the contracting to the expanding shell, resulting in a redistribution of their masses.

The results are especially simple if both shells are made of light. Their histories divide the spacetime into four sectors $A, \ldots, D$ according to the scheme

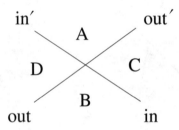

The mass parameter m in the metric function $f(r)$ of equation (7.2) takes different (constant) values $m_A, \ldots, m_D$ in these four sectors. Conservation of energy at the moment of crossing is expressed by the Dray-'tHooft-Redmount relation [16]

$$f_A(r_0) f_B(r_0) = f_C(r_0) f_D(r_0),$$

where r_0 is the crossing radius. This can be reexpressed in a number of equivalent forms, including

$$\frac{m_A - m_D}{m_C - m_B} = \frac{f_D}{f_B}, \qquad \frac{m_C - m_A}{m_B - m_D} = \frac{f_C}{f_B};$$

i.e.,

$$m'_{\text{in}} = (f_D/f_B) m_{\text{in}}, \qquad m'_{\text{out}} = (f_C/f_B) m_{\text{out}}. \tag{7.8}$$

Conservation of the total energy follows directly from (7.8):

$$m'_{\text{in}} + m'_{\text{out}} = m_{\text{in}} + m_{\text{out}}.$$

The shells may be considered to represent schematically the infalling and outgoing fluxes inside a spherical charged hole. If crossing occurs

Chapter 7. The Internal Structure of Black Holes 143

just outside the Cauchy horizon of sector B, so that $f_B(r_0)$ is negative and numerically very small, $f_D(r_0)$ negative, and $f_C(r_0)$ positive (since $m_C = m_B - m_{\text{in}} < m_B$), it follows from (7.8) that m'_{in} and m_A are greatly increased over m_{in} and m_B. (The new mass m'_{out} is then correspondingly negative, signifying that the "outgoing" shell is trapped inside the hole and now actually contracting.)

This represents a wholesale conversion of gravitational energy into material (kinetic) energy of infall. The Cauchy horizon of a black hole (like a closed universe in cosmological inflation) is a bottomless well of gravitational energy.

7.3 Is the spherical picture generic?

The key question is how far one may consider the simple toy models reviewed in the previous section as representative of conditions inside a real black hole. From the outset the stability and genericity of the spherical picture have been in doubt.

Even in the spherical case, a 1993 numerical study [17] of scalar wave tails absorbed into a charged spherical hole led to a suggestion that the Cauchy horizon might be destroyed and replaced by an $r = 0$ spacelike singularity. But this was not confirmed by subsequent analytical studies [15] and more refined numerical work [18].

The nonspherical case is much more difficult and we are still far from a comprehensive solution of this problem, although a considerable amount of work has been done [19,20,21].

In arguing the case for nongenericity of singular Cauchy horizons, a common line of reasoning is the following. We already know a class of singularities—the BKL oscillatory (mixmaster) singularities [22]—which are functionally generic in the sense that they depend on 8 physically arbitrary functions of 3 variables (6 components of intrinsic metric plus 6 components of extrinsic curvature for an initial spatial hypersurface, less 4 coordinate degrees of freedom. Imposition of specific [e.g., vacuum] field equations would subject these 8 functions to 4 constraints.). The BKL singularities are *spacelike*. Presumably they exhaust the set of generic singularities. If so, then the set of lightlike singularities would be less generic and could not arise from generic initial data. Generic perturbations should drive a lightlike singularity into a spacelike one—a phenomenon which is indeed observed in colliding plane-wave spacetimes [23].

Ori and Flanagan [24] have recently exposed a loophole in this argument. They explicitly constructed a class of *weak* lightlike singularities which *are* functionally generic. One would indeed expect the Einstein equations (which are quasi-linear) to propagate weak discontinuities and mild singularities along characteristic (i.e., lightlike) hypersurfaces.

Thus it is not possible to rule out mild lightlike singularities inside real black holes purely on the grounds that they are nongeneric. Of course, this is not a proof that such singularities actually do form inside black holes.

7.4 Covariant double-null dynamics

Although there will not be space here to go into much detail on the analysis of the nonspherical case, I shall take the opportunity of presenting the outlines of a general formalism [25] which was developed for that purpose and also proves useful in a variety of other situations where the physics singles out particular lightlike surfaces or directions. This is a (2+2)-imbedding formalism adapted to a double foliation of spacetime by a net of two intersecting families of lightlike hypersurfaces. It yields a simple and geometrically transparent decomposition of the Einstein field equations.

A number of such formalisms are extant [26]. All have the same content but they differ widely in appearance. The present version is two-dimensionally covariant and thus very compact.

Let the two families of lightlike hypersurfaces be denoted by Σ^0 (with equations $u^0 = $ const.) and Σ^1 (given by $u^1 = $ const.), where $u^A(x^\alpha)$ $(A, B, \ldots = 0, 1; \alpha, \beta, \ldots = 1, \ldots, 4)$ are a given pair of scalar fields over spacetime, with lightlike gradients:

$$\nabla u^A \cdot \nabla u^B = e^{-\lambda} \eta^{AB},$$

where the matrix

$$\eta^{AB} = \eta_{AB} = \begin{pmatrix} 0 & -1 \\ -1 & 0 \end{pmatrix}$$

will be used to raise and lower upper-case Latin indices and $\lambda(x^\alpha)$ is a scalar function. The generators $\ell^{(A)}_\alpha$ of Σ^A are conveniently defined as

$$\ell^{(A)} = e^\lambda \nabla u^A.$$

Two hypersurfaces Σ^0 and Σ^1 intersect in a 2-surface S, with parametric equations

$$x^\alpha = x^\alpha(u^A, \theta^a) \qquad (a, b, \ldots = 2, 3),$$

where (θ^2, θ^3) are intrinsic coordinates of S. Both generators $\ell^{(A)}$ are orthogonal to S.

Holonomic basis vectors $e_{(a)}$ and the intrinsic metric of S may now be defined:

$$e^\alpha_{(a)} = \frac{\partial x^\alpha}{\partial \theta^a}, \qquad g_{ab} = e_{(a)} \cdot e_{(b)}.$$

The matrix g_{ab} and its inverse g^{ab} are used to lower and raise lower-case Latin indices, so that $e^{(a)} = g^{ab} e_{(b)}$ are the dual basis vectors tangent to S.

Chapter 7. The Internal Structure of Black Holes

Two-dimensional shift vectors s_A^a are defined by

$$s_A^a = \frac{\partial x^\alpha}{\partial u^A} e_\alpha^{(a)} = -\ell_{(A)}^\alpha \frac{\partial \theta^a}{\partial x^\alpha}.$$

As in the Arnowitt-Deser-Misner formalism, the shift vector s_A^a measures how much one has to deviate from the normal direction $\ell_{(A)}$ to connect points on different 2-surfaces having the same intrinsic coordinates θ^a. An infinitesimal four-dimensional displacement dx^α can be decomposed as

$$dx^\alpha = \ell_{(A)}^\alpha \, du^A + e_{(a)}^\alpha (d\theta^a + s_A^a \, du^A).$$

Together with the completeness relation

$$g_{\alpha\beta} = e^{-\lambda} \eta_{AB} \ell_\alpha^{(A)} \ell_\beta^{(B)} + g_{ab} e_\alpha^{(a)} e_\beta^{(b)}$$

for the basis $(\ell^{(A)}, e_{(a)})$, this implies that the spacetime metric is decomposable as

$$g_{\alpha\beta} \, dx^\alpha \, dx^\beta = e^\lambda \eta_{AB} \, du^A \, du^B + g_{ab}(d\theta^a + s_A^a \, du^A)(d\theta^b + s_B^b \, du^B). \quad (7.9)$$

Associated with its two normals $\ell_{(A)}$, a 2-surface S has two extrinsic curvatures defined by

$$K_{Aab} = (\nabla_\beta \ell_{(A)\alpha}) e_{(a)}^\alpha \, e_{(b)}^\beta$$

and easily shown to be symmetric in a, b. (Since we are free to rescale the null vectors $\ell_{(A)}$, a certain scale arbitrariness is inherent in this definition.)

A further basic geometrical property of the double foliation is given by the Lie bracket of $\ell_{(0)}$ and $\ell_{(1)}$. One finds

$$[\ell_{(B)}, \ell_{(A)}] = \epsilon_{AB} \omega^a e_{(a)}, \quad (7.10)$$

where

$$\omega^a = \epsilon^{AB}(\partial_B s_A^a - s_B^b s_{A;b}^a).$$

The semicolon indicates two-dimensional covariant differentiation associated with metric g_{ab}, and ϵ_{AB} is the two-dimensional permutation symbol.

The geometrical significance of the "twist" ω^a can be read off from (7.10): the curves tangent to the generators $\ell_{(0)}, \ell_{(1)}$ mesh together to form 2-surfaces (orthogonal to the surfaces S) if and only if $\omega^a = 0$. In this case, it would be consistent to allow the coordinates θ^a to be dragged along both sets of generators and thus to gauge both shift vectors to zero.

I shall denote by D_A the two-dimensionally invariant operator associated with differentiation along the normal direction $\ell_{(A)}$. Acting on any two-dimensional geometrical object $X_{b...}^{a...}$, D_A is formally defined by

$$D_A X_{b...}^{a...} = (\partial_A - \mathcal{L}_{s_A^d}) X_{b...}^{a...}.$$

Here, ∂_A is the partial derivative with respect to u^A and $\mathcal{L}_{s_A^d}$ the Lie derivative with respect to the 2-vector s_A^d. As an example:

$$D_A g_{ab} = \partial_A g_{ab} - 2s_{A(a;b)} = 2K_{Aab}.$$

Geometrically, $D_A X^a_{b\cdots}$ is the projection onto S of the Lie derivative with respect to $\ell_{(A)}$ of the equivalent tangential 4-tensor $X^\alpha_{\beta\cdots}$.

The objects K_{Aab}, ω^a, and D_A are all simple projections onto S of four-dimensional geometrical objects. Consequently, they transform very simply under two-dimensional coordinate transformations. Under the arbitrary reparametrization

$$\theta^a \to \theta^{a'} = f^a(\theta^b, u^A) \tag{7.11}$$

(which leaves u^A and hence the surfaces Σ^A and S unchanged), ω_a and K_{Aab} transform cogrediently with

$$e_{(a)} \to e'_{(a)} = e_{(b)} \partial \theta^b / \partial \theta^{a'}.$$

By contrast, the shift vectors s_A^a undergo a more complicated gaugelike transformation, arising from the u-dependence in (7.11).

This geometrical groundwork is already sufficient to allow me to display the simple form that the Ricci components take in this formalism. (Notation for the tetrad components is typified by $R_{aA} = R_{\alpha\beta} e^\alpha_{(a)} \ell^\beta_{(A)}$.)

The results are

$$^{(4)}R_a^b = \frac{1}{2}{}^{(2)}R\delta_a^b - e^{-\lambda}(D_A + K_A)K^{Ab}_a - \frac{1}{2}e^{-2\lambda}\omega_a\omega^b - \frac{1}{2}\lambda_{,a}\lambda^{,b},$$

$$R_{AB} = -D_{(A}K_{B)} - K_{Aab}K_B^{ab} + K_{(A}D_{B)}\lambda$$
$$\quad - \frac{1}{2}\eta_{AB}[(D^E + K^E)D_E\lambda - e^{-\lambda}\omega^a\omega_a + (e^\lambda)^{;a}{}_a],$$

$$R_{Aa} = K^b_{Aa;b} - \partial_a K_A - \frac{1}{2}\partial_a D_A \lambda + \frac{1}{2}K_A \partial_a \lambda$$
$$\quad + \frac{1}{2}\epsilon_{AB}e^{-\lambda}[(D^B + K^B)\omega_a - \omega_a D^B \lambda],$$

where $^{(2)}R$ is the curvature scalar associated with the 2-metric g_{ab} and $K_A \equiv K^a_{Aa}$.

The economy and geometrical transparency of these formulae are self-evident. In particular, the shift vectors, which are largely an artifact of the choice of coordinates θ^a, make no explicit appearance.

A formalism of this kind has many applications [25, 26]—for instance, the characteristic initial-value problem, the dynamics of apparent horizons, and collisions at ultra-relativistic energies. In section 7.5 I shall indicate, without entering into formal detail, how it has been used to unravel the structure of the singularity at a generic Cauchy horizon.

7.5 Internal structure of a generic rotating black hole: remarks on the characteristic initial-value problem

There is room here only for some brief general comments on the very difficult (and basically still unsolved) problem of determining the inner evolution of a generic rotating hole.

To capture the physical essence of this problem, it should be sufficient to examine the effect of introducing, near the event horizon of an initially stationary (i.e., Kerr) black hole, an initially small perturbation representing an infalling packet of gravitational waves decaying with advanced time according to Price's power law:

$$\delta g \sim h(\theta, \varphi_+) v_+^{-\frac{1}{2}p+1} \qquad (r \approx r_+).$$

The function h describing the initial amplitude is regular [27] provided its argument φ_+ is chosen as the Eddington-Kerr advanced angular coordinate, constant along ingoing principal null rays and along the generators of the Cauchy and future event horizons—not (as is sometimes assumed) the Boyer-Lindquist coordinate φ. (These two coordinates differ by an r-dependent winding translation which becomes singular at horizons.) Small-amplitude wave packets on the Kerr background propagate inwards with constant angular momentum but accelerating angular velocity because of frame dragging; after scattering from the inner potential barrier their transmitted parts must approach CH as highly blueshifted "gravitons" moving parallel to its generators, since this is the only nonspacelike direction tangent to CH.

Determination of the back reaction is appropriately formulated as a characteristic initial-value problem. Initial data are set on an intersecting pair of lightlike hypersurfaces Σ^v and Σ^u of constant advanced and retarded time respectively, with Σ^v projecting inwards into the hole. The data on Σ^v are fixed by situating it wholly in the initial stationary (Kerr) sector of the geometry.

In the proper formulation of the problem, Σ^v and Σ^u intersect at or above the event horizon, and Σ^u lies on or above it with initial data specified as the Kerr geometry perturbed by an infalling wave tail.

These initial data now evolve in accordance with Einstein's equations. To help in visualizing this, one can broadly distinguish four stages in the evolution of a perturbation which originates on a segment of the event horizon at some late advanced time $v = v_1$:

(a) The disturbance propagates inward at constant advanced time. Part of it is scattered "outward" to cross CH transversely. The transmitted part continues in free propagation and descent along $v = v_1$ until eventually it sidles in close to CH. When it reaches a layer very near CH where blueshift and back reaction are becoming appreciable, this first phase of the evolution is deemed to be over. Up to this point, the perturbation can be treated, at least to a first approximation, as a test field propagating on a fixed Kerr background.

(b) With further descent along $v = v_1$ and deeper immersion in this layer, there is a transition from incipient to extreme blueshift and to a regime of fully developed back reaction (c).

(c) This regime appears to be the most extended part of the evolution, now parallel to the generators of CH. Experience with the spherical and planar cases, as well as the general analyses described below, suggest that the strong blueshift regime is relatively stable, changing with retarded time on the timescale associated with the slow deformation of the intrinsic geometry of CH by the (weak and unblueshifted) transverse flux arising from backscatter. The 4-geometry at CH itself will already have been dominated by this regime for the entire past history of CH (CH has no past endpoint).

(d) Finally, focusing of generators by the transverse flux leads to formation of caustics and the breakup of CH. The evolution terminates in a stronger singularity, presumably spacelike and (classically) of BKL mixmaster type.

Of these four phases, (b) and (d) are the most difficult to analyze and the least understood, while (a) is, relatively speaking, the easiest. But even (a) poses formidable problems, because the wave equation on a Kerr background is separable only for a special (harmonic or exponential) time dependence, not ideally adapted to handle power-law tails.

The detailed studies to date [19,20,21] have been largely restricted to a simplified formulation which allows one to bypass detailed consideration of phases (a) and (b). In this formulation, the initial hypersurface Σ^u, whose proper placing is along or above the event horizon, is assumed to lie inside the hole. It now intersects CH and its future end is therefore immersed in a layer of strong blueshift and back reaction where initial data are not known a priori. The strategy then has to be to make a plausible guess about the initial conditions in this layer and then try to check that the guess is self-consistent by showing that the general character of the assumed conditions near CH is preserved by the evolution. In this manner one has been able to construct a fairly plausible description of phase (c) as a self-sustaining large-blueshift regime. But the evidence that this is what *must* emerge from phases (a) and (b) still falls well short of being compelling.

The most complete analysis so far published along these lines is the work of Brady and Chambers [21], building in part on an earlier study by Ori [19], and employing an earlier variant, due to S. A. Hayward [26], of the double-null formalism described in section 7.4.

Chapter 7. The Internal Structure of Black Holes

These studies lead to a self-consistent description of phase (c) which is qualitatively similar to the spherical (section 7.2) and planar [20] pictures. The metric of two-dimensional cross sections of CH is unaffected by blueshift and slowly varying; the transverse 2-metric orthogonal to these sections is exponentially compressed; i.e., the exponent λ in (7.9) decreases linearly with external advanced time v as $v \to \infty$.

Two general observations are worth making. One is that, since $|g^{uv}| \to \infty$ at CH, the Einstein equations are actually dominated by the highest derivative terms at large blueshift, so we are not dealing with a singular boundary-value problem. Secondly, it follows from the Einstein field equations and the assumed (initial) regularity of the intrinsic geometry of CH in phase (c) that the leading divergence in λ can depend on v only:

$$\lambda = -\kappa_0 v + \text{(less divergent terms)} \qquad (v \to \infty),$$

where κ_0 is a constant. This condition already appears as a constraint on the initial data on Σ^u. A hint that it does indeed emerge from the previous evolution comes from the constancy of the surface gravity of the Kerr inner horizon, which implies that a condition of precisely this form holds for the blueshift in the unperturbed phase (a). But increasing blueshift is inseparable from back reaction, and the actual outcome will depend on how the geometry evolves through phase (b), a difficult question which remains unexplored.

7.6 Summary and concluding remarks

Analytical and numerical studies have given us a fairly complete picture of the effect of wave tails on the interior of a spherical (charged) black hole. In the nonspherical (rotating) case partial analyses and general considerations suggest a qualitatively similar picture, but evidence for this is still fragmentary.

Knowledge of the classical geometry at late advanced times near the Cauchy horizon serves as a launchpad which provides initial conditions for the subsequent quantum phase of evolution. Studies of this quantum regime [28] using spherical models are still provisional but have so far given little evidence that quantum effects will tame the classical singularity.

Nonspinning, uncharged collapse terminates in an all-enveloping, crushing spacelike singularity inside a black hole. The work I have been describing suggests some modification and softening of this (classical) scenario in the presence of spin or charge. Only a finite segment of the spacelike singularity survives; it is joined to an infinitely long lightlike precursory singularity which is milder.

Analytic continuation of the stationary Kerr and Reissner-Nordström geometries into the regions beyond their inner horizons reveals the presence of timelike singularities in those regions. Predictability in these regions is lost because unpredictable influences can emanate from the singularities. It was the desire to avoid this embarrassment that motivated strong cosmic censorship, which Penrose [11] formulated as the hypothesis that generic singularities inside black holes are achronal, i.e., nontimelike. (He was careful not to say "spacelike.") The provisional picture emerging from recent analyses of black hole interiors conforms to strong cosmic censorship in this strict sense of the term.

We thus have a provisional working picture (with conspicuous gaps and uncertainties) of the black hole interior up to the stage where curvatures are approaching Planck levels near the Cauchy horizon. Beyond that lies a sea of ignorance. The situation is well summed up by Chandra's quotation [2] of a passage from Eddington. Reviewing the state of knowledge of stellar energy sources in 1925 (before anything was known about quantum tunneling or neutrons), Eddington wrote,

> I should have liked to have closed these lectures by leading up to some great climax. But perhaps it is more in accordance with the true conditions of scientific progress that they should fizzle out with a glimpse of the obscurity which marks the frontiers of present knowledge. I do not apologize for the lameness of the conclusion, for it is not a conclusion. I wish I could feel confident that it is even a beginning.

Acknowledgments

My thanks to Robert Wald for hospitality at the University of Chicago and for helpful comments on the manuscript. This work was supported by the Canadian Institute for Advanced Research and by NSERC of Canada.

References

[1] S. Chandrasekhar, *The Mathematical Theory of Black Holes*. Clarendon Press (Oxford, 1983).

[2] S. Chandrasekhar, *Truth and Beauty*. Univ. of Chicago Press (Chicago, 1987), 154, 139.

[3] S. Chandrasekhar and B. C. Xanthopoulos, Proc. Roy. Soc. **A408**, 175 (1986); ibid. **A410**, 311 (1987).

[4] S. Chandrasekhar and J. B. Hartle, Proc. Roy. Soc. **A384**, 301 (1982).

[5] J. L. Synge, Proc. Roy. Irish Acad. **53A**, 83 (1950).

[6] R. Penrose, in *Battelle Rencontres*, ed. C. M. DeWitt and J. A. Wheeler. W. A. Benjamin (New York, 1968), 222.

Chapter 7. The Internal Structure of Black Holes   151

[7] M. Simpson and R. Penrose, Int. J. Theor. Phys. **7**, 183 (1973); Y. Gürsel, V. D. Sandberg, I. D. Novikov, and A. A. Starobinsky, Phys. Rev. **D19**, 413 (1979); ibid. **D20**, 1260 (1979); R. A. Matzner, N. Zamorano, and V. D. Sandberg, Phys. Rev. **D19**, 2821 (1979); A. Ori, Phys. Rev. **D55**, 4860 (1997); F. J. Tipler, Phys. Rev. **D16**, 3359 (1977).

[8] M. Curé and N. Zamorano, in *Relativity, Cosmology, Topological Mass and Supergravity*, ed. C. Aragone. World Scientific (Singapore, 1982), 219.

[9] R. H. Price, Phys. Rev. **D5**, 2419 (1972); C. Gundlach, R. H. Price, and J. Pullin, Phys. Rev. **D49**, 883 (1994).

[10] S. Droz, W. Israel, and S. M. Morsink, Physics World **9**, 34 (1996).

[11] R. Penrose, in *General Relativity: An Einstein Centenary Survey*, ed. S. W. Hawking and W. Israel. Cambridge Univ. Press (Cambridge, 1979), 581.

[12] W. A. Hiscock, Physics Letters **83A**, 110 (1981).

[13] E. Poisson and W. Israel, Phys. Rev. **D41**, 1796 (1990).

[14] A. Ori, Phys. Rev. Letters **67**, 789 (1991).

[15] A. Bonanno, S. Droz, W. Israel, and S. M. Morsink, Phys. Rev. **D50**, 755 (1994); Proc. Roy. Soc. **A450**, 553 (1995).

[16] C. Barrabès, W. Israel, and E. Poisson, Class. Quantum Grav. **7**, L273 (1990).

[17] M. L. Gnedin and N. Y. Gnedin, Class. Quantum Grav. **10**, 1083 (1993).

[18] P. R. Brady and J. D. Smith, Phys. Rev. Letters **75**, 1256 (1995).

[19] A. Ori, Phys. Rev. Letters **68**, 2117 (1992).

[20] A. Bonanno, S. Droz, W. Israel, and S. M. Morsink, Can. J. Phys. **72**, 755 (1995); erratum, Can. J. Phys. **73**, 251 (1996).

[21] P. R. Brady and C. M. Chambers, Phys. Rev. **D51**, 4177 (1995).

[22] V. A. Belinskii, E. M. Lifshitz, and I. M. Khalatnikov, Advances in Physics **19**, 525 (1970); C. W. Misner, Phys. Rev. Letters **22**, 1071 (1969).

[23] U. Yurtsever, Class. Quantum Grav. **10**, 117 (1993).

[24] A. Ori and E. Flanagan, Phys. Rev. **D53**, R1754 (1996).

[25] P. R. Brady, S. Droz, W. Israel, and S. M. Morsink, Class. Quantum Grav. **13**, 2211 (1996).

[26] R. Geroch, A. Held, and R. Penrose, J. Math. Phys. **14**, 874 (1973); R. A. d'Inverno and J. Smallwood, Phys. Rev. **D22**, 1233 (1980); J. Smallwood, J. Math. Phys. **24**, 599 (1983); C. G. Torre, Class. Quantum Grav. **3**, 773 (1986); S. A. Hayward, Class. Quantum Grav. **10**, 779 (1993); R. J. Epp, The symplectic structure of general relativity in the double-null (2+2) formalism, preprint, Univ. Calif., Davis (1995), gr-qc/9511060.

[27] W. Krivan, P. Laguna, and P. Papadopoulos, Phys. Rev. **D54**, 4728 (1996).

[28] R. Balbinot, P. R. Brady, W. Israel, and E. Poisson, Physics Letters **A161**, 223 (1991); R. Balbinot and E. Poisson, Phys. Rev. Letters **70**, 13 (1993); W. J. Anderson, P. R. Brady, W. Israel, and S. M. Morsink, Phys. Rev. Letters **70**, 1041 (1993); I. Oda, Mass inflation in quantum gravity, preprint, Edogawa University, Chiba, Japan.

Part II

8

Black Holes and Thermodynamics

Robert M. Wald

Abstract

We review the remarkable relationship between the laws of black hole mechanics and the ordinary laws of thermodynamics. It is emphasized that—in analogy with the laws of thermodynamics—the validity of the laws of black hole mechanics does not appear to depend upon the details of the underlying dynamical theory (i.e., upon the particular field equations of general relativity). It also is emphasized that a number of unresolved issues arise in "ordinary thermodynamics" in the context of general relativity. Thus, a deeper understanding of the relationship between black holes and thermodynamics may provide us with an opportunity not only to gain a better understanding of the nature of black holes in quantum gravity, but also to better understand some aspects of the fundamental nature of thermodynamics itself.

8.1 Introduction

Undoubtedly, one of the most remarkable developments in theoretical physics to have occurred during the past twenty five years was the discovery of a close relationship between certain laws of black hole physics and the ordinary laws of thermodynamics. It appears that these laws of "black hole mechanics" and the laws of thermodynamics are two major pieces of a puzzle that fit together so perfectly that there can be little doubt that this "fit" is of deep significance. The existence of this close relationship between these laws may provide us with a key to our understanding of the fundamental nature of black holes in a quantum theory of gravity, as well as to our understanding of some aspects of the nature of thermodynamics itself. The aim of this article is to review the nature of the relationship between the black hole and thermodynamics laws. Although some notable progress has been made, many mysteries remain.

It was first pointed out by Bekenstein [1] that a close relationship might exist between certain laws satisfied by black holes in classical general relativity and the ordinary laws of thermodynamics. The area theorem of classical general relativity [2] states that the area, A, of a black hole can never decrease in any process:

$$\Delta A \geq 0. \tag{8.1}$$

Bekenstein noted that this result is closely analogous to the statement of the ordinary second law of thermodynamics: The total entropy, S, of a closed system never decreases in any process:

$$\Delta S \geq 0. \tag{8.2}$$

Indeed, Bekenstein proposed that the area of a black hole (times a constant of order unity in Planck units) should be interpreted as its physical entropy.

A short time later, the analogy between certain laws of black hole physics in classical general relativity and the laws of thermodynamics was developed systematically by Bardeen, Carter, and Hawking [3]. They proved that in general relativity, the surface gravity, κ, of a stationary black hole (defined by eq. (8.19) below) must be constant over the event horizon of the black hole. They noted that this result is analogous to the zeroth law of thermodynamics, which states that the temperature, T, must be uniform over a body in thermal equilibrium. Finally, Bardeen, Carter, and Hawking proved the "first law of black hole mechanics". In the vacuum case, this law states that the differences in mass, M, area, A, and angular momentum, J, of two nearby stationary black holes must be related by

$$\delta M = \frac{1}{8\pi}\kappa\delta A + \Omega\delta J, \tag{8.3}$$

where Ω denotes the angular velocity of the event horizon (defined by eq. (8.18) below). Additional terms may appear on the right-hand side of eq. (8.3) when matter fields are present. They noted that this law is closely analogous to the ordinary first law of thermodynamics, which states that differences in energy, E, entropy, and other state parameters of two nearby thermal equilibrium states of a system are given by

$$\delta E = T\delta S + \text{``work terms''}. \tag{8.4}$$

If we compare the zeroth, first, and second laws of ordinary thermodynamics with the corresponding laws of black hole mechanics, we see that the analogous quantities are, respectively, $E \longleftrightarrow M$, $T \longleftrightarrow \alpha\kappa$, and $S \longleftrightarrow A/8\pi\alpha$, where α is an undetermined constant. Even in the context of classical general relativity, a hint that this relationship might be of physical significance arises from the fact that E and M represent the same physical

Chapter 8. Black Holes and Thermodynamics   157

quantity, namely, the total energy of the system. However, in classical general relativity, the physical temperature of a black hole is absolute zero, so there can be no physical relationship between T and κ. Consequently, it also would be inconsistent to assume a physical relationship between S and A. For this reason, at the time the paper of Bardeen, Carter, and Hawking appeared, most researchers (with the notable exception of Bekenstein) viewed the analogy between the black hole and thermodynamical laws as a mathematical curiosity, devoid of any physical significance.

That view changed dramatically with Hawking's discovery [4] that, due to quantum particle creation effects, a black hole radiates to infinity all species of particles with a perfect blackbody spectrum, at temperature (in units with $G = c = \hbar = k = 1$)

$$T = \frac{\kappa}{2\pi}. \tag{8.5}$$

Thus, $\kappa/2\pi$ truly is the *physical* temperature of a black hole, not merely a quantity playing a role mathematically analogous to temperature in the laws of black hole mechanics. This left little doubt that the laws of black hole mechanics must correspond physically to the laws of thermodynamics applied to a system consisting of a black hole. As will be discussed further below, it also left little doubt that $A/4$ must represent the physical entropy of a black hole in general relativity.

Thus, Hawking's calculation of particle creation effectively gave a resoundingly positive answer to the question of whether there exists any physical significance to the mathematical relationship between the laws of black hole mechanics and the laws of thermodynamics. This conclusion is particularly intriguing, since, as I shall review in sections 8.2 and 8.3, the derivations of the laws of black hole mechanics are so different in nature from those of the laws of thermodynamics that it is hard to see how it is possible that these laws could really be "the same". As I will discuss further in section 8.4, this conclusion also raises a number of new questions and issues, most prominently that of providing a physical explanation for the origin of black hole entropy.

8.2 The nature of the laws of ordinary thermodynamics

The ordinary laws of thermodynamics are not believed to be fundamental laws in their own right, but rather to be laws which arise from the fundamental "microscopic dynamics" of a sufficiently complicated system when one passes to a macroscopic description of it. The great power and utility of the laws of thermodynamics stems mainly from the fact that the basic form of the laws does not depend upon the details of the underlying microscopic dynamics of particular systems and, thus, the laws have a "universal" validity—at least for a very wide class of systems.

The analysis of how the laws of ordinary thermodynamics arise applies to a classical or quantum system with a large number of degrees of freedom, whose time evolution is governed by Hamiltonian dynamics. It is important to emphasize that it is crucial here, at the outset, that there be a well defined notion of "time translations" (to which the Hamiltonian is conjugate), and that the Hamiltonian, H, (and, thus, the dynamics) be invariant under these time translations. It then follows that the total energy, E, of the system (i.e., the value of H) is conserved.

We now shall focus on the case of a classical dynamical system. By the previous remark, it follows that each dynamical orbit in phase space is confined to its "energy shell" Σ_E, i.e., the hypersurface in phase space defined by the equation $H(x) = E = $ constant. The crucial assumption needed for the applicability of thermodynamical laws to such a system is that "generic" orbits in phase space behave "ergodically" in the sense that they come arbitrarily close to all points of Σ_E, spending "equal times in equal volumes"; equivalently, the total energy of the system is the only nontrivial constant of motion for generic orbits. (By a slight modification of these arguments, the laws also can accommodate the presence of a small number [compared with the number of degrees of freedom] of additional constants of motion—such as the angular momentum of a rotationally invariant system.) Of course, even when such ergodic behavior occurs, it would take a dynamical orbit an infinite amount of time to completely "sample" Σ_E. The degree of "sampling" which is actually needed for the applicability of thermodynamics depends upon what observable (or collection of observables) of the dynamical system is being measured—or, in more common terminology, the amount of "coarse graining" of phase space that one does. For a "fine-grained" observable (corresponding to measuring detailed information about the "microscopic degrees of freedom" of the system), the sampling of the energy shell must be extremely good, and the amount of time needed for this sampling will be correspondingly long, thereby making the laws of thermodynamics inapplicable or irrelevant for the system. However, for the types of "macroscopic, coarse-grained" observables, $\mathcal{O}$, usually considered for systems with a huge number of degrees of freedom, the "sampling" need only be quite modest and the required "sampling time" (= the timescale for the system to "reach thermal equilibrium") is correspondingly short. One may then get from the laws of thermodynamics considerable predictive power about the values of $\mathcal{O}$ that one would expect to observe for the system under various physical conditions.

The *statistical entropy*, $S_\mathcal{O}$, of a classical dynamical system relative to a macroscopic, coarse-grained observable (or collection of observables), $\mathcal{O}$, is defined to be the observable whose value at point x on Σ_E is the logarithm of the volume of the region of the energy shell at which $\mathcal{O}$ takes the same value as it does at x, i.e.,

$$\mathcal{S}_{\mathcal{O}}(x) = \ln[\text{vol}(\mathcal{R}_x)], \tag{8.6}$$

where

$$\mathcal{R}_x = \{y \in \Sigma_E | \mathcal{O}(y) = \mathcal{O}(x)\}. \tag{8.7}$$

For the types of coarse-grained observables $\mathcal{O}$ which are normally considered, the largest region, $\mathcal{R}^{\max}$, of the form (8.7) will have a volume nearly equal to that of the entire energy shell Σ_E. If dynamical orbits sample Σ_E and spend "equal times in equal volumes", then we would expect $\mathcal{S}_{\mathcal{O}}$ to increase in value until it reaches its maximum possible value, namely, $\ln[\text{vol}(\mathcal{R}^{\max})] \simeq \ln[\text{vol}(\Sigma_E)]$. Subsequently, $\mathcal{S}_{\mathcal{O}}$ should remain at that value for an extremely long time. During this extremely long period, the value of $\mathcal{O}$ remains unchanged, so no change would be perceived in the system, and the system would be said to have achieved "thermal equilibrium".

The *thermodynamic entropy*, S, of the system is defined by

$$S = \ln[\text{vol}(\Sigma_E)]. \tag{8.8}$$

Unlike $\mathcal{S}_{\mathcal{O}}$, S is not an observable on phase space but rather a function on a low dimensional *thermodynamic state space* comprised by the total energy, E, of the system, and any parameters (such as, for example, an external magnetic field) appearing in the Hamiltonian which one might contemplate varying, together with any additional constants of motion for the system (such as the total angular momentum for a rotationally invariant system). These variables characterizing thermodynamic state space are usually referred to as *state parameters*. The temperature, T, is defined by

$$\frac{1}{T} = \frac{\partial S}{\partial E}, \tag{8.9}$$

where the remaining state parameters are held fixed in taking this partial derivative. Like S, the temperature, T, is a function on thermodynamic state space, not an observable on phase space.

It follows from the above discussion that when a system is in thermal equilibrium, its statistical entropy, $\mathcal{S}_{\mathcal{O}}$, equals its thermodynamic entropy, S. Similarly, suppose a system consists of weakly interacting subsystems, so that each subsystem can be viewed as an isolated system in its own right. Suppose that $\mathcal{O}$ is comprised by a collection of observables, $\mathcal{O}_i$, for each subsytem, and suppose, in addition, that each subsystem (viewed as an isolated system) is in thermal equilibrium—although the entire system need not be in thermal equilibrium. Then $\mathcal{S}_{\mathcal{O}}$ will equal the sum of the thermodynamic entropies, S_i, of the subsystems:

$$\mathcal{S}_{\mathcal{O}} = \sum_i S_i. \tag{8.10}$$

We now are in a position to explain the origin of the laws of thermodynamics. As argued above, if $\mathcal{S}_O$ is less than its maximum possible value (for the given value of E and the other state parameters), we should observe it to increase until thermal equilibrium is reached. In particular, in the case where the total system consists of subsystems and these subsystems are individually in thermal equilibrium at both the beginning and the end of some process (but not necessarily at the intermediate stages), we should have

$$\sum_i (S_i)_1 \geq \sum_i (S_i)_0, \tag{8.11}$$

where $(S_i)_0$ and $(S_i)_1$ denote, respectively, the initial and final thermodynamic entropies of the ith subsystem. This accounts for the second law of thermodynamics, eq. (8.2).

It should be noted that the time asymmetry present in this formulation of the second law arises from the implicit assumption that, commonly, $\mathcal{S}_O$ is initially below its maximum possible value. (Otherwise, a more relevant formulation of the second law would merely state that only very rarely and/or briefly would we expect to observe $\mathcal{S}_O$ to fluctuate below its maximum possible value.) The fact that we do commonly observe systems with $\mathcal{S}_O$ below its maximum possible value shows that the present state of our universe is very "special".

The zeroth law of thermodynamics is an immediate consequence of the fact that if the subsystems appearing in eq. (8.10) are at different temperatures, then $\mathcal{S}_O$ can be increased by transferring energy from a subsystem of high temperature to a subsystem of low temperature. (This fact follows immediately from the definition, (8.9), of T.) Thus, for a thermal equilibrium state—where, by definition, $\mathcal{S}_O$ achieves its maximum value—it is necessary that T be uniform.

Finally, since S is a function on thermodynamic state space, its gradient can be written as

$$dS = \frac{1}{T}dE + \sum_j X_j d\alpha_j, \tag{8.12}$$

where α_j denotes the state parameters other than E, and $X_j \equiv \partial S/\partial \alpha_j$ (where E and the state parameters other than α_j are held fixed in taking this partial derivative). Using Liouville's theorem, one can argue that S should be constant when sufficiently slow changes are made to parameters appearing in the Hamiltonian. This fact gives TX_j the interpretation of being the "generalized force" conjugate to α_j (at least for the case where α_j is a parameter appearing in the Hamiltonian), and it gives $TX_j d\alpha_j$ the interpretation of being a "work term". This accounts for the first law of thermodynamics, eq. (8.4).

Thus far, our discussion has been restricted to the case of a classical dynamical system. However, as discussed in more detail in [5], a completely

Chapter 8. Black Holes and Thermodynamics

parallel analysis can be made for a quantum system. In this analysis, the classical coarse-grained observable $\mathcal{O}$ on phase space is replaced by a self-adjoint operator $\hat{\mathcal{O}}$ acting on the Hilbert space of states with energy between E and $E + \Delta E$. The spectral decomposition of $\hat{\mathcal{O}}$ takes the form

$$\hat{\mathcal{O}} = \sum \lambda_m \hat{P}_m, \qquad (8.13)$$

where the $\{\hat{P}_m\}$ are a family of orthogonal projection operators, and it is assumed—in correspondence with the assumptions made about coarse graining in the classical case—that the degeneracy subspaces of $\hat{\mathcal{O}}$ are large. The statistical entropy, $\hat{S}_{\hat{\mathcal{O}}}$, is then defined to be the quantum observable

$$\hat{S}_{\hat{\mathcal{O}}} = \sum \ln(d_m) \hat{P}_m, \qquad (8.14)$$

where d_m is the dimension of the mth degeneracy subspace, i.e.,

$$d_m = \mathrm{tr}(\hat{P}_m). \qquad (8.15)$$

Again, it is assumed that the maximum value of d_m is essentially the dimension of the entire Hilbert space of states of energy between E and $E + \Delta E$.

The corresponding definition of the thermodynamic entropy, S, of a quantum system is

$$S = \ln n, \qquad (8.16)$$

where n denotes the dimension of the Hilbert space of states between E and $E + \Delta E$; i.e., n is proportional to the density of quantum states per unit energy. Again, S is not a quantum observable, but rather a function on thermodynamic state space. Arguments for the zeroth, first, and second laws of thermodynamics then can be made in parallel with the classical case.

It should be noted that, thus far, I have made no mention of the third law of thermodynamics. In fact, there are two completely independent statements which are referred to as the "third law". The first statement consists of the rather vague claim that it is physically impossible to achieve $T = 0$ for a (sub)system. To the extent that it is true, I would view this claim as essentially a consequence of the second law, since it always is highly entropically favorable to take energy away from a subsystem at finite temperature and add that energy to a subsystem whose temperature is very nearly 0. The second statement, usually referred to as "Nernst's theorem", consists of the claim that $S \to 0$ (or a "universal constant") as $T \to 0$. This claim is blatantly false in classical physics – it fails even for a classical ideal gas—but it holds for many quantum systems (in particular, for standard examples of boson and fermion ideal gases). Clearly "Nernst's theorem" is actually a claim about the behavior of the density of states, $n(E)$, as the total energy of the system goes to its minimum possible value. Indeed, more precisely, as explained in section 9.4 of [6], it should be viewed as a statement about the extrapolation to minimum energy of the continuum approximation to $n(E)$. Elsewhere, I shall give some examples of quantum ideal gas systems which violate "Nernst's theorem" [7] (see also footnote 4 of section 8.4 below). Thus, while "Nernst's theorem" holds empirically for systems studied in the laboratory, I do not view it as a fundamental aspect of thermodynamics. In particular, I do not feel that the well known failure of the analog of "Nernst's theorem" to hold for black hole mechanics—where there exist black holes of finite area with $\kappa = 0$—should be viewed as indicative of any breakdown of the relationship between thermodynamics and black hole physics.

The above discussion explains the nature and origin of the laws of thermodynamics for "ordinary" classical and quantum systems. However, as I shall now briefly describe, when general relativity is taken into account, a number of new issues and puzzles arise.

In the first place, it should be noted that general relativity is a field theory and, as such, it ascribes infinitely many degrees of freedom to the spacetime metric/gravitational field. If these degrees of freedom are treated classically, no sensible thermodynamics should be possible. Indeed, this situation also arises for the electromagnetic field, where a treatment of the statistical physics of a classical electromagnetic field in a box yields the Rayleigh-Jeans distribution and its associated "ultraviolet catastrophe". As is well known, this difficulty is cured by treating the electromagnetic field as a quantum field. I see no reason to doubt that similar difficulties and cures will occur for the gravitational field. However, it should be emphasized at the outset that one should not expect any thermodynamic laws to arise from a statistical physics treatment of classical general relativity; a quantum treatment of the degrees of freedom of the gravitational field should be essential.[1]

A much more perplexing issue arises from the fact that, as emphasized above, the arguments for the validity of thermodynamics for "ordinary systems" are based upon the presence of a well defined notion of "time translations", which are symmetries of the dynamics. Such a structure is present when one considers dynamics on a background spacetime whose metric possesses a suitable one-parameter group of isometries, and when the Hamiltonian is invariant under these isometries. However, such a structure is absent in general relativity, where no background metric is present. Furthermore, when the degrees of freedom of the gravitational field are excited, one would not expect the dynamical spacetime metric to possess

---

[1] However, it should be noted that in the Euclidean approach to quantum gravity [8] (see also [9] and references cited therein) one derives the formula $S_{bh} = A/4$ in the "zero loop" approximation, so, in this sense, the entropy of a black hole appears to arise at the classical level.

Chapter 8. Black Holes and Thermodynamics

a time translation symmetry. The absence of any "rigid" time translation structure in general relativity can be viewed as being responsible for making notions like the "energy density of the gravitational field" ill defined in general relativity. Notions like the "entropy density of the gravitational field" are not likely to fare any better. It may still be possible to use structures like asymptotic time translations to define the notion of the total entropy of an (asymptotically flat) isolated system. (As is well known, total energy can be defined for such systems.) However, for a closed universe, it seems unlikely that any meaningful notion will exist for the "total entropy of the universe" (including gravitational entropy). If so, it is far from clear how the second law of thermodynamics is to be formulated for a closed universe. This issue appears worthy of further exploration.

Another important issue that arises in the context of general relativity involves ergodic behavior. As discussed above, ordinary thermodynamics is predicated on the assumption that generic dynamical orbits sample the entire energy shell, spending "equal times in equal volumes". However, gross violations of such ergodic behavior occur in classical general relativity on account of the irreversible tendency for gravitational collapse to produce singularities—from which one cannot then evolve back to "uncollapsed" states. Interestingly, however, there are strong hints that ergodic behavior could be restored in quantum gravity. In particular, the quantum phenomenon of black hole evaporation (see section 8.4 below) provides a means of evolving from a collapsed state back to an uncollapsed configuration.

Finally, as noted above, the fact that we commonly observe the increase of entropy shows that the present state of the universe is very "special". As has been emphasized by Penrose [10], the "specialness" of the present state of the universe traces back, ultimately, to extremely special initial conditions for the universe at the big bang. This specialness of the initial state of the universe should have some explanation in a complete, fundamental theory. However, at present, it remains a matter of speculation as to what this explanation might be.

The above comments already give a clear indication that there are deep and fundamental issues lying at the interface of gravitation and thermodynamics. As we shall see in sections 8.3 and 8.4, the theory of black holes gives rise to very significant further relationships between gravitation and thermodynamics.

8.3 The nature of the laws of classical black hole mechanics

As indicated in the introduction, it appears overwhelmingly likely that the laws of classical black hole mechanics must arise, in a fundamental quantum theory of gravity, as the classical limit of the laws of thermodynamics applied to a system comprised by a black hole. However, as we shall see in this section, the present derivations of the laws of classical black hole mechanics could hardly look more different from the arguments for the corresponding laws of thermodynamics, as given in section 8.2. Nevertheless, as I shall emphasize here, the derivations of the laws of black hole mechanics appear to share at least one important feature with the thermodynamic arguments: There appears to be a "universality" to the laws of black hole mechanics in that the basic form of the laws appears to be independent of the details of the precise Lagrangian or Hamiltonian of the underlying theory of gravity—in a manner analogous to the "universality" of the form of the laws of ordinary thermodynamics.

In this section, we will consider theories of gravity which are much more general than general relativity, but we shall restrict attention to geometric theories, wherein spacetime is represented by a pair (M, g_{ab}), where M is a manifold and g_{ab} is a metric of Lorentzian signature. Other matter fields also may be present on spacetime. For definiteness, I will assume that M is 4-dimensional, but all results below generalize straightforwardly to any dimension $n \geq 2$. For our discussion of the first and second laws of black hole mechanics, it will be assumed, in addition, that the field equations of the theory have been obtained from a diffeomorphism covariant Lagrangian.

In physical terms, a black hole is a region of spacetime where gravity is so strong that nothing can escape. In order to make this notion precise, one must have in mind a region of spacetime to which one can contemplate escaping. For an asymptotically flat spacetime (representing an isolated system), the asymptotic portion of the spacetime "near infinity" is such a region. The *black hole* region, $\mathcal{B}$, of an asymptotically flat spacetime (M, g_{ab}) is defined as

$$\mathcal{B} \equiv M - I^-(\mathcal{I}^+), \tag{8.17}$$

where $\mathcal{I}^+$ denotes future null infinity and I^- denotes the chronological past. The *event horizon*, $\mathcal{H}$, of a black hole is defined to be the boundary of $\mathcal{B}$.

If an asymptotically flat spacetime (M, g_{ab}) contains a black hole $\mathcal{B}$, then $\mathcal{B}$ is said to be *stationary* if there exists a one-parameter group of isometries on (M, g_{ab}) generated by a Killing field t^a which is unit timelike at infinity. The black hole is said to be *static* if it is stationary and if, in addition, t^a is hypersurface orthogonal—in which case there exists a discrete "time reflection" isometry about any of the orthogonal hypersurfaces. The black hole is said to be *axisymmetric* if there exists a one-parameter group of isometries which correspond to rotations at infinity. A stationary, axisymmetric black hole is said to possess the "t-ϕ orthogonality property" if the 2-planes spanned by t^a and the rotational Killing field ϕ^a are orthogonal to a family of 2-dimensional surfaces. In this case, there exists a discrete "t-ϕ" reflection isometry about any of these orthogonal 2-dimensional surfaces.

Chapter 8. Black Holes and Thermodynamics

For a black hole which is static or is stationary-axisymmetric with the t-ϕ orthogonality property, it can be shown [11] that there exists a Killing field ξ^a of the form

$$\xi^a = t^a + \Omega \phi^a \tag{8.18}$$

which is normal to the event horizon, $\mathcal{H}$. The constant Ω defined by eq. (8.18) is called the *angular velocity of the horizon*. (For a static black hole, we have $\Omega = 0$.) A null surface whose null generators coincide with the orbits of a one-parameter group of isometries is called a *Killing horizon*, so the above result states that the event horizon of any black hole which is static or is stationary-axisymmetric with the t-ϕ orthogonality property must always be a Killing horizon. A stronger result holds in general relativity, where, under some additional assumptions, it can be shown that the event horizon of any stationary black hole must be a Killing horizon [12]. From this result, it also follows that in general relativity, a stationary black hole must be nonrotating (from which staticity follows [13], [14]) or axisymmetric (though not necessarily with the t-ϕ orthogonality property).

Now, let $\mathcal{K}$ be any Killing horizon (not necessarily required to be the event horizon, $\mathcal{H}$, of a black hole), with normal Killing field ξ^a. Since $\nabla^a(\xi^b \xi_b)$ also is normal to $\mathcal{K}$, these vectors must be proportional at every point on $\mathcal{K}$. Hence, there exists a function, κ, on $\mathcal{K}$, known as the *surface gravity* of $\mathcal{K}$, which is defined by the equation

$$\nabla^a(\xi^b \xi_b) = -2\kappa \xi^a. \tag{8.19}$$

It follows immediately that κ must be constant along each null geodesic generator of $\mathcal{K}$, but, in general, κ can vary from generator to generator. It is not difficult to show (see, e.g., [15]) that

$$\kappa = \lim(Va), \tag{8.20}$$

where a is the magnitude of the acceleration of the orbits of ξ^a in the region off of $\mathcal{K}$ where they are timelike, $V \equiv (-\xi^a \xi_a)^{1/2}$ is the "redshift factor" of ξ^a, and the limit as one approaches $\mathcal{K}$ is taken. Equation (8.20) motivates the terminology "surface gravity". Note that the surface gravity of a black hole is defined only when it is "in equilibrium" (i.e., stationary), analogous to the fact that the temperature of a (sub)system in ordinary thermodynamics is defined only for thermal equilibrium states.

In the context of an arbitrary metric theory of gravity, the zeroth law of black hole mechanics may now be stated as the following theorem [11], [16]: *For any black hole which is static or is stationary-axisymmetric with the t-ϕ orthogonality property, the surface gravity, κ, must be constant over its event horizon $\mathcal{H}$.* The key ingredient in the proof of this theorem is the identity [16]

$$\xi_{[a} \nabla_{b]} \kappa = -\frac{1}{4} \epsilon_{abcd} \nabla^c \omega^d, \tag{8.21}$$

which holds on an arbitrary Killing horizon, where $\omega_a \equiv \epsilon_{abcd}\xi^b \nabla^c \xi^d$ denotes the twist of the Killing field ξ^a. For a static black hole, we have $\omega_a = 0$, and the constancy of κ on $\mathcal{H}$ follows immediately. Further arguments [16] similarly establish the constancy of κ for a stationary-axisymmetric black hole with the t-ϕ orthogonality property. It should be emphasized that this result is "purely geometrical" and involves no use of any field equations.

A stronger version of the zeroth law holds in general relativity. There it can be shown [3] that if Einstein's equation holds with the matter stress-energy tensor satisfying the dominant energy condition, then κ must be constant on any Killing horizon. In particular, one need not make the additional hypothesis that the t-ϕ orthogonality property holds.

An important consequence of the zeroth law is that if $\kappa \neq 0$, then in the "maximally extended" spacetime representing the black hole, the event horizon, $\mathcal{H}$, comprises a branch of a "bifurcate Killing horizon". (A precise statement and proof of this result is given in [16]. Here, a *bifurcate Killing horizon* is comprised by two Killing horizons, $\mathcal{H}_A$ and $\mathcal{H}_B$, which intersect on a spacelike 2-surface, $\mathcal{C}$, known as the *bifurcation surface*.) As stated above, the event horizon of any black hole which is static or is stationary-axisymmetric with the t-ϕ orthogonality property necessarily is a Killing horizon and necessarily satisfies the zeroth law. Thus, the study of such black holes divides into two cases: "degenerate" black holes (for which, by definition, $\kappa = 0$) and black holes with bifurcate horizons. Again, this result is purely geometrical—involving no use of any field equations—and, thus, it holds in any metric theory of gravity.

We turn, now, to the consideration of first law of black hole mechanics. For this analysis, it will be assumed that the field equations of the theory arise from a diffeomorphism covariant Lagrangian 4-form, **L**, of the general structure

$$\mathbf{L} = \mathbf{L}\left(g_{ab}; R_{abcd}, \nabla_a R_{bcde}, ...; \psi, \nabla_a \psi, ...\right), \qquad (8.22)$$

where ∇_a denotes the derivative operator associated with g_{ab}, R_{abcd} denotes the Riemann curvature tensor of g_{ab}, and ψ denotes the collection of all matter fields of the theory (with indices suppressed). An arbitrary (but finite) number of derivatives of R_{abcd} and ψ are permitted to appear in **L**. Here and below we use boldface letters to denote differential forms, and we will suppress their indices. We also shall denote the complete collection of dynamical fields, (g_{ab}, ψ), by ϕ (thereby suppressing the indices of g_{ab} as well). Our treatment will follow closely that given in [17]; much of the mathematical machinery we shall use also has been extensively employed in analyses of symmetries and conservation laws of Lagrangian systems (see, e.g., [18] and references cited therein).

The Euler Lagrange equations of motion, $\mathbf{E} = 0$, are obtained by writing

Chapter 8. Black Holes and Thermodynamics

the variation of the Lagrangian in the form

$$\delta \mathbf{L} = \mathbf{E}(\phi)\delta\phi + d\boldsymbol{\theta}(\phi, \delta\phi), \tag{8.23}$$

where no derivatives of $\delta\phi$ appear in the first term on the right-hand side. Usually, the manipulations yielding eq. (8.23) are performed under an integral sign, in which case the second term on the right-hand side becomes a "boundary term", which normally is discarded. In our case, however, our interest will be in the mathematical structure provided to the theory by $\boldsymbol{\theta}$. Indeed, the precise form of $\boldsymbol{\theta}$ and the auxilliary structures derived from it will play a crucial role in our analysis, whereas the precise form of the field equations, $\mathbf{E} = 0$, will not be of interest here.

The symplectic current 3-form, $\boldsymbol{\omega}$, on spacetime—which is a local function of a field configuration, ϕ, and two linearized perturbations, $\delta_1\phi$ and $\delta_2\phi$, off of ϕ—is obtained by taking an antisymmetrized variation of $\boldsymbol{\theta}$:

$$\boldsymbol{\omega}(\phi, \delta_1\phi, \delta_2\phi) = \delta_2\boldsymbol{\theta}(\phi, \delta_1\phi) - \delta_1\boldsymbol{\theta}(\phi, \delta_2\phi). \tag{8.24}$$

The (pre)symplectic form, Ω—which is a map taking field configurations, ϕ, together with pairs of linearized perturbations off of ϕ, into the real numbers—is obtained by integrating $\boldsymbol{\omega}$ over a Cauchy surface, Σ:

$$\Omega(\phi, \delta_1\phi, \delta_2\phi) = \int_\Sigma \boldsymbol{\omega}. \tag{8.25}$$

(This integral is independent of choice of Cauchy surface when $\delta_1\phi$ and $\delta_2\phi$ satisfy the linearized field equations.) The (pre)symplectic form, Ω, provides the structure needed to define the phase space of the theory [19]. It also provides the structure needed to define the notion of a Hamiltonian, H, conjugate to an arbitrary vector field, η^a, on spacetime:[2] H is a function on phase space satisfying the property that about any solution, ϕ, the variation of H satisfies

$$\delta H = \Omega(\phi, \delta\phi, \mathcal{L}_\eta\phi), \tag{8.26}$$

where $\mathcal{L}_\eta$ denotes the Lie derivative with respect to the vector field η^a. (Eq. (8.26) can be rewritten in the more familiar form of Hamilton's equations of motion by solving it for $\mathcal{L}_\eta\phi$, thus expressing the "time derivative" of ϕ in terms of functional derivatives of H.)

On account of the diffeomorphism covariance of $\mathbf{L}$, the infinitesimal diffeomorphism generated by an arbitrary vector field, η^a, is a local symmetry of the theory. Hence, there is an associated, conserved *Noether current* 3-form, $\mathbf{j}$, defined by

$$\mathbf{j} = \boldsymbol{\theta}(\phi, \mathcal{L}_\eta\phi) - \eta \cdot \mathbf{L}, \tag{8.27}$$

---

[2] As discussed in [19], it will, in general, be necessary to choose η^a to be "field dependent", in order that it "project" to phase space.

where the "·" denotes the contraction of the vector field η^a into the first index of the differential form $\mathbf{L}$. One can show [20] that $\mathbf{j}$ always can be written in the form

$$\mathbf{j} = d\mathbf{Q} + \eta^a \mathbf{C}_a, \tag{8.28}$$

where $\mathbf{C}_a = 0$ when the equations of motion hold; i.e., $\mathbf{C}_a$ corresponds to "constraints" of the theory. Equation (8.28) defines the *Noether charge* 2-form $\mathbf{Q}$, which is unique up to

$$\mathbf{Q} \to \mathbf{Q} + \eta \cdot \mathbf{X}(\phi) + \mathbf{Y}(\phi, \mathcal{L}_\eta \phi) + d\mathbf{Z}(\phi, \eta), \tag{8.29}$$

where $\mathbf{X}$, $\mathbf{Y}$, and $\mathbf{Z}$ are arbitrary forms which are locally constructed from the fields appearing in their arguments (and with $\mathbf{Y}$ being linear in $\mathcal{L}_\eta \phi$ and $\mathbf{Z}$ being linear in η). Here the term $\eta \cdot \mathbf{X}$ arises from the ambiguity $\mathbf{L} \to \mathbf{L} + d\mathbf{X}$ in the choice of Lagrangian, the term $\mathbf{Y}(\phi, \mathcal{L}_\eta \phi)$ arises from the ambiguity $\boldsymbol{\theta} \to \boldsymbol{\theta} + d\mathbf{Y}$ in eq. (8.23), and the term $d\mathbf{Z}$ arises directly from eq. (8.28).

The first law of black hole mechanics is a direct consequence of the variational identity

$$\delta \mathbf{j} = \boldsymbol{\omega}(\phi, \delta\phi, \mathcal{L}_\eta \phi) + d(\eta \cdot \boldsymbol{\theta}), \tag{8.30}$$

which follows directly from eqs.(8.23), (8.24), and (8.27) above. One immediate consequence of this identity, together with eq. (8.26), is that if a Hamiltonian, H, conjugate to η^a exists, it must satisfy

$$\delta H = \int_\Sigma [\delta \mathbf{j} - d(\eta \cdot \boldsymbol{\theta})]. \tag{8.31}$$

From eq. (8.28) it follows, in addition, that "on shell"—i.e., when the equations of motion hold and, hence, $\mathbf{C}_a = 0$—we have

$$\delta H = \int_\Sigma d[\delta \mathbf{Q} - \eta \cdot \boldsymbol{\theta}], \tag{8.32}$$

so, on shell, H is given purely by "surface terms". In the case of an asymptotically flat spacetime, the surface term, H_∞, arising from infinity has the interpretation of being the total "canonical energy" (conjugate to η^a) of the spacetime.

Now, let ϕ be any solution to the field equations $\mathbf{E} = 0$ with a Killing field ξ^a, and let $\delta\phi$ be any solution to the linearized field equations off ϕ (not necessarily satisfying $\mathcal{L}_\xi \delta\phi = 0$). It follows immediately from eq. (8.30) (with $\eta^a = \xi^a$) together with the variation of eq. (8.28) that

$$d[\delta \mathbf{Q} - \xi \cdot \boldsymbol{\theta}] = 0. \tag{8.33}$$

We apply this equation to a spacetime containing a black hole with bifurcate Killing horizon, with ξ^a taken to be the Killing field (8.18) normal to

Chapter 8. Black Holes and Thermodynamics

the horizon, $\mathcal{H}$. (As mentioned above, the assumption of a bifurcate Killing horizon involves no loss of generality [16] if the zeroth law holds and $\kappa \neq 0$.) We integrate this equation over a hypersurface, Σ, which extends from the bifurcation surface, $\mathcal{C}$, of the black hole to infinity. The result is

$$\delta H_\infty = \delta \int_\mathcal{C} \mathbf{Q}, \qquad (8.34)$$

where the fact that $\xi^a = 0$ on $\mathcal{C}$ has been used.

We now evaluate the surface terms appearing on each side of eq. (8.34). As noted above, H_∞ has the interpretation of being the canonical energy conjugate to ξ^a. For ξ^a of the form (8.18), we have (see [13])

$$\delta H_\infty = \delta M - \Omega \delta J + \cdots \qquad (8.35)$$

where the ellipsis denotes possible additional contributions from long range matter fields. On the other hand, it is possible to explicitly compute $\mathbf{Q}$ and thereby show that [17] (see also [21], [22], [23])

$$\delta \int_\mathcal{C} \mathbf{Q} = \frac{\kappa}{2\pi} \delta S_{\text{bh}}, \qquad (8.36)$$

where

$$S_{\text{bh}} \equiv -2\pi \int_\mathcal{C} \frac{\delta L}{\delta R_{abcd}} n_{ab} n_{cd}. \qquad (8.37)$$

Here n_{ab} is the binormal to $\mathcal{C}$ (normalized so that $n_{ab} n^{ab} = -2$), L is the Lagrangian (now viewed as a scalar density rather than a 4-form), and the functional derivative is taken by formally viewing the Riemann tensor as a field which is independent of the metric in eq. (8.22). Combining eqs.(8.34), (8.35), and (8.36), we obtain

$$\delta M = \frac{\kappa}{2\pi} \delta S_{\text{bh}} + \Omega \delta J + \cdots, \qquad (8.38)$$

which is the desired first law of black hole mechanics. Indeed, this result is actually stronger than the form of the first law stated in the introduction, since eq. (8.38) holds for nonstationary perturbations of the black hole, not merely for perturbations to other stationary black hole states.

For the case of vacuum general relativity, where $L = R\sqrt{-g}$, a simple calculation yields

$$S_{\text{bh}} = A/4. \qquad (8.39)$$

However, if one considers theories with non-minimally-coupled matter or "higher derivative" theories of gravity, additional curvature contributions will appear in the formula for S_{bh}. Nevertheless, as eq. (8.37) explicitly shows, in all cases, S_{bh} is given by an integral of a "local, geometrical expression" over the black hole horizon.

The above analysis also contains a strong hint that the second law of black hole mechanics may hold in a wide class of theories. Consider a stationary black hole with bifurcate Killing horizon, but now let us normalize the Killing field, ξ^a, normal to the horizon by the local condition that $\nabla_a \xi_b = n_{ab}$ on $\mathcal{C}$ (or, equivalently, $\nabla_a \xi_b \nabla^a \xi^b = -2$ on $\mathcal{H}$), rather than by the asymptotic behavior (8.18) of ξ^a at infinity. With this new normalization, eq. (8.36) (with the δs removed) becomes

$$S_{\text{bh}} = 2\pi \int_{\mathcal{C}} \mathbf{Q}[\xi^a]. \tag{8.40}$$

It is not difficult to show that for a stationary black hole, this equation continues to hold when $\mathcal{C}$ is replaced by an arbitrary cross-section, σ, of $\mathcal{H}$ (see [23]).

Now, consider a process in which an initially stationary black hole evolves through a nonstationary era and then "settles down" to another stationary final state. Let ξ^a be any vector field which coincides with the Killing field normal to $\mathcal{H}$ (with the above, new normalization) in the two stationary regimes. Then, by eqs.(8.40) and (8.28) (together with $\mathbf{C}_a = 0$), we obtain

$$\begin{aligned}\Delta S_{\text{bh}} &= 2\pi \int_{\sigma_1} \mathbf{Q}[\xi^a] - 2\pi \int_{\sigma_0} \mathbf{Q}[\xi^a] \\ &= 2\pi \int_{\mathcal{H}} \mathbf{j}[\xi^a], \end{aligned} \tag{8.41}$$

where σ_0 and σ_1 are, respectively, cross-sections of $\mathcal{H}$ in the initial and final stationary regimes. Equation (8.41) states that the change in black hole entropy is proportional the net flux of Noether current (conjugate to ξ^a) through $\mathcal{H}$. Now, in many circumstances, the Noether current conjugate to a suitable time translation can be interpreted as the 4-density of energy-momentum. Thus, eq. (8.41) suggests that the second law of black hole mechanics, $\Delta S_{\text{bh}} \geq 0$, may hold in all theories which have suitable positive energy properties (such as, perhaps, a positive "Bondi energy flux" at null infinity). However, I have not as yet obtained any general results along these lines. Nevertheless, it has long been known that the second law holds in general relativity [2], provided that the matter present in spacetime satisfies the following positive energy property (known as the "null energy condition"): for any null vector k^a, the matter stress-energy tensor, T_{ab}, satisfies $T_{ab} k^a k^b \geq 0$. The second law also has been shown to hold in a class of higher derivative gravity theories, where the Lagrangian is a polynomial in the scalar curvature [24].

8.4 Quantum black hole thermodynamics

In section 8.3, we used a purely classical treatment of gravity and matter fields to derive analogs of the laws of thermodynamics for black holes. However, as already noted in the introduction, in classical physics, these laws of black hole mechanics cannot correspond physically to the laws of thermodynamics. It is only when quantum effects are taken into account that these subjects appear to merge.

The key result establishing a physical connection between the laws of black hole mechanics and the laws of thermodynamics is, of course, the thermal particle creation effect discovered by Hawking [4]. This result is derived in the context of "semiclassical gravity", where the effects of gravitation are still represented by a classical spacetime (M, g_{ab}) but matter fields are now treated as quantum fields propagating in this classical spacetime. In its most general form, this result may be stated as follows (see [25] for further discussion): Consider a black hole formed by gravitational collapse, which "settles down" to a stationary final state. By the zeroth law of black hole mechanics, the surface gravity, κ, of this stationary black hole final state will be constant over its event horizon. Consider a quantum field propagating in this background spacetime, which is initially in any (nonsingular) state. *Then, at asymptotically late times, particles of this field will be radiated to infinity as though the black hole were a perfect blackbody at the Hawking temperature, eq. (8.5).* Thus, a stationary black hole truly is a state of thermal equilibrium, and $\kappa/2\pi$ truly is the physical temperature of a black hole. It should be noted that this result relies only on the analysis of quantum fields in the region exterior to the black hole. In particular, the details of the gravitational field equations play no role, and the result holds in any metric theory of gravity obeying the zeroth law.

The physical connection between the laws of black hole mechanics and the laws of thermodynamics is further cemented by the following considerations. If we take into account the "back reaction" of the quantum field on the black hole (i.e., if the gravitational field equations are used self-consistently, taking account of the gravitational effects of the quantum field), then it is clear that if energy is conserved in the full theory, an isolated black hole must lose mass in order to compensate for the energy radiated to infinity in the particle creation process. As a black hole thereby "evaporates", S_{bh} will decrease, in violation of the second law of black hole mechanics. (Note that in general relativity, this can occur because the stress-energy tensor of quantum matter does not satisfy the null energy condition—even for matter for which this condition holds classically—in violation of one of the hypotheses of the area theorem.) On the other hand, there is a serious difficulty with the ordinary second law of thermodynamics when black holes are present: One can simply take some ordinary matter and drop it into a black hole, where, classically at least, it will disappear into a spacetime singularity. In this latter process, one loses the entropy initially present in the matter, but no compensating gain of ordinary entropy occurs, so the total entropy, S, decreases. Note, however, that in the black hole evaporation process, although S_{bh} decreases, a significant amount of ordinary entropy is generated outside the black hole due to particle creation. Similarly, when ordinary matter (with positive energy) is dropped into a black hole, although S decreases, by the first law of black hole mechanics, there will necessarily be an increase in S_{bh}.

The above considerations motivated the following proposal [1], [26]. Although the second law of black hole mechanics breaks down when quantum processes are considered, and the ordinary second law breaks down when black holes are present, perhaps the following law, known as the *generalized second law*, always holds: *In any process, the total generalized entropy never decreases*:

$$\Delta S' \geq 0, \qquad (8.42)$$

where the *generalized entropy*, S', is defined by

$$S' \equiv S + S_{\text{bh}}. \qquad (8.43)$$

A number of analyses [27], [28], [29], [30] have given strong support to the generalized second law. Although these analyses have been carried out in the context of general relativity, the arguments for the validity of the generalized second law should be applicable to a general theory of gravity, provided, of course, that the second law of black hole mechanics holds in the classical theory.

The generalized entropy (8.43) and the generalized second law (8.42) have obvious interpretations: Presumably, for a system containing a black hole, S' is nothing more than the "true total entropy" of the complete system, and (8.42) is then nothing more than the "ordinary second law" for this system. If so, then S_{bh} truly is the physical entropy of a black hole.

Although I believe that the above considerations make a compelling case for the merger of the laws of black hole mechanics with the laws of thermodynamics, there remain many puzzling aspects to this merger. One such puzzle has to do with the existence of a "thermal atmosphere" around a black hole. It is crucial to the arguments for the validity of the generalized second law (see, in particular, [27]) that near the black hole, all fields are in thermal equilibrium with respect to the notion of time translations defined by the horizon Killing field ξ^a (see eq. (8.18) above). For an observer following an orbit of ξ^a just outside the black hole, the locally measured temperature (of all species of matter) is

$$T = \frac{\kappa}{2\pi V}, \qquad (8.44)$$

where $V = (-\xi^a \xi_a)^{1/2}$. Note that, in view of eq. (8.20) above, we see that

Chapter 8. Black Holes and Thermodynamics

$T \to a/2\pi$ as the black hole horizon, $\mathcal{H}$, is approached. Thus, in this limit eq. (8.44) corresponds to the flat spacetime Unruh effect [31].

Since $T \to \infty$ as the horizon is approached, this thermal atmosphere has enormous entropy. Indeed, if no cutoff is introduced, the entropy of the thermal atmosphere is divergent.[3] However, inertial observers do not "see" this thermal atmosphere and would attribute the physical effects produced by the thermal atmosphere to other causes, like radiation reaction effects [27]. In particular, with respect to a notion of time translations which would be naturally defined by inertial observers who freely fall into the black hole, the entropy of quantum fields outside of a black hole should be negligible.

Thus, it is not entirely clear what the quantity "S" appearing in eq. (8.43) is supposed to represent for matter near, but outside of, a black hole. Does S include contributions from the thermal atmosphere? If so, S is divergent unless a cutoff is introduced – although changes in S (which is all that is needed for the formulation of the generalized second law) could still be well defined and finite. If not, what happens to the entropy in a box of ordinary thermal matter as it is slowly lowered toward the black hole? By the time it reaches its "floating point" [27], its contents are indistinguishable from the thermal atmosphere, so has its entropy disappeared? These questions provide good illustrations of some of the puzzles which arise when one attempts to consider thermodynamics in the framework of general relativity, as previously discussed at the end of section 8.2.

However, undoubtedly the most significant puzzle in the relationship between black holes and thermodynamics concerns the physical origin of the entropy, $S_{\rm bh}$, of a black hole. Can the origin of $S_{\rm bh}$ be understood in essentially the same manner as in the thermodynamics of conventional systems (as suggested by the apparently perfect merger of black hole mechanics with thermodynamics), or is there some entirely new phenomenon at work here (as suggested by the radical differences in the present derivations of the laws of black hole mechanics and the laws of thermodynamics)? As already remarked near the end of section 8.2, a classical treatment (as in section 8.3), or even a semiclassical treatment (as above in this section), cannot be adequate to analyze this issue; undoubtedly, a fully quantum treatment of all of the degrees of freedom of the gravitational field will be required.

Our present understanding of quantum gravity is quite rudimentary.

---

[3] If a cutoff at the Planck scale is introduced, the entropy of the thermal atmosphere agrees, in order of magnitude, with the black hole entropy (8.39). There have been a number of attempts (see [30] for further discussion) to attribute the entropy of a black hole to this thermal atmosphere, or to "quantum hair" outside the black hole. However, if any meaning can be given to the notion of where the entropy of a black hole "resides", it seems much more plausible to me that it resides in the "deep interior" of the black hole (corresponding to the classical spacetime singularity), rather than near the "surface" of the black hole, which has no local significance.

Most approaches used to formulate the quantum theory of fields propagating in a Minkowski background spacetime either are inapplicable or run into severe difficulties when one attempts to apply them to the formulation of a quantum theory of the spacetime metric itself. However, several approaches have been developed which have some appealing features and which hold out some hope of overcoming these difficulties. The most extensively developed of these approaches is string theory, and, very recently, some remarkable results have been obtained in the context of string theory relevant to understanding the origin of black hole entropy. I will very briefly describe some of the key results here, referring the reader to the contribution of Horowitz [32] for further details and discussion.

In the context of string theory, one can consider a "low energy limit" in which the "massive modes" of the string are neglected. In this limit, string theory should reduce to a 10-dimensional supergravity theory. If one treats this supergravity theory as a classical theory involving a spacetime metric, g_{ab}, and other classical fields, one can find solutions describing black holes. These classical black holes can be interpreted as providing a description of states in string theory which is applicable at "low energies".

On the other hand, one also can consider a "weak coupling" limit of string theory, wherein the states are treated perturbatively about a background, flat spacetime. In this limit, the dynamical degrees of freedom of the theory are described by perturbative string states together with certain soliton-like configurations, known as D-branes. In the weak coupling limit, there is no literal notion of a black hole, just as there is no notion of a black hole in linearized general relativity. Nevertheless, certain weak coupling states comprised of "D-branes" can be identified with certain black hole solutions of the low energy limit of the theory by a correspondence of their energy and charges.

Now, the weak coupling states are, in essence, ordinary quantum dynamical degrees of freedom in a flat background spacetime. The ordinary entropy, S, of these states can be computed using eq. (8.16) of section 8.2. The corresponding classical black hole states of the low energy limit of the theory have an entropy, S_{bh}, given by eq. (8.39) of section 8.3. The remarkable results referred to above are the following: For certain classes of extremal ($\kappa = 0$) [33] and nearly extremal [34] black holes, the "ordinary entropy", (8.16), of the weak coupling D-brane states agrees exactly with the entropy, (8.39), of the corresponding classical black hole states occurring in the low energy limit of the theory.[4] Since these entropies have a nontrivial functional dependence on energy and charges, it is hard to imagine that this agreement could be the result of a random coincidence. Furthermore, for low energy scattering, the absorption/emission

---

[4]Note that since the D-brane states are, in essence, ordinary quantum systems, they thereby provide an example of a quantum system which violates the Nernst formulation of the third law of thermodynamics.

coefficients ("graybody factors") of the corresponding D-brane and black hole configurations also agree [35]. In particular, since the temperatures of the corresponding D-brane and black hole configurations agree (by virtue of the equality of entropies), it follows that the "Hawking radiation" from the D-brane configuration agrees with that from the corresponding black hole, at least at low energies.

Since the entropy of the D-brane configurations arises entirely from conventional "state counting", the above results strongly suggest that the physical origin of black hole entropy should be essentially the same as for conventional thermodynamic systems. However, undoubtedly, a much more complete understanding of the nature of black holes in quantum gravity when one is in neither a "weak coupling" nor a "low energy" limit will be required before a definitive answer to this question can be given.

As I have tried to summarize in this article, a great deal of progress has been made in our understanding of black holes and thermodynamics, but many unresolved issues remain. I believe that the attainment of a deeper understanding of the nature of the relationship between black holes and thermodynamics is our most promising route toward an understanding of the fundamental nature of both quantum gravity *and* thermodynamics.

Acknowledgments

This research was supported in part by NSF grant PHY 95-14726.

References

[1] J.D. Bekenstein, Phys. Rev. **D7**, 2333 (1973).

[2] S.W. Hawking, Phys. Rev. Lett. **26**, 1344 (1971).

[3] J.M. Bardeen, B. Carter, and S.W. Hawking, Commun. Math. Phys. **31**, 161 (1973).

[4] S.W. Hawking, Commun. Math. Phys. **43**, 199 (1975).

[5] R.M. Wald, Phys. Rev. **D20**, 1271 (1979).

[6] K. Huang, *Statistical Mechanics*, John Wiley & Sons (New York, 1963).

[7] R.M. Wald, "Nernst theorem" and black hole thermodynamics, Phys. Rev. D, in press (1997).

[8] G. Gibbons and S.W. Hawking, Phys. Rev. **D15**, 2752 (1977).

[9] J.D. Brown and J.W. York, Phys. Rev. **D47**, 1420 (1993).

[10] R. Penrose, in *General Relativity: An Einstein Centennary Survey*, ed. S.W. Hawking and W. Israel, Cambridge University Press (Cambridge, 1979).

[11] B. Carter, in *Black Holes*, ed. C. DeWitt and B.S. DeWitt, Gordon and Breach (New York, 1973).

[12] S.W. Hawking and G.F.R. Ellis, *The Large Scale Structure of Space-Time*, Cambridge University Press (Cambridge, 1973).

[13] D. Sudarsky and R.M. Wald, Phys. Rev. **D46**, 1453 (1992).

[14] P.T. Chrusciel and R.M. Wald, Commun. Math. Phys. **163**, 561 (1994).

[15] R.M. Wald, *General Relativity*, University of Chicago Press (Chicago, 1984).

[16] I. Racz and R.M. Wald, Class. Quant. Grav. **13**, 539 (1996).

[17] V. Iyer and R.M. Wald, Phys. Rev. **D50**, 846 (1994).

[18] M. Ferraris and M. Francaviglia, in *Mechanics, Analysis, and Geometry: 200 Years after Lagrange*, North-Holland (Amsterdam, 1991).

[19] J. Lee and R.M. Wald, J. Math. Phys. **31**, 725 (1990).

[20] V. Iyer and R.M. Wald, Phys. Rev. **D52**, 4430 (1995).

[21] T. Jacobson and R.C. Myers, Phys. Rev. Lett. **70**, 3684 (1993).

[22] M. Visser, Phys. Rev. **D48**, 5697 (1993).

[23] T. Jacobson, G. Kang, and R.C. Myers, Phys. Rev. **D49**, 6587 (1994).

[24] T. Jacobson, G. Kang, and R.C. Myers, Phys. Rev. **D52**, 3518 (1995).

[25] R.M. Wald, *Quantum Field Theory in Curved Spacetime and Black Hole Thermodynamics*, University of Chicago Press (Chicago, 1994).

[26] J.D. Bekenstein, Phys. Rev. **D9**, 3292 (1974).

[27] W.G. Unruh and R.M. Wald, Phys. Rev. **D25**, 942 (1982).

[28] W.H. Zurek and K.S. Thorne, Phys. Rev. Lett. **54**, 2171 (1985).

[29] V.P. Frolov and D.N. Page, Phys. Rev. Lett. **71**, 3902 (1993).

[30] R. Sorkin, chap. 9 of this volume.

[31] W.G. Unruh, Phys. Rev. **D14**, 870 (1976).

[32] G. Horowitz, chap. 12 of this volume.

[33] A. Strominger and C. Vafa, Phys. Lett. **B379**, 99 (1996).

[34] C. G. Callan, Jr., and J. M. Maldacena, Nucl. Phys. **B472**, 591 (1996); G. T. Horowitz and A. Strominger, Phys. Rev. Lett. **77**, 2368 (1996).

[35] J.M. Maldacena and A. Strominger, Phys. Rev. **D55**, 861 (1997).

9

The Statistical Mechanics of Black Hole Thermodynamics

Rafael D. Sorkin

Abstract

Although we have convincing evidence that a black hole bears an entropy proportional to its surface (horizon) area, the "statistical mechanical" explanation of this entropy remains unknown. Two basic questions in this connection are What is the microscopic origin of the entropy? and Why does the law of entropy increase continue to hold when the horizon entropy is included? After a review of some of the difficulties in answering these questions, I propose an explanation of the law of entropy increase which comes near to a proof in the context of the "semiclassical" approximation, and which also provides a proof in full quantum gravity under the assumption that the latter fulfills certain natural expectations, like the existence of a conserved energy definable at infinity. This explanation seems to require a fundamental spacetime discreteness in order for the entropy to be consistently finite, and I recall briefly some of the ideas for what the discreteness might be. If such ideas are right, then our knowledge of the horizon entropy will allow us to "count the atoms of spacetime".

When I first learned of the thermodynamics of black holes, and specifically of the fact that a black hole possesses an entropy proportional to its horizon area, my reaction (after thinking about it a while) was that this was just as if the horizon were divided into small "tiles" of a fixed size, with each tile carrying roughly one bit of information. To see that this would lead to the correct proportionality law, imagine a 0 or 1 engraved on each tile. If there were A tiles, then the number of possible configurations would be $N = 2 \times 2 \times 2 \times \cdots \times 2 = 2^A$, and the corresponding entropy would be $S = \log N = A \log 2$ (taking Boltzmann's constant k equal to one). In order to get the numerical coefficient right, the tile size would have to be around 10^{-65} cm; that is, it would have to be of order unity in units with $c = \hbar = 8\pi G = 1$.

Of course, no one believes (or would admit to believing) that the black hole horizon is painted with tiny 0s and 1s, but the suggestion remains that, in some less artificial manner, a cutoff occurs at around the Planck scale, and the microscopic degrees of freedom proper to the horizon carry about one bit of information per horizon "atom". In order for such an explanation to work, it would have to provide convincing answers to two principal questions: What degrees of freedom does this information capture (to what N is one really referring when one says $S = k \log N$) and why does the total entropy still increase when black holes are involved (why does the second law of thermodynamics continue to hold)?

The explanation that would resolve these twin questions is what I mean by the "statistical mechanics behind black hole thermodynamics". We can hope that in the process of arriving at such an explanation, we will learn something important about the nature of spacetime on small scales, just as the quest for the statistical mechanics of a box of gas taught us something important about the nature of ordinary matter on atomic scales, revealing the existence of atoms, their sizes, and something about their structure and quantum nature.

Equally, we may hope that the investigation of these questions will shed new light on statistical mechanics itself, and specifically on the meaning of entropy. For example, the experience so far has been that no derivation of the second law can even get started without first applying some form of coarse-graining to define the entropy, and the choice of coarse-graining seems to introduce an unwelcome element of subjectivity into the foundations of statistical mechanics. But a black hole affords an obvious objective way to coarse-grain, namely, neglect whatever is inside the horizon. Or to take another example, it is the possibility of fluctuations in the entropy that distinguishes the statistical mechanical picture from the thermodynamical one, but in practice such fluctuations are ordinarily too small to be observable. With black holes, on the other hand, one can, at least in principle, arrange for arbitrarily large fluctuations to occur [1].

Before we address the statistical mechanics of black holes as such, let me remind you very briefly of their thermodynamic properties in a little more detail. In a process like stellar collapse, a black hole is said to form when a region of spacetime develops from which signals can no longer escape. Within the context of classical general relativity the subsequent occurrence of a singularity is then inevitable, but it is expected to form inside the black hole, so that it is hidden from the view of distant astronomers (see the discussion in [2]).

The black hole's *horizon H* is by definition the three dimensional surface separating the interior of the black hole from its exterior. Formally, we have $H = \partial(\text{past}\,\mathcal{I}^+)$, where $\mathcal{I}^+$ (called "future null infinity") is defined as the set of ideal points at infinity at which outgoing light rays terminate. Thus the past of $\mathcal{I}^+$ consists of all events that can send light rays to infinity; and

H is its boundary. From this definition alone certain mathematical facts follow, including the fact that H is a continuous surface (a C^0 manifold) which, although it may not be smooth (because of caustics), nevertheless is null almost everywhere in the sense that it is a union of null geodesics which never leave H as they propagate into the future. (These ruling geodesics are photon world lines that hover on the horizon, balanced precariously between escaping to infinity and being pulled into the singularity.) When we speak of the "area of the horizon", we mean the area of the *cross section* in which H intersects some spacelike (or possibly null) hypersurface Σ on which we seek to evaluate the entropy:[1]

$$A := \text{Area}(H \cap \Sigma).$$

Now an abundance of evidence indicates that there is associated with an event horizon of this sort an entropy of

$$S_{\text{BH}} = \frac{2\pi A}{l^2}, \tag{9.1}$$

where

$$l^2 = 8\pi G\hbar \tag{9.2}$$

is the square of the "rationalized Planck length". The best known piece of evidence for this association is that the black hole radiates in precisely the right manner to be in equilibrium with a surrounding gas of thermal radiation at the temperature derived from (9.1) via the first law of thermodynamics, $dM = T\,dS$ (the Hawking radiation). But the evidence goes well beyond that, ranging from theorems in classical general relativity that can be interpreted as the zeroth through third laws of thermodynamics applied to a black hole in the $\hbar \to 0$ limit, to computations directly yielding the entropy (9.1) in the "tree level" approximation of path-integral quantum gravity.

Many of these relationships carry over to other spacetime dimensions and other gravity theories than standard general relativity, as emphasized in [4]. However, the evidence remains less complete and convincing in these other cases.[2] In particular one still lacks an analog for these other cases of the result that in standard general relativity represents the $\hbar \to 0$ limit of the second law of thermodynamics, namely, the theorem that classically the total horizon area necessarily increases (or remains constant)

---

[1] For this definition of A to make sense, it is necessary that $H \cap \Sigma$ not be too rough. I don't know whether one can prove this for every possible black hole horizon H and smooth surface Σ, but "geometric measure theory" guarantees at least that, for any given H, there exists a dense set of Σs for which A is well defined [3].

[2] In a sense it is disappointing that black hole thermodynamics appears to be so robust: the less sensitive it is to the theoretical assumptions, the less capable of guiding us to the correct underlying theory of quantum gravity!

as the hypersurface Σ on which it is evaluated moves forward in time. Because of the way that $\hbar$ enters into the expressions (9.2) and (9.1), the black hole entropy goes to infinity in the classical limit and so tends to dominate all other entropies. Therefore, this classical law of area increase can be interpreted as precisely what remains of the second law when $\hbar$ tends to zero. (Unfortunately, the proof of area increase remains incomplete even in standard general relativity, because it rests on the still unproven assumption of "cosmic censorship".)

Notice that the law of area increase, even though it is valid only in the nonquantum limit, is a fully nonequilibrium result in the sense that no requirement of stationarity is imposed on the black holes or the classical matter with which they interact. In the complementary limit of fully quantum matter interacting with a near equilibrium black hole, there exist several arguments and thought experiments suggesting the impossibility of procuring a decrease in the total entropy (black hole plus entropy of surroundings) by any process in which the black hole evolves quasi-stationarily and remains essentially classical. (Later, I will propose a completely general proof of this impossibility.) Taken together, these results in special cases strongly suggest that the second law of thermodynamics, $\Delta S \geq 0$, will continue to hold in general if we attribute to each black hole its corresponding entropy (9.1) and take for the total entropy the sum of this horizon-area entropy with the entropy of whatever else may be present *outside* the black holes:

$$S_{\text{tot}} = S_{\text{BH}} + S_{\text{outside}} \quad \text{increases with time.} \tag{9.3}$$

Remark The above discussion has defined a black hole in causal terms and correspondingly has taken the horizon to be what is more precisely referred to as the *event horizon*. A rather different concept is that of *apparent horizon*, whose definition generalizes the curvature properties of the Schwarzschild horizon rather than its causal properties. In application to time-independent black holes, the two concepts coincide, but away from equilibrium they differ. It has been proposed [5] to identify the area that enters into the formula (9.1) as the area of the apparent horizon rather than the event horizon. The apparent horizon has the apparent advantage of being definable "quasi-locally", whereas locating the event horizon requires in principle a knowledge of the entire future of the spacetime. On the other hand, it is precisely this "teleological" attribute of the event horizon that gives rise to the large entropy fluctuations alluded to above. Moreover, the concept of event horizon seems more fundamental than that of apparent horizon and therefore more robust in relation to possible discrete replacements for spacetime such as the causal set [6]. In addition, the area of the apparent horizon will often jump discontinuously, which is not how one might expect entropy to behave. For these reasons, I will stick with the identification, horizon = event horizon, but it seems prudent to bear

Chapter 9. The Statistical Mechanics of Black Hole Thermodynamics 181

the other alternative in mind as well.

Now, how would we understand the nondecreasing character of S_{tot} if a black hole were an ordinary, nonrelativistic thermodynamic object? For example, suppose the black hole were really a warm brick or, say, a hot ball of hafnium. In that case (recalling the familiar plausibility arguments, as nicely presented in [4]), we would identify N_{BH} (the subscript stands for "ball of hafnium") as the number of internal micro-states of the ball, and $S_{\text{BH}} = \log N_{\text{BH}}$ as its logarithm. If we went on to assume that the ball was *weakly coupled* to its surroundings, then we could write (with N_{out} being the number of micro-states of the surroundings),

$$N_{\text{tot}} = N_{\text{BH}} \cdot N_{\text{out}}, \quad (9.4\text{a})$$
$$S_{\text{tot}} = S_{\text{BH}} + S_{\text{out}}. \quad (9.4\text{b})$$

(On the other hand, if the coupling was not weak, then the states of the two subsystems would not be able to vary independently, and so the estimate $N_{\text{tot}} = N_{\text{BH}} \cdot N_{\text{out}}$ would be inaccurate.) Under the further assumption that the dynamical evolution of the whole system was *ergodic* (exploring all of the available state-space) and *unitary* (so preserving state-space volume as measured by number of states), we would conclude that the time spent in any region of state-space was proportional to the number of states in that region:

$$\text{dwell time} \propto N_{\text{tot}}, \quad (9.5)$$

so that the probability of a transition in which N_{tot} increased would be more likely than one in which it decreased,[3] the discrepancy being overwhelming for large entropy differences because $\Delta S \gg 1 \Rightarrow e^{\Delta S} \ggg 1$.

So what goes wrong with this reasoning if we try to apply it to a black hole? In the first place, it is at least peculiar that the number of black hole states would be proportional to e^{Area} rather than e^{Volume} as for other ther-

---

[3]The plausibility argument here could be made stronger if we assumed a form of "detailed balance" for the effective Markov process operating on the macro-states, or better "meso-states" (cf. [7]). Then it would follow immediately that the probability (per unit time, say) of a transition from state 1 to state 2, given that the system was currently in state 1, would be greater by $e^{S_2 - S_1}$ than that of the reverse transition, given that the system was currently in state 2.

It is interesting that such a Markovian model of the evolution appears easier to justify in a classical framework, where the system really is in a definite (albeit unknown) micro-state at any instant. In the quantum case, in contrast, there is no reason to believe that the state-vector normally lies within any particular eigensubspace of a macroscopic observable like the temperature distribution. Thus, although the *definition* of entropy works better in the quantum case (because "number of micro-states" really makes sense there), the argument for its *increase* seems to proceed more happily in the classical setting. To recover a Markovian picture in the quantum case as well, one might have recourse to a sum-over-histories interpretation of the formalism, or to the closely related picture in which coarse-grained histories are obtained from sequences of projection operators as in [8].

modynamic systems. This peculiarity becomes more troubling if we consider the example of the Oppenheimer-Snyder spacetime, in which a Friedmann universe *of arbitrary size* is joined onto the interior of a Schwarzschild black hole of arbitrary mass. (See the remarks in [9].) The existence of such solutions means that the number of possible *interior* states for a black hole is really infinite, which is certainly not consistent with any formula like $S = \log N$.

A second set of problems concerns *ergodicity* and *internal equilibrium*. The course of events inside a collapsing star leads classically to a singularity, and it is not at all obvious that this is consistent with an ergodic exploration of all available states, including for example the Oppenheimer-Snyder states just described. But even if quantum effects did restore ergodicity, there would remain a problem with equilibrium. Although I did not stress it above, an assumption of internal equilibrium (or partial equilibrium) is needed in order to deduce the entropy from the values of a few macroscopic variables. For example, knowing only the surface temperature of the ball of hafnium would not at all allow you to deduce its entropy unless you added the assumption of internal equilibrium; if this assumption were mistaken, you might find the apparent entropy suddenly starting to decrease, because the interior of the ball was much colder than you had assumed. In the same way, mere knowledge of the external appearance of a black hole tells you little about its interior, and since realistic black holes would seem to be far from internal equilibrium, this is another reason to doubt whether the type of state counting utilized in (9.4) and (9.5) can carry over to black holes.

A related problem with such state counting is that our earlier assumption of *weak coupling* between the subsystem and its environment is doubly wrong in the case of a black hole. The coupling from outside to inside is not weak but very strong, while the reverse coupling is not so much weak as nonexistent! Indeed, this last observation points up the fact that conditions in the interior should be irrelevant, almost by definition, to what goes on outside. And since the second law, as ordinarily formulated for black holes, makes no reference to conditions inside, it seems especially strange that it should have anything to do with counting interior states. (In contrast, the temperature distribution inside our ball of hafnium can make a big difference in its interaction with the outside world, as we have seen, and it is impossible to specify its entropy without saying something about the internal conditions, if not explicitly, then implicitly via the assumption of internal equilibrium.)

Finally, there is the vexed question of *unitarity*. Many people refer to this in terms of an "information puzzle", the puzzle being that the facts apparently belie their belief that there must be a well-defined unitary S-matrix for the *exterior* region alone; but for our purposes here, the existence of such an S-matrix is irrelevant, because all we are interested in is the sec-

Chapter 9. The Statistical Mechanics of Black Hole Thermodynamics

ond law, and that certainly imposes no such requirement.[4] Rather, what the ordinary statistical mechanical reasoning requires is *overall* unitarity (for exterior and interior regions together), and the difficulty lies in combining this overall unitarity with the mode of explanation that would locate the entropy of the black hole in the multiplicity of its interior micro-states. If the latter were correct, then a contradiction with unitarity would arise in the process of evaporation of a black hole by Hawking radiation. In fact, since the black hole loses entropy as it shrinks (its surface area decreases), its number of internal states would have to go down as well, and at some point there would no longer be enough of them to support the correlations with the radiated particles which are needed if the *overall* evolution is to remain unitary.

At least this conclusion is inevitable if we accept the semiclassical description of the radiation as consisting of correlated pairs A and $\bar{A}$, the first of which goes off to infinity while the second falls into the singularity. In this approximation of quantum field theory on a fixed, background black hole metric, the emitted particles A taken alone are described by a highly impure state of the exterior field, and the unitarity of the overall quantum evolution is restored only when the $A \leftrightarrow \bar{A}$ correlations are taken into account. (One sometimes says that the particles A are "entangled" with the particles $\bar{A}$.)

In order for these entanglement correlations to exist, however, it is necessary that the interior of the black hole support a quantum state-space of dimension at least $e^{S_{\mathrm{rad}}}$, where S_{rad} is the entropy of the emitted thermal radiation. If the number of internal states really diminished as the black hole shrank, then this would cease to be possible, and so unitarity could be maintained only if the $A \leftrightarrow \bar{A}$ correlations were transferred to ones of the form $A \leftrightarrow B$ between *emitted* particles. Ultimately, if the black hole were allowed to evaporate fully, then all of the $A \leftrightarrow \bar{A}$ correlations would have to be transferred to the outside; and thus unitary evolution of the whole system, *together with* the assumption that black hole entropy counts the number of internal states, would require the external evolution *alone* to be unitary, at least in the S-matrix sense.

Now, it is possible to imagine mechanisms by which the required $A \leftrightarrow B$ correlations could arise, but it is much less easy to imagine how the $A \leftrightarrow \bar{A}$ correlations could disappear, since they arise in an approximation which should remain good in full quantum gravity. But if they don't disappear then the quantum state would have to stay entangled, which in turn would require the number of internal states to remain much larger than allowed by the formula (9.1). Hence one has either to abandon the interpretation of black hole entropy in terms of internal states or to grant that the overall

---

[4] On the contrary, a unitary evolution of the exterior region would be inimical to the type of explanation of the second law I will propose below.

quantum evolution of the system, black hole + environment, is nonunitary (or both).

Despite the seeming inevitability of this conclusion, many workers still hope to evade it, but in seeking to do so, they are driven to rather desperate maneuvers, such as hypothesizing a new "complementarity" according to which the internal state-space would be of a different dimensionality for observers outside the black hole than for observers inside of it. Such conceptual difficulties notwithstanding, the attempt to imagine how the external region alone can possess a unitary S-matrix (or perhaps even a dynamics which remains unitary during all intermediate stages) has led to some interesting suggestions, including the suggestion [10] that at very small distances the only variables that remain are purely geometrical ones, with all other distinctions (color, flavor, generation, etc.) being washed out.

By far the greatest effort toward providing a unitary statistical mechanics for black hole thermodynamics has been exerted within the context of string theory (which is not to say that strings are necessarily tied to a unitary explanation, any more than any other approach is; cf. [11]). In fact most of the effort has been directed to only one of the two questions I emphasized at the outset, namely, the question of what degrees of freedom one must count in order to obtain the formula (9.1). The earliest calculation of this sort that I know of was performed by Steve Carlip [12] for a black hole in (2+1)-dimensional gravity. Although this calculation did not actually use a stringy version of gravity, it used string technology to count certain "gauge" degrees of freedom defined on the horizon, obtaining (9.1) with precisely the thermodynamically required coefficient. This would tend to agree with the suggestion that the entropy is localized at the horizon, rather than inside the black hole. More recently, some calculations in string theory proper have also obtained equation (9.1) for zero temperature black holes in certain higher dimensional theories of gravity coupled to special combinations of gauge fields chosen to make supersymmetric solutions possible. These more recent calculations do not actually work with black holes, rather they count certain "membrane" states in a flat-space limit and obtain a formula which can be interpreted as the analytic continuation of (9.1) to that limit. (See [13] for a review of this work.)

To the extent that the physical picture behind these stringy calculations can be extrapolated from flat spacetime to genuine black hole geometries (which assumes in particular that the "phase transition" accompanying the formation of the horizon would not interfere with analyticity), they suggest that the degrees of freedom contributing to the entropy are associated with certain types of membranes. In contrast to Carlip's calculation, they also could be taken to suggest that the relevant degrees of freedom are those of the black hole as a whole (as opposed to being localized at the horizon), but since there are no horizons in flat space, such a suggestion would have to

Chapter 9. The Statistical Mechanics of Black Hole Thermodynamics

be very tentative at best. Beyond this, the stringy calculations do not (as far as I can see) shed much light on the question that to me is most central about the entropy of black holes: Why is it finite at all? We will see later that certain identifiable contributions to the entropy that arguably must be present are infinite in the absence of a short-distance cutoff at the horizon. One might ask whether these contributions are present in the string theory picture and, if so, how the requisite cutoff arises.

Concerning the derivation of the second law and the objections raised above, string theory apparently has little to say. If, at the end of the day, it is able to produce a unitary dynamics underlying quantum gravity, and if it is able to count the states of a black hole in this framework (and also explain why a weak coupling, equilibrium approximation is really valid after all, etc.), then the derivation of the second law will be reduced to the discussion one can find in any thoughtful textbook on statistical mechanics. However, we have seen that the demand for unitarity in particular seems to drive one to a new kind of inside-outside complementarity that so far no one knows how to formulate. For this reason it seems impossible at this stage to address the question of the second law within the confines of any unitary theory that identifies the black hole entropy with the states of the black hole taken as a whole. Let us return, then, to our two main questions—Why does S increase? and What does S count?—and consider them in the order just given, hoping that an answer to the first will suggest an answer to the second.

Why does S increase?

What I want to describe now is an old proposal [14, 15] that would derive the second law by appealing directly to the property of black holes that makes them *different* from all other objects, the fact that what goes on outside the horizon takes place without any reference to what goes on inside. Classically, this independence is exact, and quantum mechanically, it can still be expected to hold to an excellent approximation, despite some quantum blurring of the horizon. But if the exterior region really has a well-defined autonomous dynamics, then it is particularly natural to "coarse-grain away" the interior region and seek to identify the entropy that enters into the second law with the entropy of some effective quantum density operator describing the exterior situation.[5]

More specifically, what we would like to do is evaluate the entropy on a

---

[5]This is not meant to imply that on philosophical grounds we must neglect the interior region because "we" cannot observe it. Rather my attitude would be that any coarse-graining at all is valid if it leads to a useful definition of entropy. The horizon is nevertheless special in this connection because, as we will see, its special causal properties make it possible to draw conclusions that could not be drawn with other types of coarse-graining.

hypersurface Σ like that discussed earlier, but with the difference that our new Σ terminates where it meets the horizon. We would like to associate an effective density operator $\hat{\rho}(\Sigma)$ with each such hypersurface and to prove that

$$S(\Sigma'') \geq S(\Sigma')$$

whenever Σ'' lies wholly to the future of Σ'. We will see, in fact, that this approach to the question leads to two proofs of entropy increase, a more fully worked out one which applies in the semiclassical approximation [15] (cf. the remarks in [16] and the related remarks in [17] and [18]) and a more sketchy one which applies in full quantum gravity [14, 15]. It also leads to the conclusions that the entropy is localized (to the extent that entropy can ever be localized) at or just outside the horizon [14, 16, 19], and that S owes its finiteness to a fundamental spacetime discreteness or "atomicity" [14, 19].

Let us take first the case of a quasi-classical, quasi-stationary black hole, by which I mean that we ignore quantum fluctuations in M and the other black hole parameters and we assume that the spacetime geometry can be well approximated at any stage by a strictly stationary metric. (In other words, we work in the framework of "quantum field theory in curved spacetime". Notice that the requirement of approximate stationarity applies only to the metric; the matter fields [among which we may include gravitons] can be doing anything they like.) In such a situation there exists for the matter fields outside the black hole a unique thermal (or "KMS") state $\hat{\rho}^0$, known as the "Hartle-Hawking state" and expressible as

$$\hat{\rho}^0 \propto e^{-\beta \hat{E}}, \qquad (9.6)$$

where $\hat{E}$ represents the energy operator for the matter fields in the external region and β is the reciprocal of the thermodynamic temperature of the black hole.[6] (For simplicity, we may as well work with a nonrotating, charge-free black hole. Alternatively, one could just reinterpret $\hat{E}$ as $\hat{E} - \omega\hat{J} - \phi\hat{Q}$.)

Now let us limit our consideration of the entropy to the hypersurfaces $\Sigma(t)$ of a foliation of the exterior region which is compatible with the Killing vector field ξ^a that expresses the stationarity of the metric. In other words, the $\Sigma(t)$ are the $t =$ constant surfaces of some coordinate system for which the metric is explicitly time independent. (Notice, however, that it would *not* be appropriate to take t to be the Schwarzschild time coordinate, for example, since then the cross section $\Sigma(t) \cap H$ would not move forward along the horizon as t increased.) To each hypersurface $\Sigma(t)$ there corresponds

---

[6]This expression is inevitably not rigorous, if only because the theory of (interacting) quantum fields is itself not rigorous. A related comment is that, in a nonstationary background, there would be difficulties with the definition of $\hat{E}$, even in the case of a free field, as Adam Helfer pointed out to me after my lecture [20].

Chapter 9. The Statistical Mechanics of Black Hole Thermodynamics 187

a quantum Hilbert space in which the field operators residing on $\Sigma(t)$ are represented (in particular the operator $\widehat{E} = \widehat{E}(t)$ of eq. (9.6)), and in which the quantum state $\widehat{\rho}(t)$ can therefore be represented as a density operator. By identifying appropriately the hypersurfaces $\Sigma(t)$ with one another (the natural identification here being that induced by the Killing vector ξ^a), we identify their attached Hilbert spaces, and the dynamical change of $\widehat{\rho}(t)$ with time becomes thereby a motion of $\widehat{\rho}(t)$ within a single Hilbert space. (Notice that this evolution of $\widehat{\rho}$ is well defined because the boundary H at which we have truncated the hypersurfaces Σ is an *event horizon*, across which no information can propagate. This property of autonomous evolution would be lost if, say, H were replaced by a timelike surface, or if we tried to evolve $\widehat{\rho}(t)$ *backward* in time.)

Now the key feature of the state (9.6) for us is that it is time independent—as every equilibrium state must be by definition. What makes this so important is that given any well-defined (*but not necessarily unitary*) evolution law for density operators $\widehat{\rho}(t)$, and given any state $\widehat{\rho}^0$ that is stationary with respect to this evolution, we can find a function $f(\widehat{\rho}(t))$, defined for *arbitrary* states $\widehat{\rho}(t)$, which is nondecreasing with time. (More precisely, this is true classically for an arbitrary Markov process, and true quantum mechanically under a certain auxiliary technical assumption.) In effect, every state "wants to evolve toward the stationary one", and f is a kind of Lyapunov functional measuring how close $\widehat{\rho}(t)$ has come to $\widehat{\rho}^0$. Now, when the stationary state has the Gibbsian form (9.6), the quantity $f(\widehat{\rho})$ turns out to be (up to an additive constant)

$$S(\widehat{\rho}) - \beta \langle \widehat{E} \rangle = \operatorname{tr} \widehat{\rho} (\log \widehat{\rho}^{-1} - \beta \widehat{E}); \qquad (9.7)$$

that is, it turns out to be the free energy up to a factor of $-1/T$.

The proof of this little known result is so remarkably simple (at least classically) that I cannot resist presenting it here, in the important special case where $\beta = 0$. To begin with, note that the function $f(x) \equiv x \ln x^{-1}$ is concave downward since $f''(x) = -1/x < 0$ in the range $0 \leq x \leq 1$. Now consider a Markov process whose probability of being in the kth state at some moment of time is p_k. The quantity (9.7), when $\beta = 0$, is nothing but the entropy

$$S(p) = \sum_k p_k \log p_k^{-1} = \sum_k f(p_k),$$

so what we have to prove is that $S(p)$ increases when the p_k get replaced at some later time by $\sum_l T_{kl} p_l$, T_{kl} being the matrix of transition probabilities. But we have

$$S(p) \;\; = \;\; \sum_k f(p_k)$$

$$\to \sum_k f\left(\sum_l T_{kl} p_l\right)$$
$$\geq \sum_{kl} T_{kl} f(p_l)$$
$$= \sum_l f(p_l),$$

where to get the inequality we used the concavity of f together with the fact that

$$\sum_l T_{kl} = 1$$

because T preserves (when $\beta = 0$) the totally random state, $p_k \equiv 1$; and in the last step we used that

$$\sum_k T_{kl} = 1$$

(conservation of probability). When the stationary state p_k^0 is not uniform, the proof is almost as simple, only S gets replaced by

$$\sum_k p_k^0 f(p_k/p_k^0),$$

which for a thermal $p_k^0 = e^{-\beta E_k}$ is just

$$\sum p_k^0 \frac{p_k}{p_k^0} \log \frac{p_k^0}{p_k} = \sum p_k(-\beta E_k - \ln p_k) = S - \beta \langle E \rangle,$$

which is indeed the classical form of (9.7).

Returning to the black hole situation, it is now simple to see that the nondecreasing character of (9.7) entails that of the total entropy in our semiclassical, quasi-stationary approximation. To that end, consider the small change of state that occurs as the hypersurface Σ' moves slightly forward in time to Σ'', and write now S_{out} for the entropy $S(\widehat{\rho})$ of the exterior matter, in order to distinguish it from the entropy S_{BH} of the black hole itself. For the latter we have (from the "first law of black hole thermodynamics") that

$$dS_{\text{BH}} = \beta \, dM, \tag{9.8}$$

M being the mass of the hole. As part of the semiclassical approximation, we also assume that the mass of the black hole adjusts itself in response to the quantum mechanical *expectation value* of the energy it exchanges

Chapter 9. The Statistical Mechanics of Black Hole Thermodynamics 189

with the matter fields. Therefore (changing "$\widehat{E}$" to "$\widehat{E}_{\text{out}}$" for notational consistency) we have[7]

$$\langle \widehat{E}_{\text{out}} \rangle + M = \text{constant}, \tag{9.9}$$

whence $dM = -d\langle \widehat{E}_{\text{out}} \rangle$ and $dS_{\text{BH}} = -\beta \, d\langle \widehat{E}_{\text{out}} \rangle$. Putting this together with the fact that the expression (9.7) must not decrease in the process (and observing that β in that expression represents a *fixed* parameter of the stationary metric), we can write for the infinitesimal change $\Sigma' \to \Sigma''$,

$$\begin{aligned}
d(S_{\text{out}} - \beta \langle \widehat{E}_{\text{out}} \rangle) &\geq 0, \\
dS_{\text{out}} - \beta \, d\langle \widehat{E}_{\text{out}} \rangle &\geq 0, \\
dS_{\text{out}} + dS_{\text{BH}} &\geq 0, \\
d(S_{\text{out}} + S_{\text{BH}}) &\geq 0.
\end{aligned} \tag{9.10}$$

Thus we have proved the second law in the semiclassical approximation, for arbitrary processes in which the black hole geometry changes sufficiently slowly. This class of situations includes all that I know of for which thought experiments have been done to check the second law (see [4] for some of them); however, it does not cover black holes that are far from equilibrium, for example, ones just formed by the coalescence of two neutron stars and still in the process of settling down to a stationary state. Notice also that in this proof we are not *deriving* the value of the horizon entropy, but only showing that the second law holds *if* we use the value (9.1) provided by thermodynamic arguments.

Finally, it should be added that the matter entropy $S(\widehat{\rho})$ we have been working with is actually infinite, due to the entanglement between values of the quantum fields just inside and just outside the horizon. In a little while we will consider this near-horizon entanglement entropy as a possible source of the black hole entropy itself, but for now it is just a nuisance since the divergence in S_{out} means that (9.7) diverges as well. Thus, making our proof rigorous would require showing that *changes* in (9.7) are nevertheless well defined and conform to the temporal monotonicity we derived for that quantity. This probably could be done by introducing a high-frequency cutoff on the Hilbert space (using as high a frequency as needed in any

---

[7] One might worry that the two terms in this equation refer to two different energies, M to the total energy as defined at infinity and $\langle \widehat{E} \rangle$ to an energy defined with respect to the stationary background geometry. That this is not really a problem can be seen, for example, by imagining an infinitesimal mass m falling into the black hole from infinity. On one hand, this augments the mass M of the black hole by m; on the other hand, by conservation of the conserved energy current $T_b^a \xi^b$ in the stationary background, the exterior energy E_{out} decreases by the same amount m as the mass passes through the horizon. A more direct proof that (classically) $dS_{\text{BH}} = -\beta \, dE_{\text{out}}$ may be extracted from the derivation following eq. (13) in the first reference of [21].

given situation) and showing that the evolution of $\widehat{\rho}$ remained unaffected because the high-frequency modes remained unexcited.[8]

To what extent can we expect to generalize these considerations to the case of greatest interest, that of a fully dynamical, quantum black hole (or holes)? In one respect, the situation actually simplifies, because the technical hypothesis of complete positivity is no longer needed. The complication, of course, is that the continued validity of some of our other assumptions will depend on the structure of the quantum gravity theory in which the proof is to be realized. Without knowing that structure in advance we can do no better than to list the features that would be needed in order for the proof to go through. In writing down this list, I will assume that the hypersurfaces Σ to which the entropy is being referred can be specified by some generally covariant prescription that continues to make sense in quantum gravity, at least in some sufficiently great subset of the situations for which we would want to formulate a second law. (For example, we might take advantage of the fact that a "box" is needed in order for thermodynamic equilibrium to be possible and specify Σ as the boundary of the future of a freely chosen cross section of the box, regarded as a timelike "world tube" of fixed geometry. [More precisely, Σ would be that part of the boundary lying inside of the box and outside of any future horizons that were present.] Such a Σ would be "achronal", though not strictly spacelike, and the temporal relationship of two Σs specified in this way would follow directly from that of their cross sections.) With respect to some such family of hypersurfaces the needed assumptions are these:

- There is defined a Hilbert space $\mathcal{H}$, and for each surface Σ, an effective density operator $\widehat{\rho}(\Sigma)$ acting in $\mathcal{H}$.

- $\widehat{\rho}$ evolves autonomously in the sense that $\widehat{\rho}(\Sigma'')$ is determined by $\widehat{\rho}(\Sigma')$ whenever Σ'' lies to the future of Σ'.

- There exists an operator $\widehat{E}$ defined at the boundary of the system (or in any case without direct reference to the region inside the black hole) and yielding the total conserved energy.

---

[8]Alternatively, perhaps one could show that the additive constant, $-\log\mathrm{tr}\, e^{-\beta H}$, omitted from (9.7) canceled the divergence in a well-defined sense. In order to make the proof rigorous, one would also have, for example, to specify an observable algebra for the exterior fields and a representation of that algebra in which the operators $\widehat{\rho}$ and $\widehat{E}$ were well defined (which in particular might raise the issue of boundary conditions near the horizon). In addition, one would have to prove that the autonomous evolution of $\widehat{\rho}$ were "completely positive", this being a condition for the applicability of the theorem that guarantees the nondecreasing nature of (9.7) in the quantum case. This probably could be done by expressing the autonomous evolution as a unitary transformation followed by a tracing out of interior degrees of freedom. See [15] for references and some further discussion of these points.

- The operator $\widehat{\rho}_{E_0} = \theta(E_0 - \widehat{E})$ is preserved by the autonomous evolution.

- For all E_0, $\dim(\mathcal{H}_{\widehat{E}<E_0}) < \infty$, where $\mathcal{H}_{\widehat{E}<E_0}$ is the subspace of $\mathcal{H}$ with total energy less than E_0.

Depending on your expectations, hopes, or fears for quantum gravity, you may find different ones of these assumptions more or less plausible or palatable. For me all are at least plausible, albeit the entire framework of Hilbert space and operator observables is unlikely to be reproduced in an exact form in quantum gravity. In any case, if we accept these assumptions, then it follows easily that the quantity

$$S_{\text{tot}} \equiv \operatorname{tr} \widehat{\rho} \log \widehat{\rho}^{-1} \tag{9.11}$$

is a nondecreasing function of Σ. (The underlying mathematical theorem is proved in [22] and also in [15], where some further discussion of the above assumptions may be found as well.) This, then, will establish the second law (9.3) if we can show in addition that S_{tot} is a sum of two terms, one associated with the horizon and one with the exterior region.

Why is S a sum?

So why would $S_{\text{tot}} = \operatorname{tr} \widehat{\rho} \log \widehat{\rho}^{-1}$ take the form of the sum $A/l^2 + S_{\text{surroundings}}$ when $\widehat{\rho}$ describes the exterior state including all fields, both gravitational and nongravitational? For this to occur, there would need to be a great many degrees of freedom just outside or "contained in" the horizon (one might expect it to be thickened due to quantum effects), making their own identifiable contribution to S_{tot}. And of course this contribution would have to be proportional to the horizon area in order to agree with equation (9.1). Three ideas that have been proposed for what the horizon degrees of freedom might be are:

(a) the geometrical shape of the horizon itself,

(b) the modes of quantum fields propagating just outside the horizon,

(c) the fundamental degrees of freedom of the substratum,

where by "substratum", I mean the underlying structure(s) of whatever turns out to be the true theory of quantum gravity. Of course these possibilities are not necessarily exclusive of each other. All three types of contribution might be present, and they might also overlap significantly (for example, substratum degrees of freedom might show up in an effective description as geometrical variables describing the horizon shape). In concluding this article, I would like to say a few words about each of these possibilities and how they bear on the question of spacetime discreteness.

An attractive feature of the first proposal [14, 23] is that it offers a geometrical explanation for the very geometrical relationship (9.1). Unfortunately, it is not easy to estimate the magnitude of the quantum fluctuations in the horizon shape, but there are indications [21] that they become significant, not at the Planck scale l (as one might have expected), but at the much larger scale $\lambda_0 \sim (Ml^2)^{1/3}$. Such fluctuations would in effect spread the horizon into a shell of thickness λ_0 harboring a potentially unlimited source of entropy. In fact, the density of fluctuation modes diverges as their transverse wave number goes to infinity. This means that, without any cutoff, the entropy residing in the shape degrees of freedom would presumably be infinite. On the other hand, the very rapidity of the modes' growth rate also means that most of the modes with wavelength greater than any specified lower bound λ_{min} have $\lambda \sim \lambda_{min}$. In consequence, a result like (9.1), (9.2) emerges automatically (at least in crudely estimated order of magnitude) if one does assume a cutoff of magnitude $\lambda_{min} \sim l$.

The second proposal leads to very similar conclusions. Here, if we work in the approximation of a fixed background geometry, and if we limit ourselves to free quantum fields, then we can actually compute the entropy $S_{out} = -\operatorname{tr} \widehat{\rho} \log \widehat{\rho}$ of the field, and we obtain [14, 23] the result $S \sim A/\lambda_{min}^2$, where λ_{min} is the cutoff. In this case, we recognize the area law explicitly, and we see again that we must choose $\lambda_{min} \sim l$ in order to recover an entropy of the correct order of magnitude.

Unlike for proposal (a), where the effect of shape fluctuations on the density operator $\widehat{\rho}(\Sigma)$ is not completely clear, a contribution of type (b) must necessarily be present in the entropy (9.11). In conjunction with the type (a) contribution, this leads to an argument for the inevitability of spacetime discreteness: under the assumption of a true continuum, the type (b) entropy could fail to be infinite only if it provided its own cutoff by inducing, and thereby coupling to, horizon fluctuations of type (a), but then these would take over and provide an entropy that therefore would still be infinite.[9]

Another facet of proposal (b) which should be mentioned here is the apparent difficulty that, since each field makes its own contribution, the black hole entropy would seem to depend on the number of species of fundamental fields in nature. However, this conclusion is inevitable only if we ignore the coupling of field fluctuations to horizon shape and also hold fixed the value of the cutoff λ_{min}. But at fixed cutoff, not only S but also the Planck length (9.2) will be affected by the addition of a new species, and it appears [25] that this dependence is just what is needed to maintain the relationship (9.1) unchanged.

Of course the need for a cutoff bespeaks an underlying spacetime discreteness, and so, to my mind, the most intriguing possibilities of type

---

[9]For a partly different interpretation see [24].

(c) are those in which the substratum has a discrete character. In such a theory one could hope to derive the entropy, not just by counting discrete quantum states of physically continuous variables, but by counting certain discrete physical elements themselves (compare the fact that the entropy of a box of gas is, up to a logarithmic factor, just the number of molecules it contains). For example, in causal set theory, one might count the number of causal links crossing the horizon (near Σ), or in canonical quantum gravity in the loop representation, one might count the number of loops cut by the horizon (within Σ). In any such case, the result would be something like the horizon area in units set by the fundamental discreteness scale. Thus, to reduce the evaluation of the entropy (9.1) to a counting exercise of this sort would be to open up a direct path to learning the value of the fundamental length. And if that were to occur, then the quest for the statistical mechanics of black hole thermodynamics would certainly have led us to something of interest: it would have led us to the atoms of spacetime itself.

Acknowledgments

This research was partly supported by NSF grant PHY-9600620.

References

[1] R.D. Sorkin and D. Sudarsky, in preparation.

[2] R. Penrose, chap. 5 of this volume.

[3] F. Morgan, *Geometric Measure Theory: A Beginner's Guide*, 2d ed. (Academic Press, 1988), sec. 4.11. I thank G.J. Galloway for this information.

[4] R.M. Wald, chap. 8 of this volume, and references therein; Entropy and black hole thermodynamics, *Phys. Rev. D* **20**, 1271–1282 (1979).

[5] S.A. Hayward, General laws of black-hole dynamics, *Phys. Rev. D* **49**, 6467–6474 (1994).

[6] L. Bombelli, J. Lee, D. Meyer, and R.D. Sorkin, Spacetime as a causal set, *Phys. Rev. Lett.* **59**, 521–524 (1987).

[7] R.D. Sorkin, Stochastic evolution on a manifold of states, *Ann. Phys. (New York)* **168**, 119–147 (1986).

[8] C.J. Isham, Quantum logic and the histories approach to quantum theory, *J. Math. Phys.* **35**, 2157–2185 (1994); gr-qc/9308006.

[9] R.D. Sorkin, R.M. Wald, and Zh.-J. Zhang, Entropy of self-gravitating radiation, *Gen. Rel. Grav.* **13**, 1127–1146 (1981).

[10] G. 't Hooft, The black hole interpretation of string theory, *Nuclear. Phys. B* **335**, 138–154 (1990).

[11] S.W. Hawking, chap. 11 of this volume.

[12] S. Carlip, Statistical mechanics of the (2+1)-dimensional black hole, *Phys. Rev. D* **51**, 632–637 (1995); gr-qc/9409052.

[13] G. Horowitz, chap. 12 of this volume.

[14] R.D. Sorkin, On the entropy of the vacuum outside a horizon, in B. Bertotti, F. de Felice, and A. Pascolini (eds.), *Tenth International Conference on General Relativity and Gravitation*, held Padova, July 4–9, 1983, *Contributed Papers*, Vol. 2 (Consiglio Nazionale Delle Ricerche, 1983), 734–736.

[15] R.D. Sorkin, Toward an explanation of entropy increase in the presence of quantum black holes, *Phys. Rev. Lett.* **56**, 1885–1888 (1986).

[16] W.H. Zurek and K.S. Thorne, Statistical mechanical origin of the entropy of a rotating, charged black hole, *Phys. Rev. Lett.* **54**, 2175 (1985).

[17] R.D. Sorkin, On the meaning of the canonical ensemble, *Int. J. Theor. Phys.* **18**, 309–321 (1979).

[18] R.D. Sorkin, A simplified derivation of stimulated emission by black holes, *Class. Quant. Grav.* **4**, L149–L155 (1987).

[19] G 't Hooft, On the quantum structure of a black hole, *Nuclear Phys. B* **256**, 727–745 (1985).

[20] A.D. Helfer, The stress-energy operator, *Class. Quant. Grav.* **13**, L129–L134 (1996).

[21] R.D. Sorkin, Two topics concerning black holes: extremality of the energy, fractality of the horizon, in S.A. Fulling (ed.), *Proceedings of the Conference on Heat Kernel Techniques and Quantum Gravity*, held Winnipeg, Canada, August 1994, Discourses in Mathematics and Its Applications, No. 4 (University of Texas Press, 1995) 387–407, gr-qc/9508002; How wrinkled is the surface of a black hole? in D. Wiltshire (ed.), *Proceedings of the First Australasian Conference on General Relativity and Gravitation*, held February 1996, Adelaide, Australia (University of Adelaide, 1996), 163–174, gr-qc/9701056; A. Casher, F. Englert, N. Itshaki, S. Massar, and R. Parenti, Black hole horizon fluctuations, hep-th/9606106.

[22] Woo Ching-Hung, Linear stochastic motions of physical systems, University of California Preprint, UCRL-10431 (1962).

[23] L. Bombelli, R.K. Koul, J. Lee, and R.D. Sorkin, A quantum source of entropy for black holes, *Phys. Rev. D* **34**, 373–383 (1986).

[24] J.D. Bekenstein, Do we understand black hole entropy? in R.T. Jantzen, G. Mac Keiser, and R. Ruffini (eds.) *The Seventh Marcel Grossmann Meeting on Recent Developments in Theoretical and Experimental General Relativity, Gravitation and Relativistic Field Theories*, proceedings of the MG7 meeting, held Stanford, July 24–30, 1994 (World Scientific, 1996); gr-qc/9409015.

[25] L. Susskind and J. Uglum, Black hole entropy in canonical quantum gravity and superstring theory, *Phys. Rev. D* **50**, 2711 (1994), hep-th/9401070; T. Jacobson, Black hole entropy and induced gravity, gr-qc/9404039.

10

Generalized Quantum Theory in Evaporating Black Hole Spacetimes

James B. Hartle

Abstract

Quantum mechanics for matter fields moving in an evaporating black hole spacetime is formulated in fully four-dimensional form according to the principles of generalized quantum theory. The resulting quantum theory cannot be expressed in a 3+1 form in terms of a state evolving unitarily or by reduction through a foliating family of spacelike surfaces. That is because evaporating black hole geometries cannot be foliated by a nonsingular family of spacelike surfaces. A four-dimensional notion of information is reviewed. Although complete information may not be available on every spacelike surface, information is not lost in a spacetime sense in an evaporating black hole spacetime. Rather, complete information is distributed about the four-dimensional spacetime. Black hole evaporation is thus not in conflict with the principles of quantum mechanics when suitably generally stated.

10.1 Introduction

The early 1980s were a memorable time to be at Chicago when Chandra was writing *The Mathematical Theory of Black Holes* [1]. His method was that of an explorer in many ways—voyaging through the complex landscape of equations that the classical theory of black holes presents—discovering novel perspectives, relationships, and hidden symmetries. We discussed these many times in long walks near the lakefront in Hyde Park on often very cold Sunday afternoons. It is therefore a special pleasure for me to contribute to this commemoration of Chandra's work

In the prologue to the *Mathematical Theory*, Chandra sums up his views on black holes in a sentence: "The black holes of nature are the most perfect macroscopic objects there are in the universe: the only elements in their construction are our concepts of space and time." I note that Chandra used the word "macroscopic" to qualify the black holes that exhibit perfection. That was either prudent or prescient, for now in theoretical physics we are engaged with the question of whether black holes are or are not a blot on the perfection of quantum theory—the organizing principle of *microscopic* physics. This essay puts forward the thesis that black hole evaporation is not inconsistent with the principles of quantum mechanics provided those principles are suitably generalized from their usual flat space form.

10.2 Black holes and quantum theory

10.2.1 Fixed background spacetimes

The usual formulations of quantum theory rely on a fixed, globally hyperbolic[1] background spacetime geometry as illustrated in figure 10.1. In these usual formulations, complete information about a physical system is available on any spacelike (Cauchy) surface and is summarized by a state vector associated with that surface. This state vector evolves through a foliating family of spacelike surfaces $\{\sigma\}$, either unitarily between spacelike surfaces

$$|\Psi(\sigma'')\rangle = U|\Psi(\sigma')\rangle, \qquad (10.1)$$

or by reduction on them

$$|\Psi(\sigma)\rangle \to \frac{P|\Psi(\sigma)\rangle}{||P|\Psi(\sigma)\rangle||}. \qquad (10.2)$$

Usual formulations of quantum theory thus depend crucially on a background spacetime exhibiting nonsingular foliations by spacelike surfaces to define the notion of quantum state and its evolution.

Nowhere does the close connection between spacetime structure and quantum theory emerge so strikingly as in the process of black hole evaporation. Black holes and quantum theory have been inextricably linked since Hawking's 1974 [3] discovery of the tunneling radiation from black holes that bears his name. That radiation when analyzed in the approximation that its back reaction on the black hole can be neglected requires nothing new of quantum theory.

The familiar story is summarized in the Penrose diagram in figure 10.2. The entire region of spacetime outside the horizon is foliable by a nonsingular family of spacelike surfaces. States of matter fields defined on a

---

[1] A spacetime is *globally hyperbolic* if it admits a surface Σ, no two points of which can be connected by a timelike curve, but such that every inextendable timelike curve in the spacetime intersects Σ. Such a surface is called a *Cauchy surface*. Classically, data on a Cauchy surface determine the entire future and past evolution of the spacetime. Globally hyperbolic spacetimes have topology $\mathbf{R} \times \Sigma$ and can be foliated by a one-parameter family of Cauchy surfaces defining a notion of time. For more details see [2].

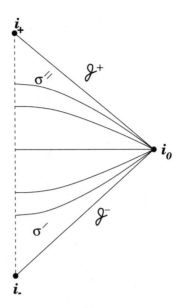

Figure 10.1: The Penrose diagram for a globally hyperbolic spacetime like flat Minkowski space.[3] A globally hyperbolic spacetime may be smoothly foliated by a family of spacelike surfaces a few of which are shown. When a spacetime can be so foliated quantum evolution can be described by a state vector that evolves unitarily between surfaces or by reduction on them.

spacelike, precollapse surface σ' in the far past evolve unitarily to states on later spacelike surfaces like σ''. The complete information represented by the state is available on any spacelike surface. The evolution defines a pure state $|\Psi(H, \mathcal{I}^+)\rangle$ on the surface $H \cup \mathcal{I}^+$ consisting of the horizon and future null infinity. The Hilbert space of states is the product $\mathcal{H}(H) \otimes \mathcal{H}(\mathcal{I}^+)$ of field states on the horizon and on future null infinity. Measurements of observers at infinity probe only $\mathcal{H}(\mathcal{I}^+)$. The probabilities of their outcomes may therefore be predicted from the density matrix

$$\rho(\mathcal{I}^+) = \text{tr}_H \left[|\Psi(H, \mathcal{I}^+)\rangle \langle \Psi(H, \mathcal{I}^+)| \right] \qquad (10.3)$$

that results from tracing over all the degrees of freedom on the horizon. For

---

[3]Spacetime has been rescaled so that it is contained in the interior of the triangle and so that radial light rays move on 45° lines. The dotted line is radius = 0. The remaining boundaries are the various parts of infinity. Future null infinity, $\mathcal{I}^+$, consists of the endpoints of light rays that escape to infinity. The points i_0, i_+, and i_- are respectively spacelike infinity—where spacelike surfaces that reach infinity end—and future and past timelike infinity, which are the endpoints of timelike curves. These conventions are used in the other figures in this paper. For more on the construction of Penrose diagrams see [2].

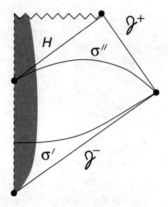

Figure 10.2: The Penrose diagram for a black hole when the back reaction of the Hawking radiation is neglected. The shaded region in this diagram represents spherically symmetric matter collapsing to form a black hole with horizon H and eventually a spacelike singularity represented by the jagged horizontal line. Light rays emitted from a point inside the horizon move on 45° lines, end at the singularity, and never escape to infinity. A state of a field in this background on an initial surface σ' can evolve unitarily through an interpolating family of spacelike surfaces to a surface like σ''. By pushing σ'' forward the state can be evolved unitarily to the boundary of the region exterior to the black hole consisting of $\mathcal{I}^+$ and the horizon H. The state $|\Psi(\mathcal{I}^+, H)\rangle$ on $\mathcal{I}^+ \cup H$ exhibits correlations (entanglements) between $\mathcal{I}^+$ and H. Thus complete information cannot be recovered far from the black hole on $\mathcal{I}^+$. Indeed, for late times the predictions of $|\Psi(\mathcal{I}_+, H)\rangle$ for observations on $\mathcal{I}^+$ are indistinguishable from those of a thermal density matrix at the Hawking temperature. This does not mean that information is lost in quantum evolution; it is fully recoverable on the whole surface $\mathcal{I}^+ \cup H$.

Chapter 10. Generalized Quantum Theory ...

observations at late times this represents disordered, thermal radiation.

Of course, information is missing from $\rho(\mathcal{I}^+)$ so that complete information is not available on $\mathcal{I}^+$. Specifically, the information in $|\Psi(H, \mathcal{I}^+)\rangle$ concerning correlations between observables on H and $\mathcal{I}^+$ is missing [4, 5]. However, it is only necessary to consider observables on both H and $\mathcal{I}^+$ to recover it. Usual quantum mechanics with its notion of a state carrying complete information evolving unitarily through families of spacelike surfaces is adequate to discuss the Hawking radiation when its back reaction is neglected.

Only in the complete evaporation of a black hole does one find a hint that the usual framework may need to be modified. Spacetime geometries representing a process in which a black hole forms and evaporates completely have a causal structure summarized by the Penrose diagram in figure 10.3. Let us consider for a moment the problem of quantum mechanics of matter fields in a fixed geometry with this causal structure. (We shall return later to the fluctuations in spacetime geometry that must occur in a quantum theory of gravity that is necessary to fully describe the evaporation process.) Any pure initial state $|\Psi(\sigma')\rangle$ leads to a state of disordered radiation on a hypersurface A after the black hole has evaporated, so we have

$$|\Psi(\sigma')\rangle\langle\Psi(\sigma')| \to \rho(A) \ , \qquad (10.4)$$

where $\rho(A)$ is the mixed density matrix describing the radiation. This cannot be achieved by unitary evolution. Indeed, it is difficult to conceive of *any* law for the evolution of a density matrix $\rho(\sigma)$ through a family of spacelike surfaces that would result in (10.4) because there is no nonsingular family of spacelike surfaces that interpolates between σ' and A. Even classically there is no well defined notion of evolution of initial data on σ' to the surface A because of the naked singularity N.

I shall not attempt to review the discussion this situation has provoked.[4] What is clear is that a generalization of quantum mechanics is needed for a quantum theory of fields in geometries such as this.

Evaporating black hole geometries are not the only backgrounds whose causal structure requires a generalization of usual quantum mechanics for a quantum theory of matter fields. Consider a spacetime with a compact region of closed timelike curves such as occur in certain wormhole geometries [7]. (See fig. 10.4.) Given a state on a spacelike surface σ' before the nonchronal region of closed timelike curves, how do we calculate the probabilities of field measurements inside the nonchronal region or indeed on any spacelike surface like σ'' such that the nonchronal region is contained between it and σ'? Certainly not by evolving a state or density matrix through an interpolating family of spacelike surfaces, whether by a unitary or nonunitary rule of evolution. No such foliating family of spacelike

---
[4]For reviews of this discussion see [6].

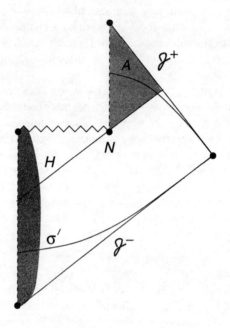

Figure 10.3: The Penrose diagram for an evaporating black hole spacetime. The black hole is assumed here to evaporate completely, giving rise to a naked singularity N, and leaving behind a nearly flat spacetime region (lightly shaded above). A spacelike surface like A is complete and data on this surface completely determine the evolution of fields to its future. Yet complete information about a quantum matter field moving in this spacetime is not available on A. Even if the initial state of the matter field on a surface like σ' is pure, the state of the disordered Hawking radiation on A would be represented by a density matrix. A pure state cannot evolve unitarily into a density matrix, so the usual formulation of quantum evolution in terms of states evolving through a foliating family of spacelike surfaces breaks down. The geometry of evaporating black hole spacetimes suggests why. There is no smooth family of spacelike surfaces interpolating between σ' and A and even classically there is not a well defined notion of evolution of initial data on σ' to A. The usual notion of quantum evolution must therefore be generalized to apply in spacetime geometries such as this.

Chapter 10. Generalized Quantum Theory ...

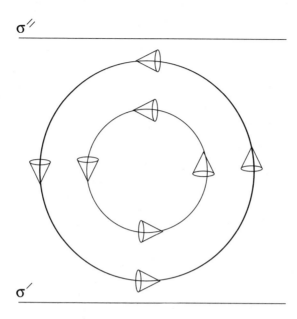

Figure 10.4: A spacetime with a compact region of closed timelike curves (CTCs). As a consequence of the CTCs there is no family of spacelike surfaces connecting an initial spacelike surface σ' with a final one σ'', both outside the CTC region. The quantum evolution of a matter field therefore cannot be described by a state evolving through a foliating family of spacelike surfaces. The notion of quantum evolution must be generalized to apply to spacetimes such as this.

surfaces exists! A generalization of usual quantum mechanics is required.

Spacetimes exhibiting spatial topology change, as in the "trousers" spacetime of figure 10.5, are another class of backgrounds for which quantum field theory requires a generalization of usual quantum mechanics. Given an initial state on a spacelike surface σ', one could think of calculating the probabilities of alternatives on spacelike surfaces A or B. But because such spacetimes are necessarily singular [8], there is no smooth family of surfaces interpolating between σ' and A and B. A generalization of usual quantum dynamics is again required.[5]

10.2.2 Quantum gravity

The evaporating black hole spacetimes illustrated in figure 10.3 are singular. Spacetimes that are initially free from closed timelike curves but evolve them later must be singular or violate a positive energy condition [11]. Spatial topology change implies either a singularity or closed timelike

---

[5]Field theory in such spacetimes has been studied by a number of authors [9, 10].

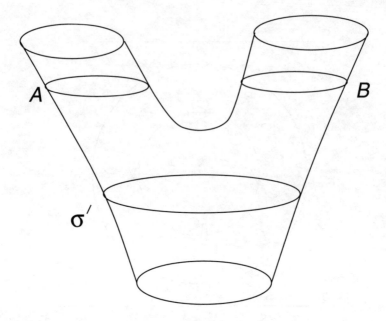

Figure 10.5: A spacetime with a simple change in spatial topology. There is no nonsingular family of spacelike surfaces between an "initial" spacelike surface σ' and "final" spacelike surfaces A and B. The usual notion of quantum evolution must therefore be generalized to apply to spacetimes such as this. Complete information about the initial state is plausibly not available on spacelike surfaces A and B separately, but only surfaces A and B together.

curves [8]. These pathologies suggest a breakdown in a purely classical description of spacetime geometry. One might therefore hope that the difficulties with usual formulations of quantum theory in such backgrounds could be resolved in a quantum theory of gravity. The recent successful calculation of black hole entropy in string theory [12] raises several questions related to this hope. First, there is the question of whether there are general principles mandating a connection between the entropy and the logarithm of a number of states in any sensible quantum theory of gravity. More important for the present discussion, however, is the question of whether these calculations mean that black hole evaporation can be described within usual quantum mechanics in string theory. It is possible that string theory will yield a unitary S-matrix between asymptotic precollapse and evaporated states. However, as with any quantum theory of gravity, the need for a generalization of the idea of unitary evolution of states through spacelike surfaces is only more acute than it is for field theory in fixed non-globally-hyperbolic spacetimes for the following reason:

We have seen how the usual quantum mechanics of fields with evolution

Chapter 10. Generalized Quantum Theory ...

defined through states on spacelike surfaces relies heavily on a fixed, globally hyperbolic, background geometry to define those surfaces. But in any quantum theory of gravity spacetime geometry is not fixed. Geometry is a quantum variable—generally fluctuating and without definite value. Quantum dynamics cannot be defined by a state evolving in a given spacetime; no spacetime is given. A generalization of usual quantum mechanics is thus needed. This need for generalization becomes even clearer if one accepts the hint from string theory that spacetime geometry is not fundamental.

In the rest of this paper we shall discuss some generalizations of quantum theory that are applicable to the process of black hole evaporation. We shall discuss these primarily for the case of field theory in a fixed background evaporating black hole spacetime. There we can hope to achieve a concreteness not yet available in quantum theories of gravity. However, there is every reason to believe that the principles of the main generalization we shall describe are implementable in quantum gravity as well [13].

10.3 The $-matrix

Hawking [14] suggested one way that the principles of quantum mechanics could be generalized to apply to an evaporating black hole. For transitions between asymptotic states in the far past and far future employ, not a unitary S-matrix mapping initial states to final states, but rather a $-matrix mapping initial density matrices to final density matrices:

$$\rho^f = \$\rho^i . \tag{10.5}$$

That way an initial pure state could evolve into a mixed density matrix as the evaporation scenario represented in figure 10.3 suggests.

Hawking gave a specific prescription for calculating the $-matrix: Use Euclidean sums over histories to calculate Euclidean Green's functions for the metric and matter fields. Continue these back to Lorentzian signature in asymptotic regions where the continuation is well defined because the geometry is flat. Extract the $-matrix elements from these Green's functions as one would for the product of two S-matrices. When the topology of the Euclidean geometries is trivial, the resulting $-matrix factors into the product of two S-matrices and (10.5) implies that pure states evolve unitarily into pure states. However, the $-matrix generally does not factor if the topology of the Euclidean geometries is nontrivial, and then pure states can evolve into mixed states. It remains an open question whether this prescription in fact yields a $-matrix. It is not evident, in particular, whether the mapping that results preserves the positivity and trace of the density matrices on which it acts.

However constructed, a $-matrix connects only asymptotic initial and final density matrices. A full generalization of quantum theory would predict the probabilities of nonasymptotic observables "inside" the spacetime.

To this end, various interpolating equations that will evolve a density matrix through a family of spacelike surfaces have been investigated [15, 16]. These typically have the form

$$i\hbar \frac{\partial \rho}{\partial t} = [H, \rho] + \begin{pmatrix} \text{additional terms} \\ \text{linear in } \rho \end{pmatrix}, \qquad (10.6)$$

where additional terms give rise to nonunitary evolution. Perhaps these could be adjusted to yield the $-matrix of Hawking's construction when evolved between the far past and far future [17]. Such equations display serious problems with conservation of energy and charge when the additional terms in (10.6) are local, although Unruh and Wald [16] have demonstrated that only a little nonlocality is enough to suppress this difficulty at energies below the Planck scale.

From the perspective of this paper, conservation or the lack of it is not the main problem with such modifications of the quantum evolutionary law. Rather, it is that the "t" in the equation is not defined. As we argued in section 10.2, we do not expect to have a nonsingular, foliating family of spacelike surfaces in evaporating black hole spacetimes through which to evolve an equation like (10.6). A generalization of quantum mechanics even beyond such equations seems still to be required.

10.4 Think four-dimensionally

In the previous section we argued that quantum theory needs to be generalized to apply to physical situations such as black hole evaporation in which quantum fluctuations in the geometry of spacetime can be expected, or situations such as field theory in evaporating black hole backgrounds where geometry is fixed but not globally hyperbolic. What features of usual quantum mechanics must be given up in order to achieve this generalization? In this section, we argue that one feature to be jettisoned is the notion of a state on spacelike surface and quantum evolution described in terms of the change in such states from one spacelike surface to another.

The basic argument for giving up on evolution by states through a foliating family of spacelike surfaces in spacetime has already been given: when geometry is not fixed, or even when fixed but without an appropriate causal structure, a foliating family of spacelike surfaces is not available to define states and their evolution.

This basic point is already evident classically. There is a fully four-dimensional description of any spacetime geometry in terms of four-dimensional manifold, metric, and field configurations. However, for globally hyperbolic geometries that four-dimensional information can be compressed into initial data on a spacelike surface. That initial data is the classical notion of state. By writing the Einstein equation in 3+1 form the four-dimensional description can be recovered by evolving the state through a family of spacelike surfaces. However, such compression is not possible in spacetimes like the evaporating black hole spacetime illustrated in figure 10.3. Only a four-dimensional description is possible.

Similarly there is a fully four-dimensional formulation of quantum field theory in background spacetime geometries in terms of Feynman's sum over field histories. Transition amplitudes between spacelike surfaces are specified directly from the four-dimensional action S by sums over field histories of the form

$$\sum_{\text{histories}} \exp[iS(\text{history})/\hbar] \,. \tag{10.7}$$

When the background geometry is globally hyperbolic, these transition amplitudes between spacelike surfaces can be equivalently calculated by evolving a quantum state through an interpolating family of spacelike surfaces. However, if the geometry is not globally hyperbolic, we cannot expect such a 3+1 formulation of quantum dynamics any more than we can for the classical theory. Following Feynman, however, we expect a fully four-dimensional spacetime formulation of quantum theory to supply the necessary generalization applicable to field theory in evaporating black hole spacetimes and to the other examples we have mentioned.[6] We shall describe this generalization and its consequences in this article.[7] Our motto is "Think four-dimensionally." When we do there is no necessary conflict between quantum mechanics and black hole evaporation.

10.5 Generalized Quantum Theory

To generalize usual quantum theory it is just as necessary to decide which features of the usual framework to retain as it is to decide which to discard. Generalized quantum theory is a comprehensive framework incorporating the essential features of a broad class of generalizations of the usual theory.[8] The full apparatus of generalized quantum theory is not essential to reach the conclusion that black hole evaporation is consistent with the principles of quantum mechanics sufficiently generally stated. Indeed, similar conclusions have been reached without invoking these general principles (see

---

[6]There are other ideas for the necessary generalization. Wald has reached similar conclusions in the algebraic approach to field theory in curved spacetime [18]. The most developed is the idea from canonical quantum gravity that states evolve, not through surfaces in spacetime, but rather through surfaces in the superspace of possible three-dimensional geometries. For lucid and, by now, classic reviews of the various aspects and difficulties with this approach see [19]. These ideas are not obviously applicable to the cases of fixed spacetime geometry discussed in this paper. For another that might be see [20], where states are defined in local regions later patched together.

[7]The application of sum-over-histories methods to gravity has a considerable history that cannot be recapitulated here. Some of the more notable early references are [21].

[8]For expositions see e.g. [22, 13].

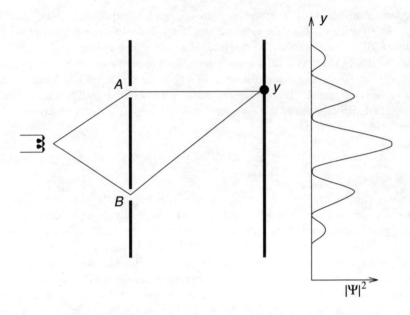

Figure 10.6: The two-slit experiment. An electron gun at left emits an electron traveling towards a screen with two slits and then on to detection at a further screen. Two histories are possible for electrons detected at a particular point on the right-hand screen defined by whether they went through slit A or slit B. Probabilities cannot be consistently assigned to this set of two alternative histories because they interfere quantum mechanically.

e.g. [18]). What is necessary is a generalization of the usual notion of unitary evolution. However, generalized quantum theory is a useful setting in which to consider generalizations because it provides basic principles and a framework for comparing different generalizations. In this section we give a brief and informal exposition of these principles.

The most general objective of a quantum theory is to predict the probabilities of the individual members of a set of alternative histories of a closed system, most generally the universe. The set of alternative orbits of the earth around the sun is an example. These are sequences of positions of the earth's center of mass at a series of times. Another example is the set of four-dimensional histories of matter fields in an evaporating black hole spacetime.

The characteristic feature of a quantum theory is that not every set of alternative histories can be consistently assigned probabilities because of quantum mechanical interference. This is clearly illustrated in the famous two-slit experiment shown in figure 10.6. There are two possible histories for an electron which proceeds from the source to a point y at the detecting screen. They are defined by which of the two slits (A or B) it passes through. It is not possible to assign probabilities to these two histories. It would be inconsistent to do so because the probability of arriving at y is not the sum of the probabilities of arriving at y *via* the two possible histories:

$$p(y) \neq p_A(y) + p_B(y) \,. \tag{10.8}$$

That is because in quantum mechanics probabilities are squares of amplitudes and

$$|\psi_A(y) + \psi_B(y)|^2 \neq |\psi_A(y)|^2 + |\psi_B(y)|^2 \tag{10.9}$$

when there is interference.

In a quantum theory a rule is needed to specify which sets of alternative histories may be assigned probabilities and which may not. The rule in usual quantum mechanics is that probabilities can be assigned to the histories of the outcomes of *measurements* and not in general otherwise. Interference between histories is destroyed by the measurement process, and probabilities may be consistently assigned. However, this rule is too special to apply in the most general situations and certainly insufficiently general for cosmology. Measurements and observers, for example, were not present in the early universe when we would like to assign probabilities to histories of density fluctuations in matter fields or the evolution of spacetime geometry.

The quantum mechanics of closed systems[9] relies on a more general rule whose essential idea is easily stated: A closed system has some initial quantum state $|\Psi\rangle$. Probabilities can be assigned to just those sets of histories for which there is vanishing interference between individual histories as a consequence of the state $|\Psi\rangle$ the system is in. Such sets of histories are said to *decohere*. Histories of measurements decohere as a consequence of the interaction between the apparatus and measured subsystem. Decoherence thus contains the rule of usual quantum mechanics as a special case. But decoherence is more general. It permits assignment of probabilities to alternative orbits of the moon or alternative histories of density fluctuations in the early universe when the initial state is such that these alternatives decohere whether or not the moon or the density fluctuations are receiving the attention of observers making measurements.

The central element in a quantum theory based on this rule is the measure of interference between the individual histories c_α, $\alpha = 1, 2, \ldots$, in a set of alternative histories. This measure is called the *decoherence functional*, $D(\alpha', \alpha)$. A set of histories decoheres when $D(\alpha', \alpha) \approx 0$ for all pairs (α', α) of distinct histories.

The decoherence functional for usual quantum mechanics is defined as follows: A set of alternative histories can be specified by giving sequences

---

[9]See e.g. [23] for an elementary review and references to earlier literature.

of sets of yes/no alternatives at a series of times $t_1, \ldots, t_n$. For example, alternative orbits may be specified by saying whether a particle is or is not in certain position ranges at a series of times. Each yes/no alternative in an exhaustive set of exclusive alternatives at time t_k is represented by a Heisenberg picture projection operator $P^k_{\alpha_k}(t_k)$, $\alpha_k = 1, 2, \ldots$, where α_k labels the different alternatives in the set. An individual history α is a particular sequence of alternatives $\alpha \equiv (\alpha_1, \ldots, \alpha_n)$ and is represented by the corresponding chain of Heisenberg picture projections

$$C_\alpha = P^n_{\alpha_n}(t_n) \cdots P^1_{\alpha_1}(t_1) \,. \tag{10.10}$$

If the initial state vector of the closed system is $|\Psi\rangle$, (nonnormalized) branch state vectors corresponding to the individual histories α may be defined by

$$C_\alpha |\Psi\rangle \,, \tag{10.11}$$

and the decoherence functional for usual quantum mechanics is

$$D(\alpha', \alpha) = \langle \Psi | C^\dagger_\alpha C_{\alpha'} | \Psi \rangle \,. \tag{10.12}$$

This is the usual measure of interference between different histories represented in the form (10.11).

The essential properties of a decoherence functional that are necessary for quantum mechanics may be characterized abstractly [22], and (10.12) is only one of many other ways of satisfying these properties. Therein lie the possibilities for generalizations of usual quantum mechanics.

10.6 Spacetime generalized quantum mechanics

Feynman's sum-over-histories ideas may be used with the concepts of generalized quantum theory to construct a fully four-dimensional formulation of the quantum mechanics of a matter field $\phi(x)$ in a fixed background spacetime. We shall sketch this spacetime quantum mechanics in what follows. We take the dynamics of the field to be summarized by an action $S[\phi(x)]$ and denote the initial state of the closed field system by $|\Psi\rangle$.

The basic (fine-grained) histories are the alternative, four-dimensional field configurations on the spacetime. These may be restricted to satisfy physically appropriate conditions at infinity and at the singularities. Sets of alternative (course-grained) histories to which the theory assigns probabilities if decoherent are partitions of these field configurations into exclusive classes $\{c_\alpha\}, \alpha = 1, 2, \ldots$. For example, the alternative that the field configuration on a spacelike surface σ has the value $\chi(\mathbf{x})$ corresponds to the class of four-dimensional field configurations which take this value on σ. In flat spacetime the probability of this alternative is the probability for the initial state $|\Psi\rangle$ to evolve to a state $|\chi(\mathbf{x}), \sigma\rangle$ of definite field on σ. The history where the field takes the value $\chi'(\mathbf{x})$ on surface σ' and $\chi''(\mathbf{x})$ on a later surface σ'' corresponds to the class of four-dimensional field configurations which take these values on the respective surfaces, and so on.

The examples we have just given correspond to the usual quantum mechanical notion of alternatives at a definite moment of time or a sequence of such moments. However, more general partitions of four-dimensional field configurations are possible which are not at any definite moment of time or series of such moments. For example, the four-dimensional field configurations could be partitioned by ranges of values of their averages over a region extending over both space and time. Partition of four-dimensional histories into exclusive classes is thus a fully four-dimensional notion of alternative for quantum mechanics.

Branch state vectors corresponding to individual classes c_α in a partition of the fine-grained field configurations $\phi(x)$ can be constructed from the sum over fields in the class c_α. We write schematically

$$C_\alpha|\Psi\rangle = \int_{c_\alpha} \delta\phi \exp\Big(iS[\phi(x)]/\hbar\Big)|\Psi\rangle \ . \tag{10.13}$$

It is fair to say that the definition of such integrals has been little studied in interesting background spacetimes, but we proceed assuming a careful definition can be given even in singular spacetimes such as those discussed in section 10.2. Even then some discussion is needed to explain what (10.13) means as a formal expression. In a globally hyperbolic spacetime we can define an operator C_α corresponding to the class of histories c_α by specifying the matrix elements

$$\langle\chi''(\mathbf{x}), \sigma''|C_\alpha|\chi'(\mathbf{x}), \sigma'\rangle = \int_{[\chi' c_\alpha \chi'']} \delta\phi \exp\Big(iS[\phi(x)]/\hbar\Big) \ . \tag{10.14}$$

The sum is over all fields in the class c_α that match $\chi'(\mathbf{x})$ and $\chi''(\mathbf{x})$ on the surfaces σ' and σ'' respectively. This operator can act on $|\Psi\rangle$ by taking the inner product with its field representative $\langle\chi'(\mathbf{x}), \sigma'|\Psi\rangle$ on a spacelike surface far in the past. By pushing σ'' forward to late times we arrive at the definition of $C_\alpha|\Psi\rangle$. The same procedure could be used to define branch state vectors in spacetimes with closed timelike curves (fig. 10.4), in spacetimes with spatial topology change (fig. 10.5), and in evaporating black hole spacetimes (fig. 10.3). The only novelty in the latter two cases is that $C_\alpha|\Psi\rangle$ lives on the product of two Hilbert spaces. There are the Hilbert spaces on the two legs of the trousers in the spatial topology change case and, in the black hole case, there are the Hilbert space of states inside the horizon and the Hilbert space of states on late time surfaces after the black hole has evaporated.

The decoherence functional is then

$$D(\alpha', \alpha) = \mathcal{N}\langle\Psi|C_\alpha^\dagger C_{\alpha'}|\Psi\rangle \tag{10.15}$$

	Usual Quantum Mechanics	Generalized Quantum Mechanics
Dynamics	$e^{-iHt}|\Psi\rangle$, $P|\Psi\rangle/\||P|\Psi\rangle\|$	$\sum_{\substack{\text{histories}\\ \in c_\alpha}} e^{iS(\text{history})/\hbar}|\Psi\rangle$
Alternatives	On spacelike surfaces or sequences of surfaces	Arbitrary partitions of fine-grained histories
Probabilities assigned to	Histories of measurement outcomes	Decohering sets of histories

Table 10.1: Usual and generalized quantum mechanics compared.

where $\mathcal{N}$ is a constant to ensure the normalization condition

$$\Sigma_{\alpha\alpha'} D(\alpha', \alpha) = 1 \ . \tag{10.16}$$

A set of alternative histories decoheres when the off-diagonal elements of $D(\alpha', \alpha)$ are negligible. The probabilities of the individual histories are

$$p(\alpha) = D(\alpha, \alpha) = \mathcal{N} \|C_\alpha|\Psi\rangle\|^2 \ . \tag{10.17}$$

There is no issue of "conservation of probability" for these $p(\alpha)$; they are not defined in terms of an evolving state vector. As a consequence of decoherence, the $p(\alpha)$ defined by (10.17) obey the most general probability sum rules including, for instance, the elementary normalization condition $\sum_\alpha p(\alpha) = 1$ which follows from (10.16).

This spacetime generalized quantum theory is only a modest generalization of usual quantum mechanics in globally hyperbolic backgrounds as table 10.1 shows. The two laws of evolution (eqs. (10.1) and (10.2)) have been unified in a single sum-over-histories expression. The alternatives potentially assigned probabilities have been generalized to include ones that extend in time and are not simply the outcomes of a measurement process. These generalizations put the theory in fully four-dimensional form.

When the more general framework is restricted to globally hyperbolic backgrounds, to histories of alternatives on spacelike surfaces, and to the outcomes of measurements, usual quantum theory is recovered as a special case. In particular, one recovers the notion of a state evolving either unitarily or by reduction through a foliating family of spacelike surfaces. To see how this works consider the globally hyperbolic spacetime shown in figure 10.7 and a single alternative where the field is restricted to some set F of spatial field configurations on a spacelike surface σ_1. The class operator C_α is defined through (10.13) by integrating $\exp(iS[\phi(x)]/\hbar)$ over *all* field configurations between $\mathcal{I}^-$ and $\mathcal{I}^+$ restricted to F on σ_1. We can

Chapter 10. Generalized Quantum Theory ...

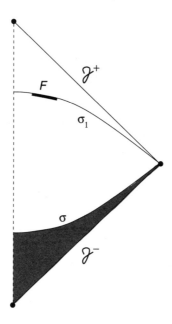

Figure 10.7: Recovery of the unitary evolution of states through spacelike surfaces from sum-over-histories quantum mechanics in a globally hyperbolic spacetime. A sum over field configurations between $\mathcal{I}^-$ and a spacelike surface σ defines a state on that spacelike surface. This sum over fields specifies the evolution of this state as σ is pushed forward in time. In a globally hyperbolic spacetime this evolution is unitary until a surface like σ_1 is encountered where there is a restriction on the spatial field configurations to a class F. This restriction is equivalent to the action of a projection on the state. Thus the two standard laws of quantum evolution in a "3 + 1" formulation are recovered from the more general four-dimensional framework. That recovery, however, requires a spacetime geometry that is smoothly foliable by spacelike surfaces.

do the integral first only over fields between $\mathcal{I}^-$ and a spacelike surface σ, and then push σ to $\mathcal{I}^+$ (see fig. 10.7). The integral over fields between $\mathcal{I}^-$ and a spacelike surface σ on which a given spatial field configuration $\chi(\mathbf{x})$ is specified defines the field representative $\langle \chi(\mathbf{x})|\Psi(\sigma)\rangle$ of a state $|\Psi(\sigma)\rangle$ on σ. As σ is pushed forward this integral defines the evolution of $|\Psi(\sigma)\rangle$. As Feynman showed, this evolution is unitary between infinitesimally separated surfaces. As a consequence $|\Psi(\sigma)\rangle$ evolves unitarily up to σ_1. There, the restriction of the sum over fields to a set F is equivalent to the projection on F acting on $|\Psi(\sigma)\rangle$. The state is reduced as in (10.2). Thus, a state evolving unitarily or by reduction is recovered from the more general sum over histories formulation in globally hyperbolic spacetimes.

In an evaporating black hole spacetime (fig. 10.8), the probabilities

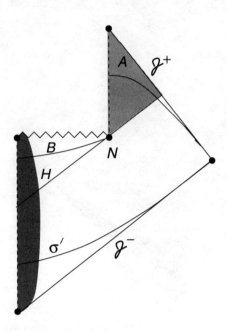

Figure 10.8: Quantum evolution in an evaporating black hole spacetime can be described four-dimensionally using a Feynman sum-over-histories. However, that evolution is not expressible in 3+1 terms by the smooth evolution of a state through spacelike surfaces. Complete information is not recoverable on surface A because of correlations (entanglements) between the field on A and the field on a surface like B. Even though information is not necessarily available on any one spacelike surface, it is not lost in an evaporating black hole geometry, but is distributed over four-dimensional spacetime.

for evolution—for decoherent alternatives on A given an initial state on σ', for instance—are generally defined four-dimensionally through (10.15) and (10.17). However, that evolution cannot be reproduced by the unitary evolution of a state on spacelike surfaces. If one attempts to define a state $|\Psi(\sigma)\rangle$ by the procedure described above for globally hyperbolic spacetimes, one finds that the surface σ cannot remain spacelike and be pushed smoothly into the lightly shaded region to the future of the naked singularity.

The best that can be done is to push the surface σ so that integration is over fields with support in the region bounded by $\mathcal{I}^-$ and the spacelike surfaces A and B shown in figure 10.8. That defines a kind of two component "state" with one component on A and the other on B. By tracing products of this object over the degrees of freedom on B and normalizing the result, a density matrix may be constructed that is sufficient for predictions of alternatives on A. However, that density matrix does not evolve

Chapter 10. Generalized Quantum Theory ...   213

by anything like an equation of the form (10.6). Indeed it evolves unitarily as the surface A is pushed forward in time. There is no problem with conservation of energy or charge because of the general arguments given in [24]. For alternatives that refer to field configurations in the interior of the spacetime even this kind of density matrix construction is in general unavailable and the general expressions (10.15) and (10.17) must be used.[10]

Similar statements also could be made for other spacetimes that are not globally hyperbolic, such as the spacetimes with closed timelike curves or spatial topology change mentioned in section 10.2, and for quantum gravity itself.

The process of black hole evaporation is thus not in conflict with the principles of quantum mechanics suitably generally stated. It is not in conflict with quantum evolution described four-dimensionally. It is only in conflict with the narrow idea that this evolution be reproduced by evolution of a state vector through a family of spacelike surfaces.

10.7 Information—where is it?

A spacetime formulation of quantum mechanics requires a spacetime notion of information that is also in fully four-dimensional form. In this section we describe a notion of the information available in *histories* and not just in alternatives on a single spacelike surface.[11] We then apply this to discuss the question of whether information is lost in an evaporating black hole spacetime.

In quantum mechanics, a statistical distribution of states is described by a density matrix. For the forthcoming discussion it is therefore necessary to generalize the previous considerations a bit and treat mixed density matrices ρ as initial conditions for the closed system as well as pure states $|\Psi\rangle$. To do this it is only necessary to replace (10.12) with

$$D(\alpha', \alpha) = \text{Tr}\left(C_{\alpha'} \rho C_\alpha^\dagger\right) . \qquad (10.18)$$

A generalization of the standard Jaynes construction [30] gives a natural definition of the missing information in a set of histories $\{c_\alpha\}$. We begin by defining the entropy functional on density matrices:

$$\mathcal{S}(\tilde\rho) \equiv -\text{Tr}\left(\tilde\rho \log \tilde\rho\right) . \qquad (10.19)$$

---

[10] It should be noted that a generalized quantum theory formulated four-dimensionally in terms of histories that extend over the whole of a background spacetime is not necessarily causal in the sense that predictions on a spacelike surface are independent of the background geometry to its future. See [25, 26, 27] for further discussion.

[11] The particular construction we use is due to Gell-Mann and the author [25]. There are a number of other ideas, e.g. [28, 29], with which the same points about information in a evaporating black hole spacetime could be made.

With this we define the *missing information* $S(\{c_\alpha\})$ in a set of histories $\{c_\alpha\}$ as the maximum of $\mathcal{S}(\tilde{\rho})$ over all density matrices $\tilde{\rho}$ that preserve the predictions of the true density matrix ρ for the decoherence and probabilities of the set of histories $\{c_\alpha\}$. Put differently, we maximize $\mathcal{S}(\tilde{\rho})$ over $\tilde{\rho}$ that preserve the decoherence functional of ρ defined in terms of the corresponding class operators $\{C_\alpha\}$. Thus, the missing information in a set of histories $\{c_\alpha\}$ is given explicitly by

$$S(\{c_\alpha\}) = \max_{\tilde{\rho}} \left[\mathcal{S}(\tilde{\rho})\right]_{\{\mathrm{Tr}(C_{\alpha'}\tilde{\rho}C_\alpha^\dagger) = \mathrm{Tr}(C_{\alpha'}\rho C_\alpha^\dagger)\}} . \tag{10.20}$$

Complete information, S_{compl}—the most one can have about the initial ρ—is found in the decoherent set of histories with the least missing information:

$$S_{\mathrm{compl}} = \min_{\substack{\text{decoherent} \\ \{c_\alpha\}}} \left[S(\{c_\alpha\})\right] . \tag{10.21}$$

In usual quantum mechanics it is not difficult to show that S_{compl} defined in this way is exactly the missing information in the initial density matrix ρ:

$$S_{\mathrm{compl}} = \mathcal{S}(\rho) = -\mathrm{Tr}(\rho \log \rho) . \tag{10.22}$$

Information in a set of histories and complete information are *spacetime* notions of information whose construction makes no reference to states or alternatives on a spacelike surface. Rather the constructions are four-dimensional, making use of histories. Thus, for example, with these notions one can capture the idea of information in entanglements in time as well as information in entanglements in space.

One can find *where* information is located in spacetime by asking what information is available from alternatives restricted to fields in various spacetime regions. For example, alternative values of a field average over a region R refer only to fields inside R. The missing information in a region R is

$$S(R) = \min_{\substack{\text{decohering}\{c_\alpha\} \\ \text{referring to } R}} S(\{c_\alpha\}) . \tag{10.23}$$

This region could be part of a spacelike surface or could extend in time; it could be connected or disconnected. When R is extended over the whole of the spacetime the missing information is complete. But it is an interesting question what *smaller* regions R contain complete information, $S(R) = S_{\mathrm{compl}}$.

When spacetime is globally hyperbolic and quantum dynamics can be described by states unitarily evolving through a foliating family of spacelike surfaces, it is a consequence of the definitions (10.23) and (10.21) that complete information is available on each and every spacelike surface:

$$S(\sigma) = S_{\mathrm{compl}} = -\mathrm{Tr}(\rho \log \rho) . \tag{10.24}$$

However, in more general cases, such as the evaporating black hole spacetimes under consideration in this paper, only incomplete information may be available on any spacelike surface, and complete information may be distributed about the spacetime. We now show why.

There is no reason to expect to recover complete information on a surface like A in figure 10.8. The analysis of the black hole spacetime without back reaction (fig. 10.2) shows that much of the information on the surface $(\mathcal{I}^+, H)$ consists of correlations, or entanglements, between alternatives on $\mathcal{I}^+$ and alternatives on H. One should rather expect complete information to be available in the spacetime region which is the union of A and a surface like B. Even though this region is not a spacelike surface there are still entanglements "in time" between alternatives on A and alternatives on B that must be considered to completely account for missing information.

The situation is not so very different from that of the "trousers" spacetime sketched in figure 10.5. Complete information about an initial state on σ' is plausibly not available on surfaces A or B separately. That is because there will generally be correlations (entanglements) between alternatives on A and alternatives on B. Complete information is thus available in spacetime even if not available on any one spacelike surface like A. The surfaces A and B of this example are similar in this respect to the surfaces A and B of the evaporating black hole spacetime in figure 10.8.

Thus, even though it is not completely available on every spacelike surface like A, information is not lost in evaporating black hole spacetimes. Complete information is distributed about the spacetime.

10.8 Conclusions

This paper has made five points:

- For quantum dynamics to be formulated in terms of a state vector evolving unitarily or by reduction through a family of spacelike surfaces, spacetime geometry must be fixed and foliable by a nonsingular family of spacelike surfaces.

- When spacetime geometry is fixed but does not admit a foliating family of spacelike surfaces or where, as in quantum gravity, spacetime geometry is not fixed, quantum evolution cannot be defined in terms of states on spacelike surfaces.

- Quantum mechanics can be generalized so that it is in fully four-dimensional, spacetime form, free from the need for states on spacelike surfaces.

- The complete evaporation of a black hole is not in conflict with the principles of quantum mechanics stated suitably generally in four-dimensional form.

- Even though complete information may not be available on every spacelike surface, four-dimensional information is not lost in an evaporating black hole spacetime. Rather, complete information is distributed over spacetime.

We have argued that, whether or not a quantum theory of gravity exhibits a unitary S-matrix between precollapse and evaporated states, some generalization of quantum mechanics will be necessary when spacetime geometry is not fixed. Black hole evaporation is not in conflict with the principles of quantum mechanics suitably generally stated. Whatever the outcome of the evaporation process generalized quantum theory is ready to describe them.

10.9 Epilogue

When I was a much younger scientist working on the physics of neutron stars, I remember reporting to Chandra how I had been criticized after a colloquium for daring to extrapolate the theory of matter in its ground state to the densities of nuclear matter and beyond. He emphatically advised me to pay no attention, as if worried that I might, saying people like my critics were never right. He might, I imagine, have been thinking of his own extrapolation of the properties of stellar matter to the densities of white dwarfs. In this article I have been engaged in extrapolating the principles of quantum mechanics to the domains of quantum gravity and black holes. One hopes Chandra would have approved.

Acknowledgments

This work was supported in part by NSF grants PHY95-07065 and PHY94-07194.

References

[1] S. Chandrasekhar, *The Mathematical Theory of Black Holes*, Oxford University Press (Oxford, 1983).

[2] S.W. Hawking and G.F.R. Ellis, *The Large Scale Structure of Spacetime*, Cambridge University Press (Cambridge, 1973).

[3] S.W. Hawking, Black hole explosions? *Nature*, **248**, 30 (1974); Particle creation by black holes, *Comm. Math. Phys.*, **43**, 199 (1975).

[4] W. Israel, Thermo-field dynamics of black holes, *Phys. Lett.*, **A57**, 107 (1976).

[5] R. Wald, On particle production by black holes, *Comm. Math. Phys.*, **45**, 9 (1975).

[6] S. Giddings, Quantum mechanics of black holes, in *Proceedings of the 1994 Trieste Summer School in High Energy Physics and Cosmology*, ed. E. Gava et al., World Scientific (Singapore, 1995), hep-th/9412138; A. Strominger, Les Houches lectures on black holes, in *Fluctuating Geometries in Statistical Mechanics and Field Theory*, Proceedings of the 1994 Les Houches Summer School LXII, ed. F. David, P. Ginsparg, and J. Zinn-Justin, Elsevier (Amsterdam, 1996), hep-th/9501071.

[7] M. Morris, K.S. Thorne, and U. Yurtsver, Wormholes, time machines, and the weak energy condition, *Phys. Rev. Lett.*, **61**, 1446 (1988); F. Echeverria, G. Klinkhammer, and K.S. Thorne, Billiard balls in wormhole spacetimes with closed timelike curves: classical theory, *Phys. Rev.*, **D44**, 1077 (1991); J.L. Friedman and M. Morris, The Cauchy problem for the scalar wave equation is well defined on a class of spacetimes with closed timelike curves, *Phys. Rev. Lett.*, **66**, 401 (1991); J. Friedman, M. Morris, I.D. Novikov, F. Echeverria, G. Klinkhammer, K.S. Thorne, and U. Yurtsever, Cauchy problem in spacetimes with closed timelike curves, *Phys. Rev.*, **D42**, 1915 (1990).

[8] R. Geroch, Topology in general relativity, *J. Math. Phys.*, **8**, 782 (1967).

[9] A. Anderson and B. DeWitt, Does the topology of space fluctuate? *Found. Phys*, **16**, 91 (1986).

[10] G. Horowitz, Topology change in classical and quantum gravity, *Class. Quant. Grav.*, **8**, 587 (1991).

[11] F. Tipler, Singularities and causality violation, *Ann. Phys.*, **108**, 1 (1977).

[12] A. Strominger and C. Vafa, Microscopic origin of the Bekenstein-Hawking entropy, *Phys. Lett.*, **B379**, 99 (1996), hep-th/9601029; G. Horowitz, The origin of black hole entropy in string theory, in *Proceedings of the Pacific Conference on Gravitation and Cosmology*, Seoul, Korea, February 1–6, 1996 (to appear), gr-qc/9604051; S.R. Das and S.D. Mathur, Comparing decay rates for black holes and D-branes, *Nucl. Phys.*, **B478**, 561 (1996), hep-th/9606185; J.M. Maldacena, Black holes in string theory, Ph.D. thesis, Princeton University, June 1996, hep-th/9607235; J. Maldacena and A. Strominger, Black hole greybody factors and D-brane spectroscopy, *Phys. Rev.*, **D55** 861 (1997), hep-th/9609204; G. Horowitz, Quantum states of a black hole, chap. 12 of this volume, gr-qc/9704072.

[13] J.B. Hartle, Spacetime quantum mechanics and the quantum mechanics of spacetime, in *Gravitation and Quantizations*, Proceedings of the 1992 Les Houches Summer School LVII, ed. B. Julia and J. Zinn-Justin, North-Holland (Amsterdam, 1995); gr-qc/9304006.

[14] S.W. Hawking, The unpredictability of quantum gravity, *Comm. Math. Phys.*, **87**, 395 (1982).

[15] J. Hagelin, J. Ellis, D. Nanopoulos, and M. Srednicki, Search for violations of quantum mechanics, *Nucl. Phys.*, **B241**, 381 (1984); T. Banks, L. Susskind, and M. Peskin, Difficulties for the evolution of pure states into mixed states, *Nucl. Phys.*, **B244**, 125 (1984); M. Srednicki, Is purity eternal? *Nucl. Phys.*, **B410**, 143 (1993), hep-th/9206056.

[16] W. Unruh and R. Wald, On evolution laws taking pure states to mixed states in quantum field theory, *Phys. Rev.*, **D52**, 2176 (1995), hep-th/9503024.

[17] S.W. Hawking, private communication.

[18] R. Wald, *Quantum Field Theory in Curved Spacetime and Black Hole Thermodynamics*, University of Chicago Press (Chicago, 1994), 178–182.

[19] K. Kuchař, Time and interpretations of quantum gravity, in *Proceedings of the 4th Canadian Conference on General Relativity and Relativistic Astrophysics*, ed. G. Kunstatter, D. Vincent, and J. Williams, World Scientific (Singapore, 1992); C. Isham, Conceptual and geometrical problems in quantum gravity, in *Recent Aspects of Quantum Fields*, ed. H. Mitter and H. Gausterer, Springer (Berlin, 1992); C. Isham, Canonical quantum gravity and the problem of time, in *Integrable Systems, Quantum Groups, and Quantum Field Theories*, ed. L.A. Ibort and M.A. Rodriguez, Kluwer (London, 1993); W. Unruh, Time and quantum gravity, in *Gravitation: A Banff Summer Institute*, ed. R. Mann and P. Wesson, World Scientific (Singapore, 1991).

[20] D. Deutsch, Quantum mechanics near closed timelike lines, *Phys. Rev.*, **D44**, 3197 (1991).

[21] C.W. Misner, Feynman quantization of general relativity, *Rev. Mod. Phys.*, **29**, 497 (1957); H. Leutwyler, Gravitational field: equivalence of Feynman and canonical quantization, *Phys. Rev.*, **B134**, 1155 (1964); B. DeWitt, Quantum gravity: the new synthesis, in *General Relativity: An Einstein Centenary Survey*, ed. S.W. Hawking and W. Israel, Cambridge University Press (Cambridge, 1979); E. Fradkin and G. Vilkovisky, S-matrix for gravitational field. II. local measure; general relations, *Phys. Rev.*, **D8**, 4241 (1973); S.W. Hawking, The path-integral approach to quantum gravity, in *General Relativity: An Einstein Centenary Survey*, ed. S.W. Hawking and W. Israel, Cambridge University Press (Cambridge, 1979); C. Teitelboim, Quantum mechanics of the gravitational field, *Phys. Rev.*, **D25**, 3159 (1983); Proper-time gauge in the quantum theory of gravitation, *Phys. Rev.*, **D25**, 297 (1983); Quantum mechanics of the gravitational field in asymptotically flat space, *Phys. Rev.* **D28**, 310 (1983).

[22] C.J. Isham, Quantum logic and the histories approach to quantum theory, *J. Math. Phys.*, **35**, 2157 (1994), gr-qc/9308006.

[23] J.B. Hartle, The quantum mechanics of closed systems, in *Directions in General Relativity*, Vol. 1, *A Symposium and Collection of Essays in Honor of Professor Charles W. Misner's 60th Birthday*, ed. B.-L. Hu, M.P. Ryan, and C.V. Vishveshwara, Cambridge University Press (Cambridge, 1993), gr-qc/9210006.

[24] J.B. Hartle, R. Laflamme, and D. Marolf, Conservation laws in the quantum mechanics of closed systems, *Phys. Rev.*, **D51**, 7007 (1995), gr-qc/9410006.

[25] J.B. Hartle, Spacetime information, *Phys. Rev.*, **D51**, 1800 (1995), gr-qc/9409005; A. Kent (*Phys. Rev. D*, to appear; gr-qc/9610075) has pointed out an error in one of the specific computations in this paper concerned with information in time-neutral formulations of quantum mechanics, but

that does not affect the general notions of spacetime information that are used here.

[26] A. Anderson, Unitarity restoration in the presence of closed timelike curves, *Phys. Rev.*, **D51**, 5707 (1995), gr-qc/9405058.

[27] S. Rosenberg, Testing causality violation on spacetimes with closed timelike curves, hep-th/9707103.

[28] C.J. Isham and N. Linden, Information-entropy and the space of decoherence functions in generalized quantum theory, quant-ph/9612035.

[29] M. Gell-Mann and J.B. Hartle, unpublished.

[30] R.D. Rosenkrantz, ed. *E.T. Jaynes: Papers on Probability Statistics and Statistical Mechanics*, Reidel (Dordrecht, 1983).

11

Is Information Lost in Black Holes?

Stephen W. Hawking

Abstract

I trace the history of black holes from Chandra's discovery of the limiting mass of white dwarfs. The no-hair theorems showed that a lot of information about the collapsed body was locked up in the black hole—inaccessible to an outside observer. When it was discovered that black holes evaporate quantum mechanically, the question arose: What happens to that information? I argue there are only two viable options: either it comes out continuously during the evaporation in the form of correlations between the different modes or it is lost forever from our region of the universe. I describe the D-brane calculations that support the first view and explain why I favor the second possibility.

Black holes are one of the few cases in science where a theory was almost completely worked out before there were any experimental data. Indeed, even now, we have observational evidence only for their existence and not for any of their more refined properties that have been predicted by theory. It was Chandra who started things off when he showed that degeneracy pressure cannot support a star of more than about one and a half solar masses. So this raised the question: What happens to a star of more than this mass when it exhausts its nuclear fuel and cools? The hostility of Eddington, Chandra's supervisor, to the idea of gravitational collapse led to Chandra's switching to other lines of research. It was not until 1939 that Oppenheimer and a graduate student, Snyder, provided the answer. They showed that a spherically symmetric star would collapse to a point of infinite density, a singularity [1].

However, no one took much notice. The war intervened, and Oppenheimer turned his attention to the atomic bomb. After the war, physicists were busy with quantum electrodynamics, which could be tested by effects like the Lamb shift, and ignored general relativity, which didn't seem to have any observable predictions at that time. It was not until the discovery of quasars in the early 1960s that interest in gravitational collapse was reawakened. Oppenheimer's work was rediscovered, but there was no significant advance in my opinion, until the Penrose singularity theorem [2] of 1965:

> **Penrose Singularity Theorem:** *Spacetime cannot be nonsingular and geodesically complete if:*
>
> *1. The weak energy condition,*
>
> $$T_{ab}k^a k^b \geq 0,$$
>
> *holds for any timelike vector k^a.*
> *2. There is a closed trapped surface.*
> *3. The spacetime admits a noncompact Cauchy surface.*

This theorem opened a new chapter. Previous work on gravitational collapse had been in particular coordinate systems, and made assumptions of symmetry, or used hand-waving approximations. But Penrose showed that once gravitational collapse had gone beyond a certain point, a singularity of spacetime was inevitable. This theorem held for any reasonable form of matter, and it applied to fully general spacetimes, without any symmetries. It was in direct contradiction to Lifshitz and Khalatnikov [3], who claimed that a general collapse would not lead to a singularity. But it is hard to argue with a mathematical theorem, so in the end they had to recant, though it took them another five years.

Penrose's theorem showed that gravitational collapse had to end in a singularity, at which general relativity broke down. We estimate that gravitational collapse occurs at least once a year in our galaxy, and at a similar rate in other galaxies. So how is it that we can predict anything? The answer is that general relativity seems to obey what Penrose called the "cosmic censorship conjecture":

> **Cosmic Censorship Conjecture:** *God abhors a naked singularity.*

The cosmic censorship conjecture says that any singularities formed in collapse from nonsingular data will be hidden behind an event horizon, another term Penrose introduced. The cosmic censorship conjecture is fundamental to all work on black holes. I therefore had a bet with Kip Thorne and John Preskill that it was true:

> Whereas Stephen W. Hawking firmly believes that naked singularities are an anathema and should be prohibited by the laws of classical physics,
> And whereas John Preskill and Kip Thorne regard naked singularities as quantum gravitational objects that might exist unclothed by horizons, for all the Universe to see,

Therefore Hawking offers, and Preskill/Thorne accept, a wager with odds of 100 pounds sterling to 50 pounds sterling, that when any form of classical matter or field that is incapable of becoming singular in flat spacetime is coupled to general relativity via the classical Einstein equations, the result can never be a naked singularity.

The loser will reward the winner with clothing to cover the winner's nakedness. The clothing is to be embroidered with a suitable concessionary message.

Stephen W. Hawking, John P. Preskill, & Kip S. Thorne, Pasadena, California, 24 September 1991

Unfortunately, I wasn't careful enough about the wording of the bet, and the similarity solutions, that represent the threshold of black hole formation, were a counterexample [4, 5, 6]. But all the evidence suggests that the conjecture is true for generic initial data [7]. The world is safe from naked singularities, at least in classical general relativity. However, I will try to convince you that the breakdown in predictability reappears in quantum theory, though in a different form.

The next major development in the field came with the unexpected discovery by Werner Israel that a static black hole had to be spherically symmetric:

Israel's Theorem: *A static, vacuum black hole with a regular event horizon has to be the Schwarzschild solution.*

My use of the term "black hole" here is an anachronism because it wasn't introduced until later, but it is a convenient shorthand. At first, many people, including Israel himself, thought this result implied that black holes would form only in collapses that were exactly spherical, which would be a set of measure zero amongst all collapses. However, Roger Penrose and John Wheeler put forward a different interpretation. This was that in gravitational collapse an asymmetric body would lose all multipole moments, except mass, angular momentum, and electric charge, which were protected by coupling to gauge fields. If this interpretation was correct, black holes could have no hair:

No Hair Theorem: *Stationary black holes are characterized by mass M, angular momentum J, and electric charge Q.*

In other words, time-independent black holes would be determined completely by their gauge charges. This was proved for Einstein-Maxwell theory by the combined work of Israel [8], Carter [9], Robinson [10], and myself [11]. For gravity coupled to Yang-Mills, Higgs, or dilaton fields, there are additional complications, but the result remains essentially true:

time-independent, stable, nonextreme black hole solutions are completely determined by their gauge charges.

The no-hair results indicated that a lot of information was lost in gravitational collapse. One ended up with a black hole that was determined by its mass, angular momentum, and electric charge, but that was otherwise independent of the nature of the body that collapsed. In a purely classical theory, there would be no information loss, because one would never actually lose sight of the collapsing body. It would appear to slow down and hover just outside the horizon, getting rapidly more redshifted, but never actually disappearing. So someone at infinity could in principle always observe the collapsing body and measure its shape and constitution. But in quantum theory, the number of photons emitted by the collapsing body, before the event horizon forms, will be finite. There will be far too few of them to carry all the information about the collapsing body. Thus an outside observer will lose information about what collapsed to form the black hole.

This loss of information didn't matter too much when it was thought that black holes went on forever. One could still say that the information was inside the black hole. It was just that it wasn't accessible to an observer at infinity. However, the situation changed when it was discovered that black holes radiate with a thermal spectrum. According to semiclassical calculations, the radiation would be in a mixed quantum state and wouldn't carry away information about what was inside the black holes. However, it would carry away energy and, hence, mass. Unless they were stabilized by something like a magnetic charge, it seemed black holes would radiate down to zero mass and disappear. What then would happen to the information they contained?

Physicists seem to have a strong emotional attachment to information. I think this comes from a desire for a feeling of permanence. They have accepted that they will die, and even that the baryons which make up their bodies will eventually decay. But they feel that information, at least, should be eternal. There have therefore been three main suggestions for preserving information in black holes:

1. Information is preserved in remnants.

2. Information emerges in the final burst of radiation.

3. Information emerges continuously during the evaporation through correlations between different modes.

The first is that the black hole doesn't evaporate completely but leaves a Planck-sized remnant containing the information. The second is that all the information in the black hole comes out in the final burst of radiation when the black hole gets down to the Planck size. That is where the semiclassical calculation of the radiation will break down, so information might emerge. The third is that the radiation coming out of the black hole contains subtle correlations between different modes, which are not seen in the semiclassical calculation, and which encode the information.

I think one can dismiss the first two suggestions. The remnant hypothesis would not be CPT invariant if black holes could form but never disappear completely. And one might expect more than the cosmological density of remnants, left over from the evaporation of black holes in the very early universe. The final burst hypothesis is in trouble because it takes energy to carry information, and there's very little energy left in the final stages. The only possibility is for the information to trickle out very slowly. So this possibility is similar to the remnant one.

In my opinion, the only way of preserving information that has not been ruled out is if there are correlations in the outgoing radiation that carry the information. For years, those who wish to make the world safe for information have been claiming there must be correlations. But they hadn't come up with a credible mechanism, until earlier this year. It is based on work started by Strominger and Vafa [12] and followed by a number of others. They showed that a collection of D-branes and strings with the same gauge charges as a near-extreme black hole had the same number of quantum states as the amount of information in the black hole. That is, the number of quantum states of the D-branes and strings is $e^{A/4}$, where A is the area of the horizon of a black hole with the same mass and gauge charges as the D-branes and strings. I obtained this formula more than twenty years ago. It is nice to see it confirmed at last.

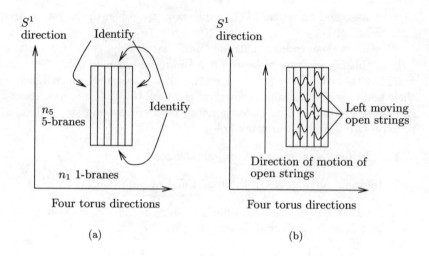

Figure 11.1: The configuration of 1-branes and 5-branes and strings in the internal five dimensions which corresponds to a five dimensional black hole with finite horizon area. The wrapping of the branes is shown in (a), and the motion of the strings is shown in (b).

By now, there are a number of different derivations of the number of internal states of a black hole. They look very different, but they all agree on the answer, $e^{A/4}$, which maybe isn't surprising, since they knew the target that they were aiming at. Whether they would have found it without my formula is another matter. In each case, there are a number of objections one can make to the treatment. But in dealing with duality, when one has a number of questionable, but apparently different arguments for the same result, one tends to believe it.

For simplicity, I shall describe only one such counting scheme, for five dimensional black holes in type IIB theory. This theory can be regarded as supergravity in ten dimensions, and it admits solutions that are extended $p+1$ dimensional objects, called p-branes, where p is odd. One can wrap a p-brane around a p dimensional torus and reduce the remaining dimensions to four or five. This gives a solution of the Einstein-Maxwell dilaton theory in those dimensions, but it corresponds to a singular limit of a black hole with a horizon of zero area. To get a nonsingular horizon of finite area, one needs at least three gauge charges in five dimensions, and four gauge charges in four dimensions.

To get a five dimensional black hole with a horizon of finite area, take n_1 1-branes and n_5 5-branes and wrap them around a five-torus, with the 1-branes all being wrapped around the same circle, or S^1, in the five-torus (see fig. 11.1(a)). When reduced to five dimensions, this still gives a

Chapter 11. Is Information Lost in Black Holes?

singular limit of a black hole, with a horizon of zero area. However, these 1- and 5-branes are what are called D-branes. The charges they carry are associated with what are called Ramond-Ramond fields in the supergravity theory. This means that the 1- and 5-branes can be the endpoints of open fundamental strings. One can therefore add a number of open strings that have one endpoint on a 1-brane and the other endpoint on a 5-brane, and that are moving in the left direction around the S^1 that the 1-branes are wrapped around (see fig. 11.1(b)). The total momentum, P, of these open strings will be some integer, n, divided by R, the radius of the S^1. This is equivalent to boosting the solution with the 1- and 5-branes along the S^1 direction. The Kaluza-Klein reduction to five dimensions then gives a nonsingular extreme black hole with finite horizon area. It has three gauge charges, associated with the numbers of 1- and 5-branes and the left-moving momentum of the open strings.

The area of the horizon of the five dimensional black hole is given by a simple formula. It is 8π times the square root of the product of the charges:

$$A = 8\pi\sqrt{nn_1n_5}. \tag{11.1}$$

One can also calculate the number of quantum states of the open strings with total left-moving momentum, P. Remarkably enough, it turns out that the logarithm of the number of states is 2π times the square root of the product of the charges:

$$\ln N = 2\pi\sqrt{nn_1n_5}. \tag{11.2}$$

Thus the entropy of the system of D-branes and strings is a quarter the area of the horizon of a black hole with the same gauge charges.

So far, there is no conflict with the prediction from semiclassical theory that information will be lost, because one does not know in which of the e^S states the open strings will be. However, one could calculate the scattering of closed strings on the D-branes. When the closed strings hit the D-branes, they can split into left and right moving open strings attached to the D-branes. Left and right moving open strings can then combine to produce closed strings, which leave the D-branes and move away (see fig. 11.2).

In black hole terms, one would say that particles went into an extreme black hole and increased the mass. This would have made it slightly nonextreme and raised the temperature above zero. The black hole would then radiate thermally and return to the zero temperature extreme state. The remarkable thing, first discovered by Das and Mathur [13], is that the emission spectrums predicted by the D-brane and black hole calculations are exactly the same, at least in the cases that have been calculated. There is, however, an important difference of principle between the two calculations. The D-brane scattering is unitary, given a knowledge of the initial quantum state of the open strings. Of course, if one knows only the gauge

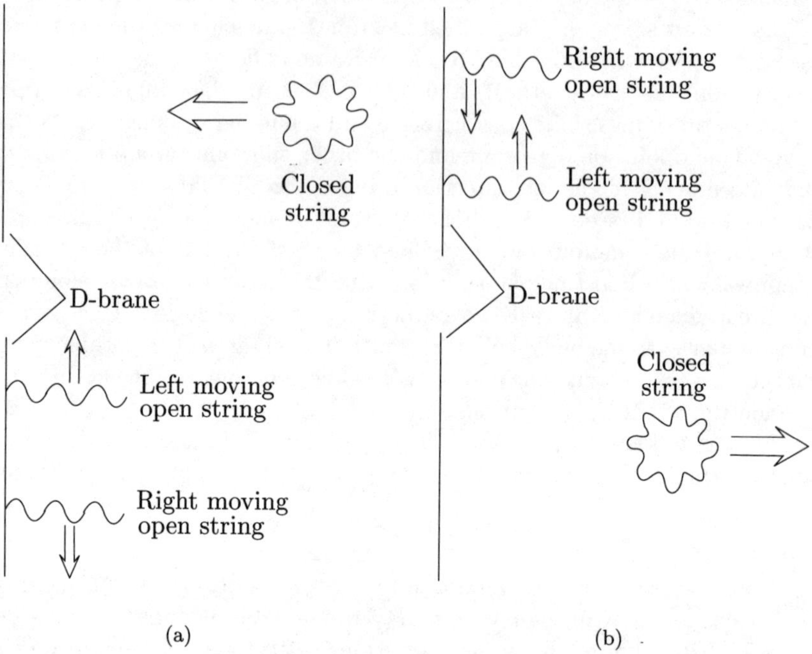

Figure 11.2: The scattering of closed strings and D-branes. In (a), a closed string hits the D-brane and splits into left and right moving open strings attached to the D-brane. The reverse process is depicted in (b).

charges of the D-brane system, it could be in any one of $e^{A/4}$ states. But with repeated scatterings, it seems that one could measure the quantum state. At one time, it seemed that the D-brane system might lose its gauge charges through the emission of massive, charged, closed string states. But Marika Taylor Robinson and I showed that such emission would be suppressed compared to neutral emission, except when one gauge charge is much greater than the others [14]. Thus for D-brane systems with roughly equal charges, it would seem that it ought to be possible to do enough scatterings to measure the quantum state before the system loses a significant fraction of its gauge charges. After that, one should be able to predict the quantum state of the outgoing radiation completely.

By contrast, according to the semiclassical treatment of black holes, the quantum state of the outgoing radiation is correlated to the quantum state inside the horizon, which is not observed. Thus, no matter how many scatterings one does, one can never predict the quantum state of the outgoing radiation. Instead, all that one can do is calculate the probabilities

Chapter 11. Is Information Lost in Black Holes?

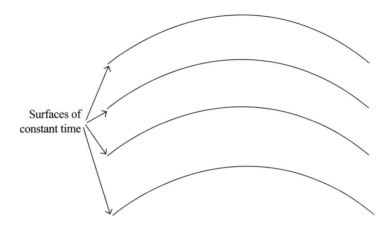

Figure 11.3: A foliation of spacetime by a family of non-intersecting surfaces of constant time.

for the outgoing radiation to be in different quantum states, depending on the quantum state inside the black hole. In other words, the outgoing radiation is in a mixed quantum state, rather than a pure quantum state, as the D-brane calculations would indicate. One can regard this as a quantum manifestation of the breakdown of predictability at singularities, because the outgoing radiation can be thought of as having quantum-tunneled out from the singularity inside the black hole.

The reason for this difference between the D-brane and black hole calculations is that in the D-brane scattering, one ignores the curvature around the D-branes and regards them as surfaces in flat space. One can foliate such a spacetime with a family of non-intersecting surfaces of constant time (see fig. 11.3). One can then evolve forward in time with the Hamiltonian and get a unitary transformation from the initial state to the final state. A unitary transformation would be a one-to-one mapping from the initial Hilbert space to the final Hilbert space. This would imply that there was no loss of information or quantum coherence.

However, if one looks closely, one finds that the presence of a D-brane warps spacetime and changes the causal structure. In flat space, the D-brane would be a timelike boundary (see fig. 11.4(a)). However, when curvature is taken into account, one gets a pair of null boundaries (see fig. 11.4(b)), like the past and future horizons of a black hole, though in general the boundary surfaces are singular. If the boundary were timelike, as in flat space, there would be natural boundary conditions that would cause reflection and preserve information and unitarity. However, with

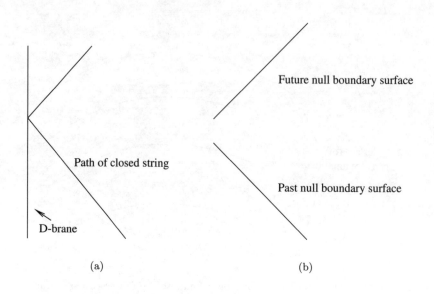

Figure 11.4: (a) The causal structure of a D-brane in flat space. (b) The causal structure of a D-brane taking curvature into account.

a pair of null boundaries, there is no natural boundary condition that preserves information, because there is no canonical map from the future boundary to the past boundary. Thus there is no natural way to say how information that reaches the future boundary will reappear at the past boundary. In the case of scattering off D-branes of a single type, all the information gets reflected before it reaches the future boundary, because a D-brane of a single type corresponds to a black hole of zero area. That is why it is a good approximation to regard D-branes as surfaces in flat space. However, in the case of the system of 1- and 5-branes and open strings, the gauge charges have been carefully chosen to balance, so that the past and future null boundaries are regular null surfaces. In this case, there is no reason why what crosses the future boundary need be related in a simple way to what comes out of the past boundary.

The presence of internal null boundaries or horizons means that data at past or future infinity do not completely determine the quantum state of fields on such a background. This in turn means that the quantum field theory on the background does not obey the axiom of asymptotic completeness. This says that the Hilbert space of the quantum field theory is not isomorphic to the initial and final Hilbert spaces. This is the necessary and sufficient condition for the map from the initial to final Hilbert spaces to be unitary.

Chapter 11. Is Information Lost in Black Holes?

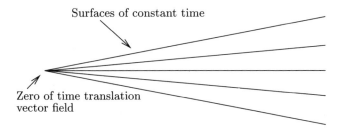

Figure 11.5: A family of constant time surfaces corresponding to the case where the time translation vector field possesses a zero.

Another way of seeing that the map from the initial Hilbert space to the final one is nonunitary is that the curvature near the D-branes changes the topology of the Euclidean section of spacetime. This means that any vector field that agrees with time translations at infinity will necessarily have zeroes in the interior of the spacetime. In turn, this will mean that one cannot foliate spacetime with a family of time surfaces. If one tries, the surfaces will intersect at the zeroes of the vector field (see fig. 11.5). One therefore cannot use the Hamiltonian to get a unitary evolution from initial state to final. But if the evolution is not unitary, there will be loss of information. An initial state that is a pure quantum state can evolve to a quantum state that is mixed. That is, it is described by a density matrix, which is a two-index tensor on Hilbert space, rather than by a state vector.

One cannot just ignore topology and pretend one is in flat space, which in effect is what those who want to preserve unitarity are doing. The recent progress in duality is based on nontrivial topology. One considers small perturbations about different vacuums of the product form (flat space)×(internal space) and shows that one gets equivalent Kaluza-Klein theories. But if one can have small perturbations about product metrics, one should also consider larger fluctuations that change the topology from the product form. Indeed, such nonproduct topologies are necessary to describe pair creation or annihilation of solitons, like black holes or p-branes.

It is often claimed that supergravity is just a low energy approximation to the fundamental theory, which is string theory. However, I think the recent work on duality is telling us that string theory, p-branes, and supergravity are all on a similar footing. None of them is the whole picture, but each are valid in different, but overlapping regions. There may be some fundamental theory from which they can all be derived as different approximations. Or it may be that theoretical physics is like a manifold that can't be covered by a single coordinate patch. Instead, we may have to use a collection of apparently different theories that are valid in different regions, but that agree on the overlaps. After all, we know from Gödel's theorem that even arithmetic can't be formulated in terms of a single set of axioms. Why should theoretical physics be different?

Even if there is a single formulation of the underlying fundamental theory, we don't have it yet. What is called string theory has a good loop expansion. But it is only perturbation theory about some background, generally flat space. So it will break down when the fluctuations become large enough to change the topology. Supergravity, on the other hand, is better at dealing with topological fluctuations, but it will probably diverge at some high number of loops. Such divergences don't mean that supergravity predicts infinite answers. It is just that it cannot predict beyond a certain degree of accuracy. But in that, it is no different from string theory. The string loop perturbation series almost certainly does not converge but is only an asymptotic expansion. Thus at finite coupling, it will only have limited accuracy.

I shall take what I have just said as justification for discussing information loss in terms of general relativity or supergravity, rather than D-branes and strings. For years, I have felt that information loss would occur, not only for macroscopic black holes, but on a microscopic scale as well, because of virtual black holes that would appear and disappear in the vacuum state. Particles could fall into these virtual black holes, which would then radiate other particles in a mixed quantum state. However, I didn't know how to describe such processes because I couldn't see how a black hole could appear and disappear in a manner that was nonsingular, at least in the Euclidean regime. I now realize that my mistake was to try to picture a single virtual black hole appearing and disappearing. Instead, black holes appear and disappear in pairs, like other virtual particles. Equivalently, one can think of the virtual pair as a single black hole moving on a closed loop.

In d dimensions, a single black hole has a Euclidean section with topology $S^{d-2} \times \mathbb{R}^2$. A real or virtual loop of black holes has Euclidean topology $S^{d-2} \times S^2$ minus a point that has been sent to infinity by a conformal transformation. For simplicity, I shall consider $d = 4$, but the treatment for higher d will be similar.

On the manifold $S^2 \times S^2$ minus a point, I shall consider Euclidean metrics that are asymptotic to flat space at infinity. Such metrics can be interpreted as closed loops of virtual black holes [15]. Because they are off-shell, they need not satisfy any field equations. But they will contribute to the path integral, just as off-shell loops of particles contribute to the path integral and produce measurable effects. The effect that I will be concerned with for virtual black holes is loss of quantum coherence. This is a distinctive feature of such topological fluctuations that distinguishes them from ordinary unitary scattering, which is produced by fluctuations that do not change the topology.

Chapter 11. Is Information Lost in Black Holes?

One can calculate scattering in an asymptotically Euclidean metric on $S^2 \times S^2$ minus a point. One then weights with e^{-I}, where I denotes the action of the metric, and integrates over all asymptotically Euclidean metrics. This would give the full scattering, with all quantum corrections. However, we can neither calculate the scattering in a general metric nor integrate over all metrics. Instead, what I shall do is point out some qualitative features of the scattering, in general metrics, that indicate that quantum coherence is lost. I shall then report on some scattering calculations that Simon Ross and I are doing [16] in a specific metric, the C-metric. It is sufficient to show that quantum coherence is lost in some metrics in the path integral, because the integral over other metrics cannot restore the quantum coherence lost in our examples.

Despite twenty years of advocating the Euclidean approach to quantum gravity, I don't have much intuition for Euclidean scattering. However, if the Euclidean metric has a hypersurface orthogonal Killing vector, it can be analytically continued to a real Lorentzian metric, in which it is much easier to see what is happening. I shall therefore consider scattering in such metrics.

The Lorentzian section will contain a pair of black holes that accelerate to infinity. One might think that this is not very physical, but it is no different from a closed loop of a particle like an electron. Closed particle loops are really defined in Euclidean space. If you analytically continue them to Minkowski space, you get an electron-positron pair accelerating away from each other. This kind of behavior is typical. Any topologically nontrivial, asymptotically Euclidean metric will appear to have solitons accelerating to infinity in the Lorentzian section. But this does not mean that there are actual black holes at infinity, any more than there are runaway electrons and positrons with a virtual electron loop. One can regard the use of the Lorentzian metric, with its black holes accelerating to infinity, as just a mathematical trick to evaluate a Euclidean path integral.

To understand the structure of these accelerating black hole metrics, it is helpful to draw Penrose diagrams. Start with the Penrose diagram for Rindler space with the two acceleration horizons and $\mathcal{I}^+$ and $\mathcal{I}^-$. A uniformly accelerated particle moves on a world line that reaches $\mathcal{I}^-$ and $\mathcal{I}^+$ at the points at which they intersect the acceleration horizons (see fig. 11.6).

We now want to replace the accelerating particle, and the similar accelerating particle on the other side, with black holes. So we replace the regions of Rindler space, to the right and left of the accelerating world lines, with intersecting black hole horizons. It turns out that the two accelerating black holes are just the two sides of the same three dimensional wormhole, so one has to identify the two sides of the Penrose diagram (see fig. 11.7).

The analytically continued Euclidean Green's function will define a vacuum state, $|0_h\rangle$, which is the analogy of the so-called Hartle-Hawking state

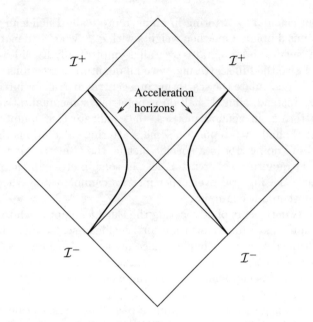

Figure 11.6: A Penrose diagram for Rindler space.

for a static black hole. I find it a bit awkward to use that term, so I shall call it the Euclidean quantum state, but it is equivalent. The Euclidean quantum state can be characterized by saying that positive frequency means positive frequency with respect to the affine parameters on the horizons. In the accelerating black hole metrics, there are two kinds of horizons, black hole and acceleration. Each kind of horizon consists of two intersecting null hypersurfaces, which I will refer to as left and right, as in the diagram. To get a Cauchy surface for the spacetime, one has to break the symmetry between left and right and choose, say, the left acceleration horizon and the right black hole horizon (see fig. 11.8). The quantum state defined by positive frequency, with respect to the affine parameters on these horizons, is the same as the quantum state defined by the other choice of horizons.

Another Cauchy surface in the future is formed by $\mathcal{I}^+$ and the future parts of the black hole horizon, as shown in figure 11.9. There is a natural notion of positive frequency on $\mathcal{I}$. On the black hole horizons, the concept of positive frequency is less well defined. One could use Rindler time, but what one observes on $\mathcal{I}$ is independent of the choice of positive frequency on the black hole horizons.

Because $\mathcal{I}^+$ and the future black hole horizons are a Cauchy surface, the quantum state of a field ϕ will be determined by data on these surfaces. This means that the Hilbert space of the spacetime, $\mathcal{H}$, will be isomorphic

Chapter 11. Is Information Lost in Black Holes?

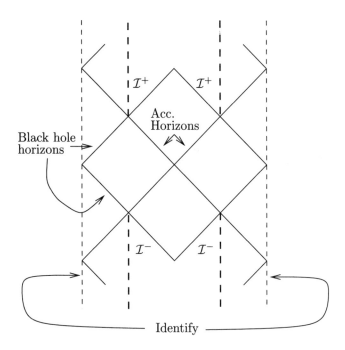

Figure 11.7: A Penrose diagram for a spacetime with accelerating black holes.

to the tensor product of the Fock space, $\mathcal{F}_{\mathcal{I}^+}$, on $\mathcal{I}^+$ and the Fock space, $\mathcal{F}_{b^+}$, on the future black hole horizons:

$$\mathcal{H} = \mathcal{F}_{\mathcal{I}^+} \otimes \mathcal{F}_{b^+}. \tag{11.3}$$

The Euclidean vacuum state, $|0_h\rangle$, can be represented as a state in this tensor product. But because of frequency mixing, it won't be the vacuum state on $\mathcal{I}$ cross the vacuum state on the black hole horizon. Rather it will be a state containing pairs of particles. Both members of the pair may go out to $\mathcal{I}^+$, or both may fall into the hole, or one go out to $\mathcal{I}^+$ and one fall in. It is this last category that can have the interesting effects. An observer on $\mathcal{I}^+$ can't measure the part of the quantum state on the black hole horizon. He or she would therefore have to trace out over all possibilities on the future black hole horizons and describe the observations by a density matrix. Thus there will be loss of quantum coherence.

The quantum state defined by the analytically continued Euclidean Green's functions will contain both incoming and outgoing radiation. Unlike the Euclidean state for static black holes, there won't be radiation to infinity at a steady rate for an infinite time. Instead, the radiation will be peaked around the points on $\mathcal{I}$ where the acceleration horizons intersect $\mathcal{I}$. The radiation will die off at early and late times on $\mathcal{I}$, and the total energy

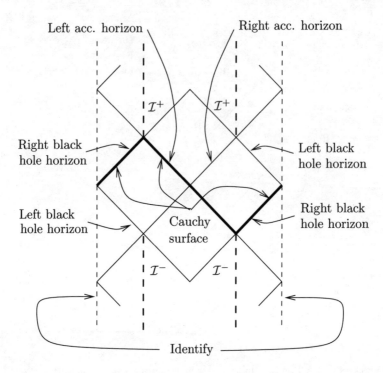

Figure 11.8: A Cauchy surface for the spacetime with accelerating black holes.

radiated in any given direction will be finite.

How should we interpret this? In the case of a static black hole, one usually imposes the boundary condition that there is no incoming radiation on $\mathcal{I}^-$. This means that one has to subtract the incoming radiation from the Euclidean state to give what is called the Unruh state. This is singular on the past horizon, but that doesn't matter, as one normally replaces this region of the metric with the metric of a collapsing body. The energy for the steady rate of outgoing radiation comes from a slow decrease of the mass of the black hole formed by the collapse. However, in the case of a virtual black hole loop, there is no collapse process to remove the singularities on the past horizons of the black holes or supply the energy of the outgoing radiation. My view, therefore, is that integrating over virtual black hole metrics will cause the amplitude to be zero, unless the energy of the outgoing particle or particles is matched by particles with the same energy falling in. One might object that one would never have exactly the combination of incoming particles that corresponded to the quantum state obtained from the Euclidean Green's functions. But I don't think that matters. All the metric cares about is the energy-momentum

Chapter 11. Is Information Lost in Black Holes?

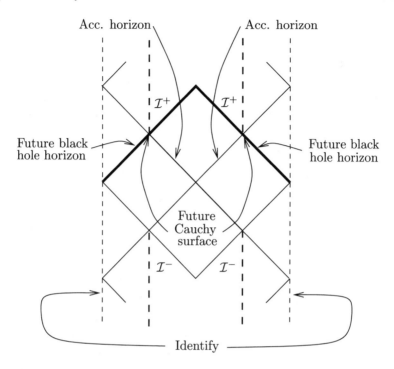

Figure 11.9: Another Cauchy surface for the spacetime with accelerating black holes.

tensor. It doesn't matter what species of particles produces it. I think all that is important is that there should be an energy-momentum balance between the particles that fall into the accelerating black holes and those that come out. The black holes don't emit a definite collection of particles. Instead, the outgoing radiation is in a mixed quantum state, with different probabilities for sending out different collections of particles. All that one does is compare the relative probabilities for those collections of particles with the same energy as the particles that fall in. Thus one can consider processes in which one or two particles fall into a virtual black hole loop, and one or two come out. These processes look like Feynman diagrams, but the difference is that the outgoing particles will be in a mixed state. There will be loss of quantum coherence.

These virtual black holes will presumably be of Planck size. The cross section for a low energy particle to fall into a Planck-size static black hole is very low unless the particle is spin zero. This suggests that the effects of virtual black holes will be small except for scalar particles. To check this, Ross and I wanted to do a scattering calculation in an accelerating black hole metric. We chose the C-metric [17]. This doesn't really qualify as a virtual black hole metric, because it has string singularities pulling the black holes apart. But it has the same structure as a virtual black hole, and it has the great advantage that the wave equation separates.

The C-metric can be written in the form

$$ds^2 = \frac{1}{A^2(x-y)^2}\left[G(y)dt^2 - \frac{dy^2}{G(y)} + \frac{dx^2}{G(x)} + G(x)d\varphi^2\right]. \quad (11.4)$$

Here G is a quartic with four real roots, ξ_1 to ξ_4:

$$\begin{aligned}G(\xi) &= (1 - \xi^2 - r_+ A\xi^3)(1 + r_- A\xi) \\ &= -r_+ r_- A^2 (\xi - \xi_1)(\xi - \xi_2)(\xi - \xi_3)(\xi - \xi_4). \end{aligned} \quad (11.5)$$

The coordinate x is an angular variable like the angle θ. The coordinate y can be thought of as a radial coordinate. The ranges of x and y are

$$\xi_3 \leq x \leq \xi_4, \quad -\infty \leq y \leq x. \quad (11.6)$$

The black hole horizon is at $y = \xi_2$, the acceleration horizon is at $y = \xi_3$, and infinity is at $y = x$, between ξ_3 and ξ_4. The coordinate t is like Rindler time, rather than Minkowski time or retarded time on $\mathcal{I}$. The Euclidean extension of the C-metric is periodic in it, with period 2π. Thus to someone moving on the orbits of the t Killing vector, the black holes would seem to be in thermal equilibrium with the acceleration radiation at a temperature of $1/2\pi$.

The scalar wave equation

$$\Box\phi = 0 \quad (11.7)$$

separates in the C-metric with the ansatz

$$\phi = A(x-y)\alpha(x)\gamma(y)e^{i\omega t}e^{im\varphi}. \quad (11.8)$$

We obtain

$$\partial_x(G(x)\partial_x\alpha(x)) - \frac{m^2\alpha(x)}{G(x)} + \left(\frac{1}{6}G''(x) + D\right)\alpha(x) = 0 \quad (11.9)$$

and

$$\partial_y(G(y)\partial_y\gamma(y)) + \frac{\omega^2\gamma(y)}{G(y)} + \left(\frac{1}{6}G''(y) + D\right)\gamma(y) = 0, \quad (11.10)$$

where D is a separation constant. The equation in the angular variable, x, has solutions that are regular on both axes only for discrete values of D. In the limit in which the accelerating black holes are small, the angular eigenfunctions will be spherical harmonics. One can then use these values of D in the radial y equation.

The black hole horizons will radiate particles according to the Bose-Einstein factor

$$\text{Emission} = \frac{T(\omega)}{e^{2\pi\omega} - 1}, \tag{11.11}$$

where ω is the frequency with respect to Rindler time. Some of these particles will be reflected back into the black holes, and some will propagate to the acceleration horizon and then on to $\mathcal{I}^+$. On $\mathcal{I}^+$, one can take Fourier transforms of the mode with respect to the affine parameter on the generators to obtain the particle content.

Ross and I are still working on this program. However, we have shown that the transmission factor between the black hole and acceleration horizons is proportional to ω^2 for small ω. This will cancel the ω^{-1} divergence in the Bose-Einstein factor and give a finite emission in each angular mode. We also find that the dominant contribution comes from the $l = 0$ mode, and that higher modes are suppressed. This suggests that the main effect of virtual black holes will be on scalar fields, and that the interaction with higher spin fields will be suppressed.

To summarize, the no-hair property of black holes means that information about the body that collapsed can't be measured from the outside. If the black hole survives forever, one could say that the information is still there, inside the black hole. However, if the black hole evaporates and disappears, the question arises: What happens to the information? I have taken it that general relativity or supergravity, rather than string theory or D-branes in a flat background, is the right arena for this problem, because of horizons and nontrivial topology. On this basis, it is clear that information is lost, and the evolution is nonunitary. One can see this in the Lorentzian regime, because $\mathcal{I}^+$ on its own is not a Cauchy surface for spacetime. One has to add the future black hole horizons. Because one cannot measure the data on the black hole horizons, one loses information and quantum coherence. In the Euclidean regime, the nontrivial topology shows that one cannot foliate with a family of time surfaces and get a unitary Hamiltonian evolution. To illustrate this, I presented calculations on virtual black holes. These indicate that quantum coherence can be lost on a microscopic scale, and that the main effect would be on scalar fields. Could this be why we have not seen the Higgs?

References

[1] J. R. Oppenheimer and H. Snyder, *Phys. Rev.* **56**, 455 (1939).

[2] R. Penrose, *Phys. Rev. Lett.* **14**, 57 (1965).

[3] E. M. Lifshitz and I. M. Khalatnikov, *Advan. in Phys.* **12**, 185 (1963).

[4] M. Choptuik, *Phys. Rev. Lett.* **70**, 9 (1993).

[5] R. S. Hamade and J. S. Stewart, *Class. Quant. Grav.* **13**, 497 (1996).

[6] D. Christodoulou, *Ann. Math.* **140**, 607 (1994).

[7] D. Christodoulou, The instability of naked singularities in the gravitational collapse of a scalar field, *Ann. Math.*, in press.

[8] W. Israel, *Phys. Rev.* **164**, 1776 (1967); *Commun. Math. Phys.* **8**, 245 (1968).

[9] B. Carter, *Phys. Rev. Lett.* **26**, 331 (1971); in *Black Holes*, ed. C. DeWitt and B. S. DeWitt (New York: Gordon & Breach, 1973).

[10] D. C. Robinson, *Phys. Rev. Lett.* **34**, 905 (1975).

[11] S. W. Hawking, *Commun. Math. Phys.* **25**, 152 (1972); S. W. Hawking and G. F. R. Ellis, *The Large Scale Structure of Space-Time* (Cambridge: Cambridge University Press, 1973).

[12] A. Strominger and C. Vafa, *Phys. Lett.* **B379**, 99 (1996); hep-th/9601029.

[13] S. Das and S. Mathur, *Nucl. Phys.* **B478**, 561 (1996); hep-th/9606185.

[14] M. T. Robinson and S. W. Hawking, Evolution of near extremal black holes, hep-th/9702045.

[15] S. W. Hawking, *Phys. Rev.* **D53**, 3099 (1996).

[16] S. W. Hawking and S. Ross, Loss of quantum coherence through scattering off virtual black holes, hep-th/970514.

[17] W. Kinnersley and M. Walker, *Phys. Rev.* **D2**, 1359 (1970).

12
Quantum States of Black Holes

Gary T. Horowitz

Abstract

I review the recent progress in providing a statistical foundation for black hole thermodynamics. In the context of string theory, one can now identify and count quantum states associated with black holes. One can also compute the analog of Hawking radiation (in a certain low energy regime) in a manifestly unitary way. Both extremal and nonextremal black holes are considered, including the Schwarzschild solution. Some implications of conjectured nonperturbative string "duality transformations" for the description of black hole states are also discussed.

12.1 Introduction

When I was a graduate student at the University of Chicago in the late 1970s, Chandra often talked about the surprises he found in his study of black holes [1]: the separation of the Dirac equation in the Kerr metric, the equality of reflection and transmission coefficients for different types of perturbations, etc. Given his well known fascination with black holes, I am sure Chandra would be interested in the unexpected results that have been discovered in the past year or two about black holes, described below.

It is by now well known that black holes have thermodynamic properties (for a recent review see [2]). In particular, they have an entropy equal to one quarter of the area of their event horizon in Planck units. This is an enormous entropy. One way to see this is to consider thermal radiation (which, of course, has the highest entropy of ordinary matter) and ask what entropy it would have were it to form a black hole. In Planck units $G = c = \hbar = k = 1$, a ball of radiation at temperature T and radius R has mass $M \sim T^4 R^3$ and entropy $S \sim T^3 R^3$. The radiation will form a black hole when $R \sim M$, which implies $T \sim M^{-1/2}$ and hence $S \sim M^{3/2}$. In contrast, the entropy of the resulting black hole is $S_{\text{bh}} \sim M^2$. So any black hole with $M \gg 1$, i.e. mass much larger than the Planck mass, has an entropy much larger than the entropy of the thermal radiation that formed it. In terms of a more fundamental description, the entropy should be a measure of the number of underlying quantum states. The problem, which has puzzled physicists for more than twenty years, is to find a more fundamental theory which contains the quantum states predicted by black hole thermodynamics.

An answer has recently been provided in the context of string theory. This is a new physical theory which came into prominence in the mid-1980s. This theory has been hailed as a "theory of everything" and scorned as a "typical end of the century phenomenon". I think the first view is much closer to the truth. String theory is a promising candidate for a consistent quantum theory of gravity and a unified theory of all forces and particles. It has not been proved that string theory can achieve these goals, but there is increasing evidence (especially over the past few years [3]) that it will.

To understand the basic idea behind the recent explanation of black hole entropy, one only needs three facts about string theory [4]. The first is that when one quantizes a string in flat spacetime there is an infinite tower of massive states. For every integer N there are states with $M^2 \sim N/l_s^2$, where l_s is a new length scale in the theory set by the string tension. These states are highly degenerate, and one can show that the number of string states at excitation level $N \gg 1$ is e^{S_s}, where

$$S_s \sim \sqrt{N} \; ; \tag{12.1}$$

i.e. the string entropy is proportional to the mass in string units. One can understand this in terms of a simple model of the string as a random walk with step size l_s.[1] As a result of the string tension, the energy in the string after n steps is proportional to its length: $E \sim n/l_s$. If one can move in k possible directions at each step, the total number of configurations is k^n, so the entropy for large n is proportional to n, i.e. proportional to the energy.

The second fact is that string interactions are governed by a string coupling constant g (which is determined by a scalar field called the dilaton). Newton's constant G is related to g and the string length l_s by $G \sim g^2 l_s^2$ in four spacetime dimensions. We will sometimes use string units where $l_s = 1$ and sometimes use Planck units where $G = l_p^2 = 1$. It is important to distinguish them, especially when g changes. Since g is in fact determined by a dynamical field, one can imagine that it changes as a result of a physical process, e.g. a wave of dilaton passing by. However, it will often be convenient to assume the dilaton is constant and treat g as just a parameter in the theory. In general, physical properties of a state can change when g is varied. But we will see that in some cases one can argue that certain properties remain unchanged.

---

[1] This model is surprisingly accurate: the typical configuration of the string in a highly excited state is indeed a random walk [5].

Chapter 12. Quantum States of Black Holes

The third fact is that the classical spacetime metric is well defined in string theory only when the curvature is less than the string scale $1/l_s^2$. This follows from the fact that, fundamentally, the metric is unified with all the other modes of the string. This is easily seen in perturbation theory where the graviton is just one of the massless excitations of the string. When the curvature is small compared to $1/l_s^2$, one can integrate out the massive modes and obtain an effective low energy equation of motion which takes the form of Einstein's equation with an infinite number of correction terms consisting of higher powers of the curvature multiplied by powers of l_s. When curvatures approach the string scale, this low energy approximation breaks down.

In section 12.2 we give a general argument which shows that string theory can explain the entropy of black holes. This argument applies to essentially all black holes and reproduces the correct dependence on the mass and charges, but it is not precise enough to check the numerical coefficient. In section 12.3 we show that for certain black holes even the numerical coefficient in the entropy can be computed and shown to agree. We also describe some recent calculations showing that the spectrum of Hawking radiation from black holes can be reproduced by string theory. In section 12.4, we consider the effect of conjectured nonperturbative "duality" symmetries on the description of black hole states. Section 12.5 contains a discussion of some open issues.

12.2 Correspondence between black holes and strings

12.2.1 Schwarzschild black holes

A few years ago, Susskind [6] suggested that there should be a one-to-one correspondence between strings and black holes. The idea was the following: Consider a highly excited string state at level N and zero string coupling. As we mentioned above, a typical configuration of such a string is a random walk with a length proportional to its energy $L \sim N^{1/2} l_s$ and hence a radius $R \sim N^{1/4} l_s$. Now imagine increasing the string coupling g and recall that $G \sim g^2 l_s^2$. Two effects take place. First, the gravitational attraction of one part of the string on the other causes the string size to decrease. Second, since G increases, the gravitational field produced by the string becomes stronger and the effective Schwarzschild radius GM increases in string units. Clearly, for a sufficiently large value of the coupling, the string forms a black hole.

Conversely, suppose one starts with a black hole and decreases the string coupling. Then the Schwarzschild radius shrinks in string units and eventually becomes smaller than the string scale. At this point the metric is approximately flat except for a small region where it is no longer well defined. Susskind suggested that the black hole becomes an excited string state. Further evidence was presented in [7].

When I first heard this, I didn't believe it. The first half of the argument sounded plausible enough, but the second half seemed to contradict the well known fact that the string entropy is proportional to the mass while the black hole entropy is proportional to the mass squared. It turns out that there is a simple resolution of this apparent contradiction [8]. Consider the familiar Schwarzschild black hole

$$ds^2 = -\left(1 - \frac{r_0}{r}\right) dt^2 + \left(1 - \frac{r_0}{r}\right)^{-1} dr^2 + r^2 \, d\Omega. \qquad (12.2)$$

The mass of the black hole is $M_{bh} = r_0/2G$. We want to equate this with the mass of a string state at excitation level N, which is $M_s^2 \sim N/l_s^2$ at zero string coupling. Now imagine increasing the string coupling, keeping the state fixed. Clearly, M_s is constant in string units where l_s is held fixed and Newton's constant is changing.[2] The analog of keeping the state fixed for the black hole is to keep the entropy fixed.[3] Thus M_{bh} is constant in Planck units where G is fixed and the string length changes. (The black hole does not know about g or l_s separately, but only the combination $G \sim g^2 l_s^2$.) In other words, the ratio M_s/M_{bh} depends on g and the masses cannot be equal for all values of the coupling. If we want to equate the masses, we have to decide at what value of the coupling they should be equal. Clearly, the natural choice is to let g be the value at which the string forms a black hole or vice versa. If we start with a black hole and decrease the coupling, the black hole description will remain valid until the horizon is of order the string scale. Setting the masses equal when $r_0 \sim l_s$ yields

$$M_{bh}^2 \sim \frac{l_s^2}{G^2} \sim \frac{N}{l_s^2}. \qquad (12.3)$$

The black hole entropy is then

$$S_{bh} \sim \frac{r_0^2}{G} \sim \frac{l_s^2}{G} \sim \sqrt{N}. \qquad (12.4)$$

So the Bekenstein-Hawking entropy is comparable to the string entropy! One cannot compare the numerical coefficients in the entropy formulas this way, since that would depend on exactly when the transition occurred. But this clearly shows that strings have enough states to reproduce the entropy of black holes. It follows from (12.3) that the transition from a string

---

[2] There is a small correction to M_s due to gravitational binding energy, but this is negligible compared to the effect discussed here.

[3] For a system with a rapidly growing density of states, a narrow band of states can be labeled by its entropy. The argument below can be reinterpreted as saying that if you insist on keeping the entropy constant through the transition from strings to black holes, the mass changes by at most a factor of order unity.

state to a black hole occurs when the string coupling is still rather small: $g \sim 1/N^{1/4} \ll 1$ for large N.

Let me emphasize that the fact that $r_0 \sim l_s$ does not mean that the black hole must be small. In fact, since the entropy is $S_{\text{bh}} \sim \sqrt{N}$, the Schwarzschild radius in Planck units is $r_0 \sim N^{1/4}$. It is probably more intuitive to describe the transition between string states and black holes in Planck units where G is fixed. Suppose we start with a solar mass black hole. As we decrease g, the string length l_s increases; i.e. the string tension decreases. Eventually, it is of order a kilometer and comparable to the horizon size. At this point, the black hole metric is no longer well defined in string theory, and the system is better described as a highly excited string state.

Starting at small coupling, the transition from a string to a black hole is analogous to the accretion of matter by a neutron star. In Planck units, the string tension increases like g^2. Thus, for fixed excitation N, the mass of a string state *increases* as g increases until it forms a black hole. For both the neutron star and the string, there is not a sudden change in mass when the black hole forms. However, there is one important difference: a neutron star has much less entropy than the resulting black hole, while a string does not.

The above counting of string states does not include the center-of-mass degree of freedom. This is appropriate since the entropy of a black hole does not include a contribution from its location in spacetime. It also does not include the possibility that a black hole is described by several strings at weak coupling. Since the string entropy is proportional to its energy, two strings with energy $E/2$ have the same entropy as one string with energy E. However, it turns out that the ratio of the total entropy of all multistring states to the entropy of a single string goes to one for large E [5]. So the entropy can be approximated by a single string.

We have seen that as one increases g a perturbative string state evolves into a black hole. In addition to the string–black hole transition, there seems to be an implicit quantum-classical transition as well. This can be made explicit as follows. Recall that for a given excitation N the transition between the black hole and string regimes occurs when $gN^{1/4} \sim 1$. For $gN^{1/4} > 1$ the Schwarzschild radius is larger than the string scale and the black hole description is valid. If $gN^{1/4} < 1$, the effective Schwarzschild radius is less than the string scale and the perturbative string description is valid. *In either case*, the classical limit is $g \to 0$, $N \to \infty$ with $gN^{1/4}$ fixed. In this limit, the Planck length $l_p \sim gl_s$ goes to zero, and the entropy $S \sim \sqrt{N}$ diverges, which agrees with the fact that a purely classical black hole indeed has infinite entropy. Quantum corrections to the black hole can be computed in terms of a string loop expansion. The classical limit on the perturbative string theory side does not yield a single classical string.

Since string theory is a "second quantized" theory of strings, the states of the first quantized string describe the classical fields of the final theory. So the $g \to 0$ limit is essentially a classical field theory with an infinite number of fields. From this viewpoint, it is interesting to note that the black hole entropy is related to the *number of fields* in the theory with a given mass. (Since many of these fields have large spin, one is actually counting the number of components of these massive fields.) In the second quantized theory, the first quantized string states simply arise as states in the "one string sector" of the usual perturbative Fock space. In the following, we will often keep N fixed and vary g. Thus the black hole will be referred to as the "strong coupling regime".

The agreement between black hole entropy and the number of string states extends to Schwarzschild black holes in higher dimensions. The easiest way to see this is to note that the entropy of a Schwarzschild black hole in any dimension can be expressed

$$S_{\text{bh}} \sim r_0 M_{\text{bh}}. \tag{12.5}$$

Clearly, when $r_0 \sim l_s$, the entropy is proportional to the mass in string units, exactly like a free string.

12.2.2 A charged black hole

The agreement between the black hole entropy and the counting of string states also extends to charged black holes. The simplest case to consider is a Kaluza-Klein black hole where the charge comes from momentum in an internal direction [9]. Given a five dimensional metric which is independent of x_5, define a four dimensional metric $g_{\mu\nu}$, gauge field A_μ, and scalar χ by

$$ds^2 = e^{-4\chi/\sqrt{3}}(dx_5 + 2A_\mu dx^\mu)^2 + e^{2\chi/\sqrt{3}} g_{\mu\nu} dx^\mu dx^\nu. \tag{12.6}$$

Then the five dimensional Einstein action is equivalent to

$$S = \frac{1}{16\pi G} \int d^4x \sqrt{-g} \left(R - 2(\nabla\chi)^2 - e^{-2\sqrt{3}\chi} F_{\mu\nu} F^{\mu\nu} \right). \tag{12.7}$$

Charged black hole solutions can be obtained by starting with the product of the four dimensional Schwarzschild metric (12.2) and a line, boosting along the line, periodically identifying x_5, and reducing to four dimensions using (12.6). The result is

$$ds^2 = -\Delta^{-1/2}\left(1 - \frac{r_0}{r}\right) dt^2 + \Delta^{1/2}\left[\left(1 - \frac{r_0}{r}\right)^{-1} dr^2 + r^2 d\Omega\right], \tag{12.8}$$

$$A_t = -\frac{r_0 \sinh 2\gamma}{4r\Delta}, \qquad e^{-4\chi/\sqrt{3}} = \Delta,$$

Chapter 12. Quantum States of Black Holes

where
$$\Delta = 1 + \frac{r_0 \sinh^2 \gamma}{r}.$$

The ADM mass and electric charge are

$$M_{\text{bh}} = \frac{r_0}{8G}(3 + \cosh 2\gamma), \tag{12.9}$$

$$Q = \frac{r_0}{4G} \sinh 2\gamma. \tag{12.10}$$

The black hole entropy is

$$S_{\text{bh}} = \frac{\pi r_0^2}{G} \cosh \gamma. \tag{12.11}$$

We want to show that this entropy is reproduced by counting string states. The electric charge on the black hole is twice the total momentum in the internal direction. A string in five dimensions with momentum P in the fifth direction has entropy $S_s \sim \sqrt{N}$, where now

$$M_s^2 - P^2 \sim \frac{N}{l_s^2} \tag{12.12}$$

and M_s is the four dimensional mass of the string. We want to match the black hole solution to the string solution, so we set $P = Q/2$. We also set $M_{\text{bh}} = M_s$ when the curvature is of order the string scale. It is the curvature of the original five dimensional solution which is important here, so the matching occurs when $r_0 \sim l_s$. The result is that

$$\frac{N}{l_s^2} \sim M_{\text{bh}}^2 - \frac{Q^2}{4} \sim \frac{l_s^2}{G^2}(5 + 3\cosh 2\gamma). \tag{12.13}$$

Comparing with (12.11) we see that up to a (γ dependent) factor of order unity

$$S_{\text{bh}} \sim \sqrt{N} \sim S_s. \tag{12.14}$$

Thus the two entropies agree for all values of γ, i.e. for all values of Q/M.

12.2.3 Correspondence principle

For the Kaluza-Klein black hole, we assumed there was an internal direction in spacetime. String theory actually predicts many internal directions. (In the above examples, the extra dimensions were taken to be a trivial torus.) These give rise to a large number of different types of charges in the effective four dimensional theory. In addition to internal momentum, one has charges associated with strings winding around each compact direction. There are also higher dimensional extended objects called D-branes (which will be discussed in the next section) which look like localized charged particles when wrapped around the internal space. Black holes can carry any of these charges. In almost all cases, the size of the black hole in string units becomes smaller when g is decreased.[4] The entropy of all of these black holes can be understood by matching onto a weak coupling description when the black hole is of order the string scale. However, there is one additional fact about string theory which must be taken into account. For many charged black holes, the dilaton field ϕ is not constant. In this case, strings couple to a metric $g^S_{\mu\nu}$ which is conformally related to the metric with the standard Einstein action $g^E_{\mu\nu}$. In D spacetime dimensions, one has $g^S_{\mu\nu} = e^{4\phi/D-2} g^E_{\mu\nu}$. Both of these metrics play a role in our discussion. The spacetime metric ceases to be well defined when the curvature of the string metric is of order the string scale, while the black hole entropy is, of course, related to the area of the event horizon in the standard Einstein metric.

The general relation between black holes and strings can now be stated in terms of the following **correspondence principle**:

(i) When the curvature at the horizon of a black hole (in the string metric) becomes greater than the string scale, the typical black hole state becomes a state of strings and D-branes with the same charges and angular momentum.

(ii) The mass changes by at most a factor of order unity during the transition.

An appropriate measure of the curvature near the horizon is given by the curvature invariants, since these control the higher order string corrections to the field equations. (The physical tidal forces felt by an infalling observer can be much larger than the curvature invariants would suggest [13].) It has been shown [8] that for a large class of charged black holes this correspondence principle correctly reproduces the Bekenstein-Hawking entropy up to factors of order unity. The two examples discussed earlier are easily seen to be special cases of this principle.

As a final example, consider the following metric describing a five dimensional black string [14]:

$$ds^2 = F\left[-\left(1 - \frac{r_0}{r}\right) dt^2 + dz^2\right] + \left(1 - \frac{r_0}{r}\right)^{-1} dr^2 + r^2 \, d\Omega, \qquad (12.15)$$

where

$$F^{-1} = 1 + \frac{r_0 \sinh^2 \gamma_1}{r}. \qquad (12.16)$$

---

[4]The one exception is black holes carrying Kaluza-Klein monopole charge, or another magnetic charge dual to string winding number. Approaches to understanding the entropy of black holes with these charges have been discussed in [10, 11, 12].

Chapter 12. Quantum States of Black Holes   249

This is not the Einstein metric, but the string metric $g^S_{\mu\nu}$. The dilaton is $e^{2\phi} = F$. The extremal limit, $r_0 \to 0$, $\gamma_1 \to \infty$ with $r_0 \sinh^2 \gamma_1$ fixed, describes the strong coupling geometry of an unexcited string wrapped around an internal direction. When this solution was discovered, it was suggested that the nonextremal black string should describe the strong coupling limit of the excited states of the wound string. We now see that this is indeed the case. In fact, (12.15) reduces to the same four dimensional black hole that we discussed earlier (12.8), and the counting of states for a string with nonzero winding number is again given by (12.12) with P replaced by the energy in the winding mode. So the number of string states with nonzero winding number agrees with the black string entropy.

12.3 Precise agreements between black holes and strings

12.3.1 Supersymmetric black holes

Using the correspondence principle to understand black hole entropy has the virtue that it can be applied to essentially any black hole, but it is not yet able to compare the precise coefficients in the entropy formulas. For a special class of black holes, as one decreases the string coupling, one has more control over the transition to a perturbative string state and even the coefficients in the entropy formulas can be compared. These more precise calculations use the fact that string theory is supersymmetric. Its low energy limit is a supergravity theory that admits black hole solutions which are invariant under some supersymmetry transformations. These supersymmetric solutions are extremally charged black holes with mass and charge related by $M = cQ$ for some constant c. In the limit of weak coupling, the mass and charge of all perturbative string states satisfy inequalities of the form $M \geq cQ$.[5] States with $M = cQ$ are called BPS states and have the special property that their mass does not receive any quantum corrections.

One can thus start with a supersymmetric black hole and decrease the string coupling. One then counts the number of BPS states at weak coupling with the same charge as the black hole. Notice that in this case the issue of when to match the mass of the black hole to the perturbative string state never arises: in both regimes the mass is completely fixed by the charge. The remarkable result is that the number of BPS states at weak coupling turns out to be precisely the exponential of the Bekenstein-Hawking entropy of the black hole at strong coupling [15]. (For a comprehensive review, see [16].)

---
[5]This is only true for certain charges in extended supersymmetry which appear in the supersymmetry algebra.

This sounds so easy, one might wonder why this calculation wasn't done years ago. There are two main reasons. The first is that most supersymmetric black holes are not really black holes at all: they have zero horizon area. The problem is that supergravity theories have scalars which couple to the gauge fields. Nonextremal charged black hole solutions exist with familiar properties, but as one approaches the extremal limit, the scalars become large at the horizon, causing it to shrink to zero size. To obtain a supersymmetric black hole with nonzero horizon area, one needs to include several charges to stabilize the scalars. This results in a second problem, since some of these charges are not carried by fundamental strings. Instead, they are carried by nonperturbative solitons. Thus, rather than simply counting states of a string at weak coupling, one must quantize the solitons and count bound states of the solitons and strings, which is much more difficult. These problems were recently solved when Polchinski discovered a new representation for these solitons called "D-branes" [17].

A D-brane has a mass proportional to $1/g$, so at weak coupling it is very massive (and hence nonperturbative). But since Newton's constant $G \sim g^2$, the gravitational field produced by the D-brane, which is proportional to GM, goes to zero as $g \to 0$. Thus there exists a flat space description of these nonperturbative states. This is obtained by adding to one's theory of closed strings a set of open strings where the endpoints of the open strings are constrained to live on a particular surface. These surfaces can have any dimension and can be viewed as generalizations of membranes. Since the endpoints of the open string satisfy Dirichlet boundary conditions in the directions normal to the surface, the surface is called a D-brane.

The excited states of a D-brane are described by quantizing the open strings. At low energies, only the massless modes contribute. The massless states of an open string include some scalars which describe the fluctuations of the brane in the surrounding spacetime and a U(1) gauge field on the brane. When two D-branes come together this is enhanced to a U(2) gauge field. The extra massless states needed to change U(1)^2 into U(2) arise from open strings that are stretched between the two D-branes. These states have a mass proportional to the separation of the branes. When this separation goes to zero, they become massless and combine with the U(1) gauge fields on each brane to produce a U(2) gauge field. Similarly, when m D-branes come together one obtains a U(m) gauge theory. In terms of the low energy gauge theory on the brane, the fact that U(m) reduces to U(1)^m when the branes are slightly separated is described by the usual Higgs mechanism.

A fascinating story is emerging from the study of D-branes at very short distances. I do not have time to develop it in detail (and it is not required to understand the recent progress in black holes), but I cannot resist mentioning it. It appears that D-branes are able to probe distances much shorter than the string scale [18]. Since the graviton is a mode of a closed string,

Chapter 12. Quantum States of Black Holes

the usual metric description cannot apply at these short distances. Consider p dimensional D-branes. When m of them come together, their low energy dynamics is governed by the reduction of a ten dimensional $U(m)$ Yang-Mills theory to p dimensions. Thus, in addition to the $U(m)$ gauge field on the brane, there are matter fields X_i, $i = p+1,\ldots,9$, which are $m \times m$ hermitian matrices with a potential $V \sim [X_i, X_j]^2$. For the ground states, $[X_i, X_j] = 0$, and one can simultaneously diagonalize these matrices. When $X_i \neq 0$ the symmetry is broken to $U(1)^m$ and the diagonal entries can be interpreted as the positions of the (now separated) branes. Therefore, there is a one-to-one relation between the position of the D-brane in spacetime and the moduli space of ground states of the gauge theory on the D-brane. For slowly moving D-branes there is a natural metric on this moduli space which controls the physics. In some cases, this metric turns out to be identical to the metric on spacetime measured at much larger distances. Perhaps the most intriguing observation is that the variables in the gauge theory which correspond to position in spacetime commute only for the ground states but in general are noncommuting! This suggests that at very short distances spacetime may be described by a form of noncommutative geometry [19].

Returning to our discussion of black holes, a single D-brane with p spatial dimensions carries a charge with respect to a $(p+2)$-form field strength F. In other words, the charge is defined by integrating $^*F$ over a sphere encircling the $p+1$ dimensional world volume of the brane. In fact, a D-brane has the minimum possible mass for this type of charge and is (in a well defined sense) a BPS state. Now suppose one compactifies spacetime on a p dimensional internal space and wraps the D-brane around this space. From the standpoint of the reduced theory, the D-brane appears to be a pointlike object carrying a charge associated with a usual two-form Maxwell field. Since a single D-brane carries only one type of charge, its strong coupling limit does not have nonzero horizon area. However, bound states of several different types of D-branes yield black holes with regular horizons.

It turns out that supersymmetric black holes with nonzero horizon area can only exist in four and five dimensions. In five dimensions one needs three different types of charges, while in four dimensions one needs four.[6] The first precise calculation of black hole entropy was performed by Strominger and Vafa [15] for an extreme nonrotating five dimensional black hole. This was quickly generalized to include rotation [22] and four dimensional extremal black holes [23, 24] (as well as small deviations from extremality which will be discussed shortly). Even the entropy of solu-

---

[6]More precisely, this is true in $N = 4$ or $N = 8$ supergravity. In $N = 2$ supergravity, there are supersymmetric black holes with one charge and nonzero horizon area [20]. However, this theory does not arise when compactifying string theory on a torus. One needs more complicated internal spaces [21].

tions which depend on arbitrary functions has been reproduced by counting states of D-branes [25].[7]

As perhaps the most interesting example of the above results, we consider the familiar extreme Reissner-Nordström solution:

$$ds^2 = -\left(1 - \frac{GM}{r}\right)^2 dt^2 + \left(1 - \frac{GM}{r}\right)^{-2} dr^2 + r^2 d\Omega. \qquad (12.17)$$

As explained above, the counting of states for this black hole is rather complicated since we need to consider four different charges. In other words, one views (12.17) as a composite of four different fundamental objects. There is a more general solution (given below) where these charges are all independent parameters, which reduces to (12.17) in a certain limit. Many of these charges are represented in weak coupling by D-branes wrapped around internal directions. There are, in fact, several different possible choices for the charges which all include (12.17) as a special case.

One way to obtain the Reissner-Nordström solution is the following [23]. One starts at weak coupling with ten dimensional flat spacetime and compactifies six dimensions on a torus, which is convenient to think of as the product of a four torus with volume $(2\pi)^4 V$ and two circles with radii R and $\tilde{R}$. One then takes Q_6 D-sixbranes wrapped around the six torus. One adds Q_2 D-twobranes wrapped around the two circles. One adds Q_5 fivebranes wrapped around the four torus and R. All of these branes lie over the same point in the three noncompact spatial directions, so this configuration describes a localized object in four spacetime dimensions. Notice that the intersection of these branes is the circle R. When R is large the low energy excitations of these branes are described by open strings moving along this circle. It turns out that these strings have $4Q_2Q_5Q_6$ massless bosonic degrees of freedom and an equal number of fermionic degrees of freedom. (This is obtained by analyzing the induced gauge theory on the branes.) Momentum along the circle is quantized in units of $1/R$. The number of states with right-moving momentum n/R (and no left-moving momentum) is e^S, where S is given by the general formula for a one dimensional gas, $S = 2\pi\sqrt{cn/6}$. The constant c receives a contribution of one for every bosonic field and one half for every fermionic field. In our case, we have $c = 6Q_2Q_5Q_6$ and hence

$$S = 2\pi\sqrt{Q_2 Q_5 Q_6 n}. \qquad (12.18)$$

I should emphasize that this calculation is done in flat spacetime. There is no event horizon present. One is simply counting states of strings on D-branes. Notice that the entropy depends only on the integer charges, and not on the continuous parameters $V, R, \tilde{R}$.

---

[7] In this case there is a small puzzle since, although the horizon area is well defined, the curvature diverges there [26].

Chapter 12. Quantum States of Black Holes

The strong coupling description of this system is found by solving the ten dimensional supergravity equations with these charges. After reducing along the six torus, the four dimensional (Einstein) metric becomes[8] [28]

$$ds^2 = -f^{-1/2}(r)\left(1 - \frac{r_0}{r}\right)dt^2$$
$$+ f^{1/2}(r)\left[\left(1 - \frac{r_0}{r}\right)^{-1} dr^2 + r^2\, d\Omega\right], \quad (12.19)$$

where

$$f(r) = \left(1 + \frac{r_0 \sinh^2 \alpha_2}{r}\right)\left(1 + \frac{r_0 \sinh^2 \alpha_5}{r}\right)$$
$$\times \left(1 + \frac{r_0 \sinh^2 \alpha_6}{r}\right)\left(1 + \frac{r_0 \sinh^2 \alpha_p}{r}\right).$$

This metric is parameterized by the five independent quantities α_2, α_5, α_6, α_p, and r_0. They are related to the integer charges by

$$Q_2 = \frac{r_0 V}{g} \sinh 2\alpha_2,$$
$$Q_5 = r_0 \tilde{R} \sinh 2\alpha_5,$$
$$Q_6 = \frac{r_0}{g} \sinh 2\alpha_6,$$
$$n = \frac{r_0 V R^2 \tilde{R}}{g^2} \sinh 2\alpha_p, \quad (12.20)$$

where we have set $l_s = 1$. The event horizon lies at $r = r_0$. The special case $\alpha_2 = \alpha_5 = \alpha_6 = \alpha_p$ corresponds to the Reissner-Nordström metric. Notice that if we set all charges except the momentum n to zero the metric (12.19) reduces to the Kaluza-Klein solution (12.8) as it should. Furthermore, since all charges enter (12.19) symmetrically, the four dimensional metric generated by any one of these charges is the Kaluza-Klein black hole. The precise relation between the four dimensional Newton constant and the string coupling is $G = g^2/(8VR\tilde{R})$ in string units ($l_s = 1$). The ADM mass of the solution is

$$M = \frac{r_0 V R \tilde{R}}{g^2}(\cosh 2\alpha_2 + \cosh 2\alpha_5 + \cosh 2\alpha_6 + \cosh 2\alpha_p), \quad (12.21)$$

and the Bekenstein-Hawking entropy is

$$S_{\text{bh}} = \frac{A}{4G} = \frac{8\pi r_0^2 V R \tilde{R}}{g^2} \cosh \alpha_2 \cosh \alpha_5 \cosh \alpha_6 \cosh \alpha_p. \quad (12.22)$$

---
[8] We follow the discussion in [27].

The extremal limit corresponds to $r_0 \to 0$, $\alpha_i \to \pm\infty$ with Q_i fixed. In this limit, the entropy becomes

$$S_{\rm bh} = 2\pi\sqrt{Q_2 Q_5 Q_6 n}, \tag{12.23}$$

in precise agreement with the string result (12.18)! The Reissner-Nordström solution is clearly just a special case of this.

Even though we have not needed the correspondence principle to reproduce the black hole entropy, one can still use it to estimate when the black hole description breaks down and must be replaced by the D-brane description. In string units, the area of the event horizon is approximately $g^2\sqrt{Q_2 Q_5 Q_6 n}$. The curvature at the horizon will be of order the string scale when this is of order one. If we assume all the charges are comparable, the transition occurs when $gQ \sim 1$.

In the extremal limit, the ADM mass becomes

$$M = \frac{R\tilde{R}}{g} Q_2 + \frac{RV}{g^2} Q_5 + \frac{R\tilde{R}V}{g} Q_6 + \frac{n}{R}, \tag{12.24}$$

which is easily seen to be just the sum of the energies of the four constituents making up the black hole.[9] In the case of the Reissner-Nordström solution when all the boost parameters are set equal, one can easily verify that each of these constituents contributes equally to the total mass. Notice that if one fixes the charges and size of the internal torus the mass (in string units) changes with the string coupling. This is consistent with the BPS bound since the constant c (relating M and Q) can depend on g.

12.3.2 Near extremal black holes

Since supersymmetry seemed to play an important role in the above discussion, one might have thought that the entropy could be calculated precisely only in this case. This turns out to be incorrect. In fact, soon after the first precise calculation of black hole entropy, it was shown that the entropy of certain slightly nonextremal black holes can also be calculated exactly [30, 31]. At present, this is well understood only in the regime where one of the constituents is much lighter than the rest. This case is clearly the simplest to consider since the maximum entropy is obtained by adding energy to the lightest degrees of freedom. For the case of Reissner-Nordström, where all the branes contribute equally to the mass, if one adds a small amount of energy, the excitation of all the branes will contribute equally to the entropy and the counting is much more difficult. It is clear from

---

[9]The energy of the fivebranes goes like $1/g^2$ rather than $1/g$ because they are not D-fivebranes but rather solitonic fivebranes [23]. Fortunately, the counting of states can still be carried out in this case. For an alternative description of the four dimensional black hole without solitonic fivebranes see [29].

Chapter 12. Quantum States of Black Holes

(12.24) that when R is large the momentum modes are much lighter than the branes. So if one adds a small amount of excess energy, it simply goes into exciting left and right moving modes. Since the left and right moving modes are largely noninteracting at weak coupling, the entropy is additive and we obtain

$$S = 2\pi\sqrt{Q_2 Q_5 Q_6}\left(\sqrt{n_\text{L}} + \sqrt{n_\text{R}}\right), \qquad (12.25)$$

where the left and right moving momenta are n_L/R and n_R/R respectively. This agrees precisely with the Bekenstein-Hawking entropy computed from the near extremal hole solution (12.22). For other choices of the parameters $V, R, \tilde{R}$, one of the branes may be much lighter than the other constituents. In this case, the entropy can again be understood by a suitable counting argument. But when several branes are light (or all contribute equally, as in Reissner-Nordström), one does not yet have a precise counting of the entropy of the near extremal solution.

At weak coupling the interactions between the left and right moving modes are small but nonzero. Occasionally, a left-moving mode can combine with a right-moving mode to form a closed string which can leave the brane. This represents the decay of an excited configuration of D-branes and is the weak coupling analog of Hawking radiation. Given the remarkable agreement between the entropy of the black hole and the counting of states of the D-branes, the next step is clearly to ask how the radiation emitted by the D-brane compares to Hawking radiation. Since the entropy as a function of energy agrees in the two cases, it is not surprising that the radiation is approximately thermal with the same temperature in both cases. What is surprising is that the overall rate of radiation agrees [32]. What is even more remarkable is that the deviations from the blackbody spectrum also agree [33]. On the black hole side, these deviations arise since the radiation has to propagate through the curved spacetime outside the black hole. This produces potential barriers which give rise to frequency-dependent graybody factors. On the D-brane side there are deviations since the modes come from separate left and right moving sectors on the D-branes. The calculations of these deviations could not look more different. On the black hole side, one solves a wave equation in a black hole background. The solutions involve hypergeometric functions. On the D-brane side, one does a calculation in D-brane perturbation theory. Remarkably, the answers agree.

To be a little more precise, the calculations were first done for the (five dimensional) near extremal black hole with $R \gg 1$. Since the black hole is near extremality, the temperature is very low and hence the wavelength of the radiation is much larger than the size of the black hole. One considers radiation by a minimally coupled scalar field. On the D-brane side, one starts with a thermal distribution of left and right moving modes with temperatures T_L and T_R respectively. The decay rate for left and right moving excitations, each with energy $\omega/2$, to produce an outgoing S-wave mode with energy ω is

$$\Gamma = \frac{g_{\text{eff}}^2 \omega}{(e^{\omega/2T_\text{L}} - 1)(e^{\omega/2T_\text{R}} - 1)} \frac{d^4k}{(2\pi)^4}. \tag{12.26}$$

The factors in the denominator are the usual thermal factors for the left and right moving modes, and g_{eff} is a frequency-independent effective coupling constant. To compare with the black hole, one computes the average left and right moving energies, which determine n_L and n_R. This fixes the total energy and momentum of the black hole. The Hawking temperature turns out to be related to T_L and T_R by

$$\frac{1}{T_\text{R}} + \frac{1}{T_\text{L}} = \frac{2}{T_\text{H}}. \tag{12.27}$$

The black hole decay rate is given by [34]

$$\Gamma = \frac{\sigma_{\text{abs}}(\omega)}{(e^{\omega/T_\text{H}} - 1)} \frac{d^4k}{(2\pi)^4}, \tag{12.28}$$

where $\sigma_{\text{abs}}(\omega)$ is the graybody factor, which equals the classical absorption cross section. One calculates $\sigma_{\text{abs}}(\omega)$ by studying solutions to the wave equation in the black hole background. For the Schwarzschild and Kerr metrics, this was extensively studied more than twenty years ago [35]. It has recently been shown that for any black hole the limit of $\sigma_{\text{abs}}(\omega)$ as $\omega \to 0$ is the area of the event horizon [36]. After a lengthy calculation in the metric analogous to (12.19) describing a five dimensional black hole with three charges, one finds [33] that for $\omega \leq T_\text{H}$

$$\sigma_{\text{abs}}(\omega) = \frac{g_{\text{eff}}^2 \omega (e^{\omega/T_\text{H}} - 1)}{(e^{\omega/2T_\text{L}} - 1)(e^{\omega/2T_\text{R}} - 1)}, \tag{12.29}$$

so the two rates (12.26) and (12.28) agree!

It is worth emphasizing that it is not just a few parameters which agree. Since the calculation is valid for $\omega \leq T_\text{H}$, the entire functional form is significant. It is as if the black hole knows that its states are described by an effective 1+1 dimensional field theory with left and right moving modes. It appears that the black hole also knows that some of these modes are fermionic. In the weak coupling description, an outgoing mode with angular momentum $\ell = 1$ arises from left and right moving fermions on the D-brane. Remarkably, when the graybody factor is computed for this case, one again finds that it factors into left and right moving thermal factors, but now they take the form $(e^{\omega/2T} + 1)$ appropriate for fermions [37]! (The overall numerical coefficient has not yet been checked for this case.) More generally, for ℓ odd, one obtains the fermionic factors, while for ℓ even, one obtains the bosonic factors as expected from the D-brane description.

It is not yet clear why a calculation of decay rates at weak coupling can be extrapolated without modification into the strong coupling regime. One possible explanation was given by Maldacena [38], who argued that at low energy the relevant interactions did not receive quantum corrections due to a supersymmetric non-renormalization theorem (see also [39]).

Attempts to extend this calculation have met with mixed success. Agreement was found for four, as well as five, dimensional black holes [40], charged scalars [33, 40], and certain non-minimally-coupled scalars [41, 42], but disagreement was found for higher energies [43, 44], other near extremal regimes, e.g. $R \sim 1$ [45, 44], and other non-minimally-coupled scalars [46]. However, in most of the cases where disagreement was found, even the weak coupling D-brane calculations are not yet completely understood.

These results have immediate implications for the well known black hole information puzzle. Hawking has argued that the radiation emitted from a black hole is independent of what falls in. Thus, if an extreme black hole absorbs a small amount of matter and then radiates back to extremality, information will be lost and unitarity will be violated. String theory provides a manifestly unitary description of a system of strings and D-branes with the same entropy and rate of radiation. If one throws a small amount of energy toward a system of D-branes, the branes become excited and then decay. The final system is in a pure state, with correlations between states of the D-branes and the radiation. If one traces over the D-brane states, the radiation is approximately thermal. One might imagine that the same thing happens for black holes. But this does not avoid the difficulties of how information gets out from inside the black hole. Suppose that one repeats this experiment many times. When the entropy in the radiation becomes greater than the entropy in the D-branes, it can no longer be the case that the radiation is thermal after tracing out the D-brane states. There will be correlations between the radiation emitted at early times and the radiation emitted at late times [47]. Since this system of strings and D-branes becomes a black hole when one increases the string coupling, it is very tempting to conclude that even in the black hole regime there will be correlations between the radiation at early and late times and the evaporation process will be unitary. In fact, the argument based on supersymmetric non-renormalization theorems [38] supports this view for very low energy quanta.

However, Hawking has stressed that the causal structure of the black hole is very different from the flat space description, since there is no analog of the event horizon. It is logically possible that the evaporation process is unitary at weak coupling and fails to be unitary at strong coupling. Before one can conclude that quantum mechanics is not violated in black hole evaporation, one needs at least a convincing explanation of where

Hawking's original arguments break down.

12.4 Duality

In sections 12.2 and 12.3, we considered the transition from a weakly coupled state of strings and D-branes to a black hole. We have seen that this occurs when $gN^{1/4} \sim 1$ or $gQ \sim 1$, which implies that the fundamental string coupling g is still rather small. During the past few years there have been a series of conjectures about the behavior of string theory when the coupling becomes much greater than one. It was previously believed that there were five fundamental string theories in ten dimensions which differed in the amount of supersymmetry and type of gauge groups they contained. The low energy limits of these theories were ten dimensional supergravity coupled to matter. In addition, there was an eleven dimensional supergravity theory which did not seem to fit into string theory at all. It is now believed that all of these theories are connected in the sense that the large coupling limit of one theory compactified on one type of internal space is equivalent to the weak coupling limit of another compactified on a possibly different internal space. This suggests that there is one universal theory with a different weak coupling limit corresponding to each of the known theories. The conjectures relating these theories are known as S-duality. The evidence for them has been accumulating for the past two years and is now rather convincing [48]. This includes the fact that there are solitons in one theory with the same properties as the fundamental strings of the other, and the spectra of BPS states in the two theories (which do not depend on the coupling) agree.

I do not have time to discuss this exciting subject in detail, but let me mention two consequences for our discussion of black holes. For simplicity, I will not discuss compactification but will consider black holes directly in ten dimensions. Since all string theories include gravity, the Schwarzschild black hole is a solution to each theory. By taking different limits, one can represent its states in different ways. Let us consider the type IIB theory which has both fundamental strings and D-strings. The D-strings have a tension $1/gl_s^2$ and hence are very heavy when $g \ll 1$. But when $g \gg 1$, the D-strings become much lighter than the fundamental strings. This limit is described by another IIB string theory, but now the role of the fundamental strings is played by the D-strings and the coupling constant is $\hat{g} = 1/g$. (In other words, the IIB theory is self-dual.) We saw in section 12.2 that when $g \ll 1$, the states of a Schwarzschild black hole could be represented by states of an ordinary string. We now see that in the limit $g \gg 1$ the same black hole can be described in terms of states of a weakly coupled D-string.

Thus one has the following picture as one increases g. One can start with an excited state of the fundamental string at level N when $g = 0$. As we have seen, when $g \sim 1/N^{1/4}$ the Schwarzschild radius is of order the

Chapter 12. Quantum States of Black Holes

string scale and the state forms a black hole. If we continue to increase the coupling, the black hole remains unchanged until $g \sim N^{1/4}$, which is when the Schwarzschild radius is of order the length scale set by the D-string tension. Beyond this point, the black hole can be described by an excited state of a weakly coupled D-string at the same level N as the initial fundamental string.

In fact, the low energy IIB string Lagrangian has an $SL(2, \mathbb{Z})$ symmetry, under which the Einstein metric is invariant, but the fundamental string is mapped into an (m, n) string which carries m units of fundamental string charge and n units of D-string charge. Starting with the Schwarzschild black hole, there are different weak coupling limits of this theory in which the states are excitations of the (m, n) strings.

It is interesting to consider the black hole information puzzle from this standpoint. If we compactify five dimensions of the IIB string theory and wrap various D-branes around the internal five torus, we can obtain extreme and near extremal five dimensional black holes. We have seen that the spectrum of radiation at infinity remains unchanged as we increase the coupling and the description of the state changes from slightly excited D-branes to near extremal black holes. The D-brane radiation is known to be unitary, yet it has been argued that the black hole radiation will not be unitary. The duality conjectures imply that if we continue to increase g the black hole can again be described by a weakly coupled dual string theory. The spectrum will remain the same, and the radiation will again be unitary. Of course the spacetime geometry in both limits $g \gg 1$ and $g \ll 1$ is flat, so there are qualitative differences from the black hole. But still it seems rather unlikely that a physical process would change from being unitary to nonunitary and back to unitary as a parameter is continuously increased.

As a second application of duality ideas to black holes, we consider the IIA string theory. This theory has a series of BPS states (bound states of D-zerobranes) with masses which are integer multiples of $1/gl_s$. This is similar to the spectrum of Kaluza-Klein states in a theory compactified on a circle of radius $R = gl_s$. Note that for weak coupling, the radius is very small, but at strong coupling, it becomes large. For this reason (and others), it is believed that the low energy limit of the strongly coupled ten dimensional IIA string theory is eleven dimensional supergravity compactified on a circle of radius $R = gl_s$. The eleven dimensional Planck length turns out to be $l_p = g^{1/3}l_s$. One can now trace out the following behavior of an excited IIA string state as the coupling is increased. One starts at zero coupling with a state at level N, so $M = \sqrt{N}/l_s$ and $S \sim \sqrt{N}$. One now increases the string coupling keeping the state (i.e. the entropy) fixed. As before, when $g \sim 1/N^{1/4}$, one forms a ten dimensional black hole with $r_0 = N^{1/16}$ in ten dimensional Planck units (since $S \sim r_0^8 \sim N^{1/2}$). When $g \sim 1$, the length of the eleventh dimension becomes greater than the eleven dimensional Planck length and so becomes physically meaningful. Since the ten dimensional IIA supergravity theory is just the dimensional reduction of eleven dimensional supergravity, the ten dimensional black hole then becomes an eleven dimensional black string; i.e. the solution becomes the product of Schwarzschild and a circle. As the coupling increases, the length of the eleventh dimension becomes larger. When it becomes of order r_0, the black string becomes unstable [49] and probably forms an eleven dimensional black hole. This occurs when $r_0 \sim R$, which implies $R \sim S^{1/9} l_p$. Since $S \sim N^{1/2}$, one has $R \sim N^{1/18} l_p$ and hence $g \sim N^{1/12}$. As g is increased further, corresponding to larger values of R, the state remains an eleven dimensional black hole. Conversely, starting with an eleven dimensional Schwarzschild solution with one dimension periodically identified, one can follow the above description in reverse as one decreases g: the eleven dimensional black hole transforms into an eleven dimensional black string, which then becomes a ten dimensional black hole, and finally a weakly coupled string in ten dimensions. This is one approach toward understanding the entropy of eleven dimensional black holes. Of course, it would be more satisfactory to have a direct explanation of the entropy of eleven dimensional black holes, one that does not require a direction to be compactified. But this requires the full eleven dimensional quantum theory (called M theory) which is not yet well understood.

12.5 Discussion

Our understanding of black hole microstates provided by string theory is progressing rapidly. To illustrate this, let me briefly list some of the highlights from 1996.[10] In January, the first calculation showing precise agreement between the entropy of an extremal five dimensional black hole and the counting of string states was performed [15]. In February, this was extended to near extremal black holes [30, 31] and extreme rotating black holes [22] (still in five dimensions). In March, the entropy of four dimensional black holes, both extremal [23, 24] and near extremal [27], was reproduced. In June, the rate of low energy radiation from a near extremal five dimensional black hole was shown to agree exactly with the rate from excited D-branes [32]. In August, this was extended to four dimensions [50]. In September, it was shown that the deviations from the blackbody spectrum agree, both in five [33] and four [40] dimensions. Finally, in December, the entropy of black holes far from extremality was understood, up to an overall coefficient, in terms of a correspondence principle [8].

Despite this enormous progress, our understanding is still far from complete. One outstanding issue is the resolution of the black hole information

---

[10]This is by no means an exhaustive list of all of the contributions to this subject; such a list would include well over a hundred papers.

Chapter 12. Quantum States of Black Holes   261

puzzle, which was discussed in sections 12.3 and 12.4. Perhaps a more modest question is whether the entropy of black holes far from extremality can be reproduced exactly. When the supersymmetric black hole calculations were first carried out, there was a strong belief that supersymmetry was playing a key role and it would be impossible to do the same thing for nonsupersymmetric configurations. But then it was found that the entropy continued to agree for near extremal black holes and also for extremal black holes which are not supersymmetric, e.g. extreme rotating four dimensional black holes [27, 51]. I now believe that the entropy of all large black holes should be computable, including the overall coefficient. The correspondence principle certainly shows that the main contribution to the entropy can be understood without supersymmetry.

One might ask whether the correspondence principle could be extended to compare the precise coefficients in the expression for the entropy. At first sight this appears to be difficult since it requires a better understanding of the string state when it is at the string scale and interactions become important. However, there are indications that things are simpler than they appear. One of these comes from studies of the near extremal threebrane. Using the correspondence principle, one can understand the entropy of all near extremal black p-branes in terms of a gas of massless strings on the brane at weak coupling [8]. However, for the threebrane, the entropy turns out to be independent of when the masses are set equal. If one takes the coupling all the way to zero and compares the entropy of a free gas with that of the black threebrane, one finds $S_{\mathrm{bh}} = (3/4)S_{\mathrm{gas}}$ [52]. The correspondence principle predicts that the transition to the black p-brane occurs when the temperature of the gas is of order the string scale, so the interactions should be important. Since the potential energy is positive, this explains why $S_{\mathrm{bh}} < S_{\mathrm{gas}}$: by ignoring the interactions, the energy of each state of the gas has been underestimated, and hence the total number of states with a given energy has been overestimated. But it is not clear why the interactions should simply produce a factor of 3/4.

We have seen that the transition from a perturbative string state to a black hole occurs when the string coupling is still rather small, $g \sim 1/N^{1/4}$ or $g \sim 1/Q$ (for near extremal black holes). So one might wonder why string perturbation theory cannot be used to directly study properties of black holes. The reason is that the effective coupling constant, due to the long string or large number of D-branes, is really $gN^{1/4}$ or gQ, which is becoming of order one. Even though string perturbation theory is known to diverge [53], there is an important difference between a coupling constant that is of order one, and a coupling constant that is small. The perturbation series has the form $\sum C_n \tilde{g}^n$, where $C_n \sim (2n)!$ and $\tilde{g}$ is the effective coupling constant [54]. This is an asymptotic series, and the best approximation to the exact answer is obtained by cutting the series off after $\tilde{g}^{-1/2}$ terms, when the individual terms become greater than one. It turns out that the error one introduces this way is of order $e^{-1/\tilde{g}}$. (In ordinary field theory, $C_n \sim n!$, so cutting the series off after $1/\tilde{g}$ terms introduces an error $e^{-1/\tilde{g}^2}$.) Clearly, when $\tilde{g} \sim 1$, no useful information can be obtained from the perturbation series. Fortunately, for $\tilde{g} \gg 1$, one has an alternative description of the system in terms of a semiclassical black hole.

For the near extremal black holes discussed in section 12.3.2, the energy above extremality is independent of the string coupling g. So one can compare the entropy as a function of energy for the black holes and strings. Since they agree, one knows that the temperature of the two systems are equal. This is not true for black holes far from extremality. We saw in section 12.2 that the mass of a state does not remain constant as one changes the string coupling g. In string units, the mass of a perturbative string state is approximately constant until it forms a black hole, and then it decreases with g. In Planck units, the mass of a string state increases with g until it forms a black hole, and then it remains constant. This means that even though the entropies of black holes and strings agree, their temperatures do not. For a highly excited string the natural temperature is a constant of order the string scale (the Hagedorn temperature) rather than the Hawking temperature.

The understanding of black hole entropy provided by the correspondence principle leads to a simple picture of the evaporation of a Schwarzschild black hole, if the string coupling is small in nature. In most of our previous discussion, we imagined varying the string coupling g. Now we suppose that $g \ll 1$ is fixed, so the string length scale is much larger than the Planck length. A large black hole will Hawking-evaporate until the curvature at the horizon reaches the string scale. At this point it turns into a highly excited string state with $N \sim 1/g^4$, since for this value of N the string has a mass comparable to that of the black hole: $M_s^2 \sim N/l_s^2 \sim 1/g^4 l_s^2 \sim l_s^2/G^2 \sim M_{\text{bh}}^2$. The entropy of the string is also comparable to the black hole entropy at this point, so the string can carry the remaining information in the black hole. The excited string will then continue to decay via string interactions. The temperature slowly increases as the black hole evaporates and reaches the string scale at the transition point. It then remains at this temperature as the excited string decays.[11] Eventually one is left with an unexcited string, i.e. an elementary particle like a photon.

The history of particle physics is full of examples where objects which were thought to be elementary were later found to be composed of more fundamental entities. The fact that the entropy of black holes has now been reproduced by counting quantum states strongly suggests that we

---

[11]The mass is likely to decay exponentially if one assumes that each segment of the long string has the same probability to break off a small loop of string. Since $M \sim L$, $dM/dt \sim -M$. (I thank L. Susskind for pointing this out.)

have finally identified the fundamental degrees of freedom and there is not another level of structure waiting to be uncovered. However, as we saw in the section 12.4, these fundamental degrees of freedom can take different forms in different weak coupling limits of the theory.

There remains the puzzling question of *why* the counting of string states turns out to reproduce the area of the event horizon of a black hole. This is undoubtably tied up with the fundamental question of what the origin of spacetime geometry in string theory is. One recent suggestion, which is motivated by the relation between position in spacetime and the moduli space of gauge theories on D-branes, starts with a quantum theory of $N \times N$ matrices in the limit of large N [55]. The description of black holes in this context is beginning to be investigated [56, 57]. At the risk of sounding too conservative, I will state my belief that spacetime itself will ultimately be made up of strings. After all, perturbations in the spacetime metric are just one mode of the string. With the possible exception of zerobranes, it is difficult to see how charged objects like D-branes can produce a neutral object like Minkowski spacetime. I believe the entropy calculations are providing a glimpse into the quantum origin of spacetime. It is tempting to turn the current calculations around and use them to try to *define* a metric or area in spacetime in terms of the number of states in a Hilbert space. Following this approach may expand our glimpse until the full picture of quantum spacetime is finally revealed.

Acknowledgments

It is a pleasure to thank A. Ashtekar, T. Jacobson, B.-L. Hu, and S. Ross for raising questions which (it is hoped) improved the clarity of this presentation. I also wish to thank J. Polchinski, A. Strominger, and L. Susskind for discussions which improved my understanding of the results discussed here. This work was supported in part by the NSF under grant PHY95-07065.

References

[1] S. Chandrasekhar, *The Mathematical Theory of Black Holes*, Oxford University Press (Oxford 1983).

[2] R. Wald, Chapter 8 of this volume; gr-qc/9702022.

[3] For a recent review see C. Vafa, hep-th/9702201.

[4] See e.g. M. Green, J. Schwarz, and E. Witten, *Superstring Theory*, Cambridge University Press (Cambridge 1987).

[5] D. Mitchell and N. Turok, *Nucl. Phys.* **B294**, 1138 (1987).

[6] L. Susskind, hep-th/9309145.

[7] E. Halyo, A. Rajaraman, and L. Susskind, *Phys. Lett.* **B392**, 319 (1997), hep-th/9605112; E. Halyo, B. Kol, A. Rajaraman, and L. Susskind, *Phys. Lett.* **B401**, 15 (1997), hep-th/9609075.

[8] G. Horowitz and J. Polchinski, *Phys. Rev.* **D55**, 6189 (1997), hep-th/9612146.

[9] H. Leutwyler, *Arch. Sci.* **13**, 549 (1960); P. Dobiasch and D. Maison, *GRG* **14**, 231 (1982); A. Chodos and S. Detweiler, *GRG* **14**, 879 (1982); D. Pollard, *J. Phys.* **A16**, 565 (1983); G. Gibbons and D. Wiltshire, *Ann. Phys.* **167**, 201 (1986), erratum *Ann. Phys.* **176**, 393 (1987).

[10] F. Larsen and F. Wilczek, *Phys. Lett.* **B375**, 37 (1996), hep-th/9511064; *Nucl. Phys.* **B475**, 627 (1996), hep-th/9604134; *Nucl. Phys.* **B488**, 261 (1997), hep-th/9609084.

[11] M. Cvetič and A. Tseytlin, *Phys. Rev.* **D53**, 5619 (1996), hep-th/9512031; A. Tseytlin, *Nucl. Phys.* **B477**, 431 (1996), hep-th/9605091.

[12] R. Dijkgraaf, E. Verlinde, and H. Verlinde, *Nucl. Phys.* **B486**, 77 (1997), hep-th/9603126; *Nucl. Phys.* **B486**, 89 (1997), hep-th/9604055.

[13] G. Horowitz and S. Ross, *Phys. Rev.* **D56**, 2180 (1997), hep-th/9704058.

[14] G. Horowitz and A. Strominger, *Nucl. Phys.* **B360**, 197 (1991).

[15] A. Strominger and C. Vafa, *Phys. Lett.* **B379**, 99 (1996), hep-th/9601029.

[16] J. Maldacena, hep-th/9607235.

[17] J. Polchinski, *Phys. Rev. Lett.* **75**, 4724 (1995), hep-th/9510017. For a recent review see J. Polchinski, *TASI Lectures on D-Branes*, ITP preprint NSF-ITP-96-145, hep-th/9611050.

[18] M. Douglas, D. Kabat, P. Pouliot, and S. Shenker, *Nucl. Phys.* **B485**, 85 (1997), hep-th/9608024.

[19] A. Connes, *Noncommutative Geometry*, Academic Press: San Diego, (1994).

[20] G. Gibbons and C. Hull, *Phys. Lett.* **109B**, 190 (1982).

[21] D. Kaplan, D. Lowe, J. Maldacena, and A. Strominger, *Phys. Rev.* **D55**, 4898 (1997), hep-th/9609204; K. Behrndt and T. Mohaupt, *Phys. Rev.* **D56**, 2206 (1997), hep-th/9611140; J. Maldacena, *Phys. Lett.* **B403**, 20 (1997), hep-th/9611163; W. Sabra, hep-th/9703101.

[22] J. Breckenridge, R. Myers, A. Peet, and C. Vafa, *Phys. Lett.* **B391**, 93 (1997), hep-th/9602065.

[23] J. Maldacena and A. Strominger, *Phys. Rev. Lett.* **77**, 428 (1996), hep-th/9603060.

[24] C. Johnson, R. Khuri, and R. Myers, *Phys. Lett.* **B378**, 78 (1996), hep-th/9603061.

[25] G. Horowitz and D. Marolf, *Phys. Rev.* **D55**, 835 (1997), hep-th/9605244; *Phys. Rev.* **55**, 846 (1997), hep-th/9606113.

[26] G. Horowitz and H. Yang, *Phys. Rev.* **D55**, 7618 (1997), hep-th/9701077.

Chapter 12. Quantum States of Black Holes

[27] G. Horowitz, D. Lowe, and J. Maldacena, *Phys. Rev. Lett.* **77** (1996) 430, hep-th/9603195.

[28] M. Cvetič and D. Youm, hep-th/9508058; hep-th/9512127.

[29] V. Balasubramanian and F. Larsen, *Nucl. Phys.* **B478**, 199 (1996), hep-th/9604189.

[30] C. Callan and J. Maldacena, *Nucl. Phys.* **B472**, 591 (1996), hep-th/9602043.

[31] G. Horowitz and A. Strominger, *Phys. Rev. Lett.* **77** (1996) 2368, hep-th/9602051.

[32] S. Das and S. Mathur, *Nucl. Phys.* **B478**, 561 (1996), hep-th/9606185.

[33] J. Maldacena and A. Strominger, *Phys. Rev.* **D55**, 861 (1997), hep-th/9609026.

[34] S. Hawking, *Commun. Math. Phys.* **43**, 199 (1975).

[35] A. Starobinsky and S. Churilov, *Sov. Phys. JETP* **38**, 1 (1974); G. Gibbons, *Commun. Math. Phys.* **44**, 245 (1975); D. Page, *Phys. Rev.* **D13**, 198 (1976); *Phys. Rev.* **D14**, 3260 (1976); W. Unruh, *Phys. Rev.* **D14**, 3251 (1976).

[36] S. Das, G. Gibbons, and S. Mathur, *Phys. Rev. Lett.* **78**, 417 (1997), hep-th/9609052.

[37] J. Maldacena and A. Strominger, hep-th/9702015.

[38] J. Maldacena, *Phys. Rev.* **D55**, 7645 (1997), hep-th/9611125.

[39] S. Das, *Phys. Rev.* **D56**, 3582 (1997), hep-th/9703146.

[40] S. Gubser and I. Klebanov, *Phys. Rev. Lett.* **77**, 4491 (1996), hep-th/9609076.

[41] C. Callan, S. Gubser, I. Klebanov, and A. Tseytlin, *Nucl. Phys.* **B489**, 65 (1997), hep-th/9610172.

[42] I. Klebanov, A. Rajaraman, and A. Tseytlin, hep-th/9704112.

[43] S. de Alwis and K. Sato, *Phys. Rev.* **D55**, 6181 (1997), hep-th/9611189.

[44] F. Dowker, D. Kastor, and J. Traschen, hep-th/9702109.

[45] S. Hawking and M. Taylor-Robinson, *Phys. Rev.* **D55**, 7680 (1997), hep-th/9702045.

[46] M. Krasnitz and I. Klebanov, *Phys. Rev.* **D56**, 2173 (1997), hep-th/9703216.

[47] D. Page, *Phys. Rev. Lett.* **71**, 1291 (1993).

[48] For a review, see J. Polchinski, *Rev. Mod. Phys.* **68**, 1245 (1996), hep-th/9607050.

[49] R. Gregory and R. Laflamme, *Phys. Rev. Lett.* **70**, 2837 (1993); *Nucl. Phys.* **B428**, 399 (1994).

[50] S. Gubser and I. Klebanov, *Nucl. Phys.* **B482**, 173 (1996), hep-th/9608108.

[51] A. Dabholkar, *Phys. Lett.* **B402**, 53 (1997), hep-th/9702050.

[52] S. Gubser, I. Klebanov, and A. Peet, *Phys. Rev.* **D54**, 3915 (1996), hep-th/9602135; A. Strominger, unpublished.

[53] D. Gross and V. Periwal, *Phys. Rev. Lett.* **60**, 2105 (1988).
[54] S. Shenker, RU-90-047, 1990.
[55] T. Banks, W. Fischler, S. Shenker, and L. Susskind, *Phys. Rev.* **D55**, 5112 (1997), hep-th/9610043.
[56] M. Li and E. Martinec, hep-th/9703211, hep-th/9704134.
[57] R. Dijkgraaf, E. Verlinde, and H. Verlinde, hep-th/9704018.

Chandra Remembered

Chandra: A Tribute

Kameshwar C. Wali

Chandra often remembered his close friend of earlier Cambridge years Edward Arthur Milne and quoted him as writing

> Posterity, in time, will give us all our true measure and assign to each of us our due and humble place; and in the end it is the judgement of posterity that really matters. He really succeeds who perseveres according to his lights, unaffected by fortune, good or bad. And it is well to remember there is no correlation between the judgement of posterity and the judgement of contemporaries.

One year after may not be a true measure of posterity. But this two-day symposium in Chandra's honor certainly marks the beginning of posterity's assignment to bestow him his due place. Posterity will also reckon other tributes, expressions of wonder and admiration for an extraordinary life.

For me personally, the years I spent working with Chandra and writing his biography were the most enjoyable and creative years in my life. I began the venture with a great deal of apprehension, but it didn't take long to feel at ease as Chandra left himself completely open to questions about himself and his life, and I had complete freedom to write whatever I chose to write. It became an overwhelmingly delightful experience to be with him and be transferred most of the time to a different world, a world in which the life of the mind dominated. Soon it became evident that Chandra, as Res Jost had said, "was a member of an ideal community of geniuses who weave and compose the fabric of our culture."

After the completion of the book, although my visits with Chandra became less frequent, our friendship continued to grow and develop. During the summer of 1994, Lalitha, Chandra, my wife, and I spent a week together at the Stratford Shakespeare Festival in Canada. "Get the best seats for the plays," Chandra had said to me while I was making the arrangements for our trip. And when I called him a few days before our scheduled meeting

---

Talk given after the Symposium Banquet, December 14, 1996.

in Stratford, he said he was rereading *Othello, Hamlet,* and *Twelfth Night,* the plays we were going to see. Along with Lalitha he was also listening to the records. Thus he came fully prepared to enjoy his rare vacation, setting aside his preoccupation with Newton at the time. We all had such a good time, seeing a new play every day, exploring new eating places, and taking sight-seeing side trips in the country surrounding Stratford.

On one of these car trips, Chandra surprised me by asking, "Which moves faster, heat or cold?" While I was racking my brain, thinking about Boltzmann and Maxwell's demon and stuff, he said, with a twinkle in his eye, "Heat, of course, cold you can catch!"

It was great to see Chandra as a full-time tourist, so lighthearted and impulsive in enjoying himself. Without the slightest hesitation, he bought a large-size painting he liked from a sidewalk exhibition. And when inadvertently we walked into an open air restaurant called Anna Banana, Chandra had to eat a veggi-burger without a table setting and without knife and fork. While I sat embarrassed at having submitted him to this ordeal, Chandra seemed to be totally at ease and to relish the experience.

I was not bugging him with questions about his life, his childhood, his days in Madras and Cambridge, his encounters with Milne and Eddington, the Yerkes Observatory, the University of Chicago, or the University of Chicago Press to which he was strongly, almost sentimentally, attached as the editor of the *Astrophysical Journal* for nearly twenty years. Over the years I had made him tell and retell these stories, and without the least annoyance, he had obliged.

The last time I talked to him on the phone was during the first week of August 1995, when I received a complimentary copy of his last book, *Newton's Principia*. It had just come out. Still he regretted that I had not received the copy sooner. I thanked him and congratulated him. We both agreed that the Oxford University Press had done an excellent job in producing the book with such virtuosity and elegance. I said to Chandra that this work of his would go down as monumental. He had his doubts, he replied. He had seen one or two critical reviews. But he accepted my compliment and said he had no longer the energy or stamina to do hard work. He complained about exhaustion and how he had to be helped to get back home when he was taking a short walk near his apartment. Those were gruelingly hot days in Chicago. I reminded him of that and said in a rather harsh tone, "Chandra, I forbid you to work hard anymore. You have done enough. You must relax and enjoy." "Yes, yes," he replied, "that is exactly what I am going to do. Just two short papers to be finished with Valeria Ferrari. I AM relaxing. Reading *Les Misérables*."

Famous last words, I said to myself. Chandra often reflected upon how some great scientists, artists, and men of letters spent their later years—what were their intimate thoughts about their lives. He had no sympathy for those who had lived their last years in misery, unable to bear the isolation, the lack of limelight and public adoration that they were used to. Chandra said he was happy that he was spending his last years in the company of Newton. I am sure, if he were alive, he would be thinking and working on Newton and Michelangelo. He would be writing about a comparison between the motivations of scientists and artists in their creative pursuits. He had once said that that is what he would want to do when he disengaged himself from serious scientific work.

Reading Chandra's essay on the series paintings of Claude Monet and the landscape of general relativity, one cannot fail to see an analogy. In Monet's series paintings, the same scene is depicted over and over again under different natural illuminations and seasonal variations. The valley, the trees and the fields, the haystacks are the same. Superficially they may appear boring and repetitive. However, the different paintings radiate totally different aesthetic content. In a similar fashion, in Chandra's hands, the seemingly same equations and solutions take on vastly different physical situations and their mathematical descriptions. In concluding that essay, Chandra says he does not know if there has been any scientist who could have said what Monet said on one occasion:

> I would like to paint the way a bird sings.

But we do know a scientist who spoke like a poet on one occasion:

> The pursuit of science has often been compared to the scaling of mountains, high and not so high. But who amongst us can hope, even in imagination, to scale the Everest and reach its summit when the sky is blue and the air is still, and in the stillness of the air survey the Himalayan range in the dazzling white of the snow stretching to infinity? None of us can hope for a comparable vision of nature and the universe around us. But there is nothing mean or lowly in standing in the valley below and awaiting the sun to rise over Kanchanjunga.

And one who closed his Halley Lecture in 1972 discussing the increasing role of general relativity and the occurrence of black holes in nature with the memorable parable:

> The parable, entitled "Not lost but gone before," is about larvae of dragonflies deposited at the bottom of a pond. A constant source of mystery for these larvae is what happens to them, when on reaching the stage of chrysalis, they pass through the surface of the pond never to return. And each larva, as it approaches the chrysalis stage and feels compelled to rise to the surface of the pond, promises to return and tell those that remain behind what really happens, and confirm or deny a rumour attributed to a frog that when a larva emerges

on the other side of their world it becomes a marvellous creature with a long slender body and iridescent wings. But on emerging from the surface of the pond as a fully-formed dragonfly, it is unable to penetrate the surface no matter how much it tries and how long it hovers. And the history books of the larvae do not record any instance of one of them returning to tell them what happens to it when it crosses the dome of their world. And the parable ends with the cry

...Will none of you in pity,
To those you left behind, disclose the secret?

With such writings and quotes from modern and ancient literature, with his Ryerson Lecture, "Shakespeare, Newton and Beethoven," and his book *Truth and Beauty,* Chandra built a bridge between what C. P. Snow calls the "two cultures," the cultures of the sciences and the humanities. As numerous obituaries, reviews, and letters testify, Chandra combined with his extraordinary success in scientific endeavors an equally extraordinary personality marked by an intensity and fervor for completeness and elegance, and above everything else, a wish to gain a personal, aesthetic perspective in his scientific work and to practice its precepts and live up to its ideals to the closest possible limit. He was a true scholar, a man of letters, a humanist and a rationalist in whose life religion and religious beliefs of any sort played no role.

What happens to a man after his death? Is there an answer for those who do not believe in simplistic answers, such as heaven and hell, the cycle of birth and rebirth and Karma, etc.? I recall the reply of an ancient sage in India when asked by his disciples. He is supposed to have said

swa karyeshu pratishthanti swa kale kalatitee cha

They stand outside time by their deeds as they did while they were living.

Our Song

Lalitha Chandrasekhar

I should like to take you back sixty years to a home in Madras, India. It was called Chandra Vilas. Chandra and I had just been married. I was entering my husband's home for the first time. And what do you expect everyone to ask the bride on an occasion such as this? "Sing us a song," they asked me. I selected a favorite song of mine to sing. But then something unexpected happened: Chandra's father was annoyed over the selection of the song. "A most improper selection," he muttered to himself and, in his irritation, walked out of the room! I, the poor bride, was bewildered. "What did I do to offend Chandra's father?" I asked myself. At the time I didn't know the answer, but on hindsight I can see that it was a bad selection for a girl just married to sing. The composer Papanasam Sivan bemoans the innumerable births and deaths that have been his fate. To put it in his own words, "There has been another conception and another birth only to be thrown into the ditch." By ditch he meant death. The composer prostrates himself before the Deity and begs him to release him from this torture of repeated births and deaths. He pleads, "Please don't brush aside this supplicant" ...But we must go back to the bride. Though Chandra's father had a point in objecting to her selection, you must admit that he was not being kind in welcoming her into her new home in this harsh way!

I didn't ask Chandra about his own reaction till later. It was only after we were at Trinity College in Cambridge, England, where we spent a month for him to pack up his things for our trip to America, that I asked him one day, "Chandra, did you dislike my song as much as you father did?" "Oh no, on the contrary I loved it, and you sang it so well and with so much feeling." "Oh Chandra, I am so glad you loved my song. I sang it because I also loved it. And now that both of us love it, does it matter what your father thought of it?" "Of course it doesn't matter." It was so soothing for me to hear him say this. The matter was ended and we never again

---

Talk given after the Symposium Banquet, December 14, 1996 (a videotape is on deposit in the Chandrasekhar archives). The song referred to in the title, "Thiruvadi Charanam" by Papanasan Sivan, has been recorded on the CD *Chellatha Mariyathu* by L. R. Eswari (Khazana Music, KH2570).

thought about it.

That song, or, shall we say, "The Song," played a very significant role in our lives. There was a parallel between the cycles of birth and death that Papanasam Sivan sang about and the cycles of scientific endeavour that have been associated with Chandra. You remember how as a young man of nineteen, on his voyage to England from India, he had discovered what seemed startling: that a star of mass more than 1.4 times the mass of the sun cannot end as a white dwarf; it must indeed contract beyond that stage! But a well-known authority in astronomy at the time, Sir Arthur Stanley Eddington, denied this possibility. He didn't deny this in a quiet way but without warning at a meeting of the Royal Astronomical Society in London where Chandra was to give his paper on his startling discovery. Eddington said after Chandra's paper, "I felt driven to the conclusion that this was almost a reductio ad absurdum of the relativistic degeneracy formula. Various accidents may intervene to save the star, but I want more protection than that. I think there should be a law of nature to prevent a star from behaving in this absurd way!"

It was devastating! Why couldn't Eddington have given a warning to Chandra when he periodically came over to Chandra's rooms at Trinity College to see how his calculations were coming along? He could have said, "Look here, Chandrasekhar, I do not agree with your conclusions. In fact, I am going to oppose them at the next R.A.S. meeting where you are going to present your paper." That would have been sportsmanlike. But he kept it a secret and attacked without warning.

Chandra knew he was right, but he could not go on fighting a world-recognized authority. All he could do was publish his results in a paper and later in a chapter of his book *An Introduction to the Theory of Stellar Structure*. After this was done, he could move on to other fields in which he could work in peace. This was the first cycle of birth and death that Chandra had to face. But there was something unnatural about this death. Chandra was at the height of his scientific investigation and was not ready to quit the subject. There were still many ideas he wanted to work on. His wings were clipped. The astonishing thing is the resilience he showed in being able to climb out of what was indeed a calamity. What is even more astonishing is that, once he searched and found another field to work on, he was able to shut his eyes to the calamity. The concentration was all on the new field. He worked on stellar atmospheres for awhile—an old love. But he didn't write a book on the subject till he took it up again in his research on radiative transfer.

It is clear that Chandra found an affinity between the composer of my song being thrown in the ditch and his being, in some ways, also thrown in the ditch by what Eddington had said at the R.A.S. meeting. But as I said before, he climbed out of the ditch to work in another field. And it was also in Chandra's generous nature to forgive, he would prefer to call it "overlook," and continue his friendship with Eddington. Eddington would take him to see tennis and cricket matches. And before Chandra left Cambridge for America, Eddington treated him to a rare privilege: he spread before him his map of England on which he had carefully traced out in black ink all the different cycling tours he had taken over the years. He and Sydney Chapman had devised a criterion for judging cycling records. The criterion was the largest number n such that one had cycled n or more miles on n different days. Eddington would include in every letter to Chandra his latest value of n. It was 77 in 1943, the year before he died.

Chandra was very impressed with how rapidly Eddington solved crossword puzzles. There was a particular puzzle in which the question asked was "What is a bagged relative?" Eddington quickly wrote down BAKING. When Chandra asked for an explanation, Eddington told him, inside the BAG was the KIN. In my kitchen cupboard, my bottle of baking powder has been labelled in Chandra's handwriting, "Powder a la bagged relative!"

There is no question that Chandra had a great respect for Eddington. This started even when he was a student at Presidency College, Madras, India. There was an essay competition on quantum theory that he entered by submitting the essay "The Study of the Fine Structure of Spectral Lines and Modern Spectroscopy." On receiving the first prize, he was asked by Professor Appa Rao, who had arranged the essay competition, what book he would like to receive as his prize. Chandra had no hesitation in saying, *"The Internal Constitution of the Stars,* by Eddington." The book played an important role in his discovery of the limiting mass. Chandra's early respect for Eddington continued all through his scientific life, as can be seen from the Centenary Lectures[1] in his memory that Chandra gave at Trinity College on October 19 and 20, 1982. In these lectures, Chandra gave a detailed account of the many fundamental contributions that Eddington had made in astrophysics. But it is astonishing that Eddington could not get over his dislike and disbelief of the limiting mass to the end of his life!

Here is another thing: It has often been said, and incorrectly, that Chandra left England to come to America because of his controversy with Eddington. Actually what happened was this: Chandra asked Eddington and Fowler what chances he had of finding a position in England after the end of his fellowship at Trinity College. They said the chances were nil. If he had secured a position, he would have stayed on in England even after he was offered positions at Yerkes and at Harvard. He had many friends in England: Eddington, Milne, Davenport, to name a few, and that is where he wanted to be. If he went to America, he would have to start life afresh there. There are so many mistaken statements made about Chandra that I am glad that I am able to correct at least one of them now!

---

[1] "Eddington: The Most Distinguished Astrophysicist of His Time" and "Eddington: The Expositor and the Exponent of General Relativity," reprinted in *Truth and Beauty* (Chicago: University of Chicago Press, 1987).

One last thing: Chandra used to tell me often, "You know it is because of Eddington that I became the sort of scientist changing my field periodically from one to another. I had to change my field after the controversy, and I continued doing that after each field was explored." Kamesh Wali in his biography of Chandra suggests that an analysis of Chandra's scientific life reveals this influence of Eddington.[2] If Kamesh felt that Chandra was unaware of this influence he is mistaken, since Chandra told me of this many years before Kamesh even started on his biography!

Each field of study, or cycle as we shall call it, took anywhere from ten to fifteen years, for the selection of the subject for investigation, study of the available scientific literature on the subject, his own research that followed, the scientific papers he wrote on the subject, and, finally, the way he gathered all the material that lay in front of him into a coherent whole that was his book on the subject. After a cycle ended and he was searching for a new field to enter, there was an interim. He referred to it as his fallow period. It was during these fallow periods that he would ask me to sing "The Song." He would refer to the composer as "Your friend." He would say, "Your friend, sing his song." I would sing it to him. It was depressing to be in a fallow period. My song gave him comfort and soothed him. Is it just seven or eight times that you sang "The Song" to him you may ask. No, many more times than that since research in a field is not one huge climb and descent. There are constant ups and downs. Sometimes a problem can last for a long time and has to be coaxed and teased, broken to pieces, and brought together again, goaded and entreated until at last it yields and reveals the answer. Whenever there was a letdown he would say to me, "Your friend, sing his song to me" and I would sing to him.

Perhaps he felt it hardest after his work on radiative transfer, which was, in his words, "the happiest period of my scientific life. The subject seems to have its own momentum. I just sat there on the driver's seat astonished to see it move forward on its own." Because of that rich experience, he felt the void that followed a little hard to accept. He turned repeatedly to my song during that transition period. It was also during this period that he was tempted to go to the attic at Yerkes to retrieve his manuscript bundled and put away! There were still ideas there he could work on. But he felt ashamed of doing this and promptly went up and put it back! It was not in his spirit to pick up crumbs. What he wanted was a whole new field spread in front of him for exploration.

Not every period of research ended in his writing a book, as for example in his work on turbulence and relativistic astrophysics. A book was published on his work in plasma physics, but that was based upon the notes taken by his student Trehan. Chandra looked at his manuscript, made some changes, and gave some suggestions. But that book was Trehan's

---

[2]Kameshwar C. Wali, *Chandra* (Chicago: University of Chicago Press, 1984).

undertaking.

When Chandra started to write a book, he had a sense of how long it would take him to finish it. He would decide on a deadline by which it should be finished, and he invariably managed to finish it by then. In the case of his book *Hydrodynamic and Hydromagnetic Stability,* he went one step further. He had set a deadline when he would hand over the manuscript to an Oxford University Press representative at Heathrow airport between our flight to London and our flight to India. There were four hours available for handing over the manuscript and having lunch together. But unfortunately, the plane was two hours late in reaching London. There was only time to hand over the manuscript but not for lunch! Chandra was also sick before finishing the book. He had to take medication to keep him going. It was altogether a very trying time.

He had a sense of duty to SCIENCE, spelt in capital letters. It is that mass of the subject that has been accumulating over the years. He felt it his duty to write his *Ellipsoidal Figures of Equilibrium* at a time when his interest had turned to relativistic astrophysics. "If I didn't write this when it is all fresh in my mind, an enormous amount of material that I worked on will become unavailable for others to work on," he told me. It was hard to lay aside what he was actively working on to do what he felt was his duty to SCIENCE. There was something ennobling in his sense of dedication.

The last book he wrote on his scientific research, *The Mathematical Theory of Black Holes,* brought him full cycle around to the subject he had to leave because of the controversy with Eddington. Chandra's prediction, which Eddington denied, that a star of mass more than 1.4 times the mass of the sun cannot end as a white dwarf but must go on contracting was proved correct. What seemed absurd to Eddington was in fact proved to be a law of nature: Black holes do exist!

There was a difference between Papanasam Sivan's cycles and Chandra's: Papanasam Sivan objected to the entire process of entry, living through, and exit from each cycle. Chandra felt dejected only after the exit. A cycle has ended. Was the effort I put in worthwhile? What am I going to do next? The quest and the search was depressing. But once he found the field he could work on, he plunged in with enthusiasm. There was the entire field lying open for his exploration, every square inch of it to be weeded and cultivated and planted with flowers. And when the whole field lay in front of him like a beautiful garden, he could move back and look at it. He called what he saw his perspective. The writing of the book was when he reduced the whole picture into a coherent whole.

Chandra had stated in his will that if there was no request forthcoming from his relatives for his ashes that they should be discarded; that is, the crematorium should discard the ashes in a wastebasket! He forgot about one relative who had other ideas about this. As it turned out, there was no request for the ashes from Chandra's relatives. His other relative, meaning myself, decided to sprinkle his ashes on the university grounds. About a year ago, I requested that President Sonnenschein and Mr. Kleinbard, vice president of University News and Community Affairs, accompany me in sprinkling Chandra's ashes on every spot of the university that he had visited: the Moore Statue, the Laboratory for Astrophysics and Space Research, the Crerar Library, the Regenstein Library, Mandell Hall, the Court Theatre, and the Seminary Bookstore, where he loved to browse through the books. I not only sprinkled them on the university grounds, but in the American Indian tradition threw them up in the air and in our Indian tradition discarded the last of the ashes in Lake Michigan to be churned by the waves into the Atlantic Ocean and from hence to all the oceans of the world. So Chandra is everywhere and because of that he is here with us now. If you feel a tickle in your throat it is because you are probably inhaling a molecule of Chandra! I hear a voice asking me, "Lalitha, won't you sing your friend's song once more for me and to all those who have gathered here in remembrance of me?" "Yes, Chandra, I will sing it once more for you, and it will also be a fitting way to end this symposium held in remembrance of you, and a fitting way to end this banquet of delicious Indian dishes that the Office of Special Events at the university and the chefs of the Quadrangle Club have so graciously provided us."

I cannot give you the voice I had sixty years ago. Two months ago I reached the ripe old age of eighty-six! But they say that practice makes one perfect; and since I have sung this song so many times to Chandra, I could dare to sing it once again to him, to all of you gathered here, and to the university he loved and which in turn treasured him.

I sang "Our Song" once more; and as I sang I thought of a room in Chandra Vilas where I had sung the same song sixty years ago. Someone left the room annoyed, but it didn't matter. "Here is our song that brought us together over the years, our song that I am singing once more for you. Let this song give you comfort wherever you are. I can imagine you sitting on top of the Himalayas, thinking, always thinking ... Do you have a new kind of vision with which you can see Black Holes that we on earth cannot see? I can't wait to join you so that you can show me what with your special vision you are able to see."